THE ENCYCLOPEDIA OF
TRANSPORT

Marshall Cavendish London & New York

The Mercury *was a stream-lined steam locomotive which operated on the New York Central line. It was designed by Henry Dreyfuss.*

Edited by Donald Clarke
Designed by Chris Lower

Published by Marshall Cavendish Publications Limited
58 Old Compton Street
London W1V 5PA

This material was first published by Marshall Cavendish Limited in *How It Works*.

First printing: 1976

Printed in Great Britain by Jarrold & Sons, Norwich

ISBN 0 85685 176 0

INTRODUCTION

The machinery of transport has fascinated mankind since ancient times; from early sailing ships to the most advanced moon rockets, the devices which take us from here to there are endlessly studied, modelled, and photographed. This book is a chronological survey of things that float, fly or run on wheels. The social significance of each innovation in transport can be traced in the historical detail. The engines which turn the wheels are also fully described, with the help of dozens of photographs and cut-away drawings, many colour-coded for easier understanding. Other important facets of transport, such as docks, airports and navigation, are also covered, making the Encyclopedia of Transport a truly comprehensive reference book.

The internal combustion engine quickly took over from the horse. This is a taxi-cab rank in London, 1929.

CONTENTS

XN 8787
YM 3505

METHODS OF PROPULSION

The invention of the wheel is an event lost in history. Apart from the use of wind in sails, the earliest challenge posed in the history of transport was how to turn the wheels without the use of muscle power. Only in modern times have metallurgy and machine-building advanced far enough to obtain the energy latent in combustion. Rising heat from a fire was used in ancient times to turn sails which rotated meat roasting on a spit, a primitive form of gas turbine. But the example was not enough; it was centuries before the technology for use in transport caught up with the theory. To propel a locomotive, steam boilers had to withstand high pressure; to power aircraft, engines such as the one on the right must have a suitable power-to-weight ratio. There could be no modern transport without the designer of the machine and the mechanic to maintain it.

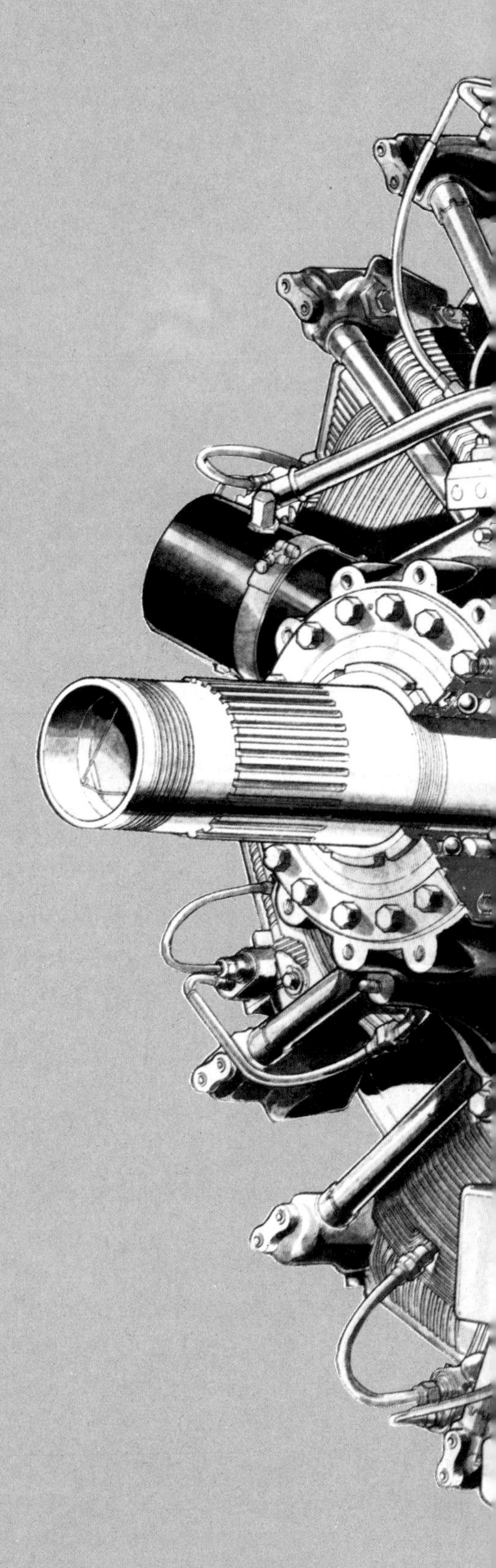

Above, top: this Sumerian mosaic dates from about 2100 BC; it shows part of a battle and victory scene. The cart shown has wheels made of two planks held together by battens.
Above: an early 19th century Yorkshire wagon. The dished wheels allowed for wider loads on the same width of track, and were at a better angle for carrying heavy, swaying loads.
Opposite page: the popularity of the bicycle was celebrated in Victorian music, as in this popular song of 1894.

The wheel

The ancient invention of the wheel made possible a greater mobility of goods and people, allowing the organization of society into urban communities to which food and other goods could be transported.

Archaeologists do not know whether the potter's wheel or the vehicle wheel was invented first; both were made of wood and have long since disappeared. The earliest evidence of either dates from about 3500 BC. The oldest surviving vehicle wheels, found in Mesopotamian tombs of 3000 to 2000 BC, were made of three planks clamped together edge to edge with cross struts, using a natural knot-hole in the central plank as a pivot. A knot-hole is very hard and wear resistant, which indicates that the wheel may have turned on a fixed axle.

The first improvement to be made was to add a wooden rim, so that wear would be even all the way around. Leather rims were also used, and later, metal ones. The spoked wheel was invented at about the same time as the rim, probably for lightness on war chariots. In the 16th century, dished wheels were invented, with the spokes arranged in a flat cone; this allowed the tops of the wheels to be angled outward, while the hub remained directly above the rim on the road. Carts could be built wider while still running on the same track, with the supporting spoke upright for greatest strength. No more technological advances were made until the 1870s and the invention of the wire wheel for bicycles.

The bicycle

The word 'bicycle' is derived from the Latin *bi* meaning 'two' and the Greek *kyklos* meaning 'circle' or 'wheel'. The physics of the various forces used in balancing a ridden bicycle is not well established. The gyroscopic effect of the rotating front wheel has little effect on the balancing, except in the case of motor cycles, which have heavy wheels. More important is the self-centring action of the front wheel. All bicycles are designed so that the steering axis, if continued downwards, would reach the ground ahead of the point of contact of the wheel with the ground. This causes the wheel to 'trail' like a castor, giving rise to the correcting effect. The rider has only to perform minor corrections to the steering, and the steering geometry takes care of the rest.

The first bicycles were made in France in the 1790s. The Célerifère, a two-wheeled wooden horse, could not be steered as the front wheel was fixed, and it was propelled by the rider pressing his feet alternately on the ground in a kind of waddling, pushing motion. The Draisienne, first shown publicly in 1818, had a wooden frame, large spoked wooden wheels with metal tyres, a saddle, and an arm rest. The front wheel was steerable, and speeds of up to ten miles per hour (16 km/h) were easily attained. The design was successful in Germany, France, the USA, and England, where it was known as the hobby-horse. Several hundred were sold and it became very fashionable to ride one, but by about 1830 the novelty had worn off and no more were made.

The first pedal-operated bicycle was invented by Scottish blacksmith Kirkpatrick Macmillan in 1839. Instead of merely modifying an existing machine he produced a new design with foot treadles driving the rear wheel. The steerable front wheel was carried on an iron fork fitted to the front of the frame, which was made of wood and divided at the other end to form a fork for the rear wheel. The treadles were fitted to the ends of two levers which pivoted on the frame near the front fork, and connecting rods transmitted the movement of the levers to cranks attached to the rear axle.

The late 1860s brought many notable developments in

FARE YOU WELL, DAISY BELL
SEQUEL TO "DAISY BELL".
Yours fin de
Harry Dacre
(Author of "DAISY BELL")
WRITTEN & COMPOSED BY
HARRY DACRE.
Copyright
Price 4/-
LONDON:
FRANCIS, DAY & HUNTER, 195, OXFORD STREET, W,
Publishers of
Smallwood's Celebrated Pianoforte Tutor. " Smallwood's 55 Melodious Exercises, Etc.
NEW YORK T. B. HARMS & Co 18 EAST 22ND ST

Right: the Singer Special Safety bicycle, made in 1896, a chain-driven machine with a diamond-shaped frame. Below: some unusual hobby-horse bicycles c1818; below these, 'ordinary' or 'penny-farthing' bicycles in a race at the Kennington Oval, London, 1874.

bicycle design. The Hanlon brothers in the USA introduced a pedal-crank to adjust the length of pedal throw, Madison and Copper produced wheels with wire spokes, and the first rubber tyres were demonstrated by J Hastings in 1868. A crude rear wheel brake appeared in the form of a spoon or shoe which was applied to the metal or rubber tyre by the rider pulling on a cord attached to the handlebars.

During the 1870s a bicycle evolved which was to become symbolic of early cycles. The 'penny-farthing' comprised a large front driving wheel of up to 50 inches (127 cm) diameter and a rear wheel of about 24 inches (60 cm) diameter. The frame was a single metal tube which ran from the steering head just above the top of the front wheel to a forked end carrying the rear wheel. The saddle was mounted high up on this tube over the front wheel and close to the handlebars. The pedals were attached to cranks on the front hub. Numerous designs were produced to avoid having the pedals directly attached to cranks on the front wheel hub. Some incorporated long levers attached to cranks on the hub, and a chain drive version appeared in 1879 with the front wheel driven by cranks placed below its centre and connected with it by a chain and sprocket arrangement. By 1885 it had reached the peak of its development, with solid rubber tyres, brakes available for both wheels, and a growing range of accessories which included lamps, toolbags and touring bags, mileage recorders, and bells, gongs and whistles to warn of the rider's approach. Long distances could be covered

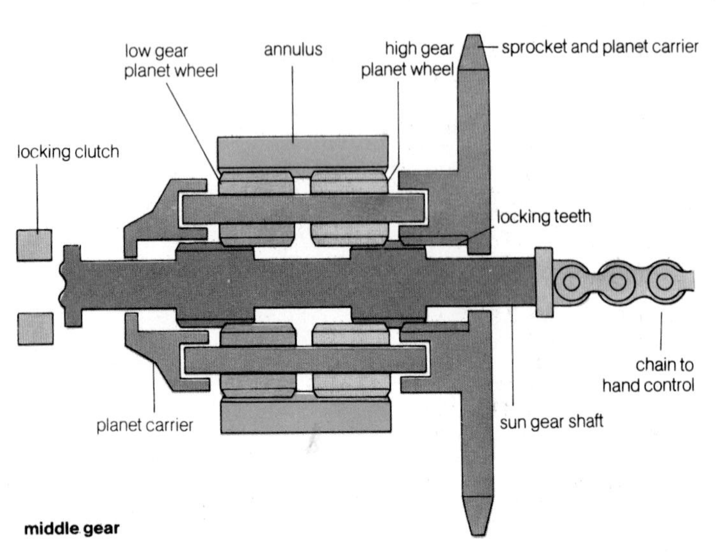

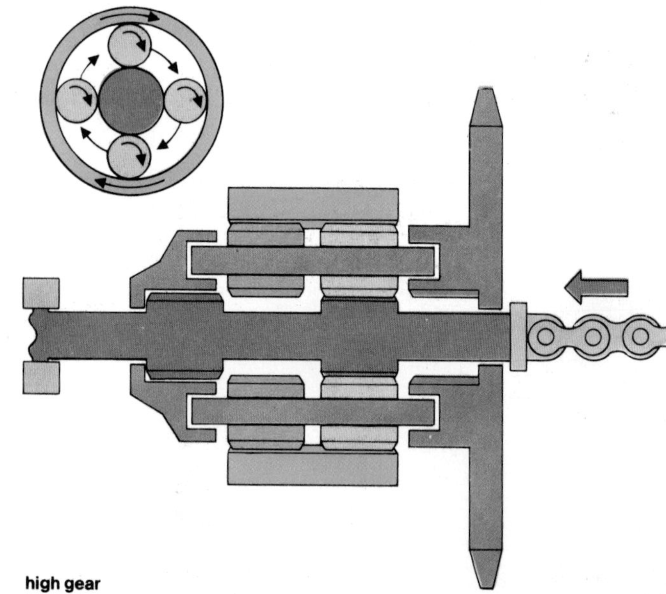

on these later cycles: in 1888 a light 37 lb (16 kg) 'Geared Facile' machine was ridden over 290 miles (465 km) in less than 24 hours, and in 1890 the journey from London to Brighton and back, about 120 miles (190 km) in all, was completed within eight hours.

The 'Safety' bicycle was the forerunner of the modern design, and the first known models were built around 1870 by various makers in France and England, notably Meyer et Cie, F W Shearing, and H J Lawson. The main features of these designs were two medium-sized wheels of approximately equal diameter, a chain drive to the rear wheel, rubber tyres, and wire-spoked wheels. The Rover safety bicycles built by J K Starley from 1884 were the first commercially successful designs, and his third version of 1888 was to determine the general configuration of bicycle frames for the next 60 years. Many other designs were produced, often incorporating springing to help absorb road shocks. Another design, the Dursley Pedersen of 1893, had a triangulated frame in which narrow section tubes were duplicated. As all the tubes were subjected to compression forces only, lightweight tubing could be used which provided a much lighter construction than an orthodox frame. The racing cycle made by Dan Albone in 1886 used the cross-frame, which never caught on at the time but made a great comeback in 1962 with Alex Moulton's design and its many imitators.

The first production bicycle to have the true diamond-shaped frame was the 1890 Humber, and by 1895 this form of frame was standard, except for ladies' cycles which had a dropped top tube.

By the late 1900s the bicycle had ceased to be merely a novel status symbol and was becoming a cheap and practical form of personal transport. Mass production of the bicycle began, and by 1914 the practical utility bicycle was being assembled from some 300 separate components, the majority of which were manufactured by specialist firms.

The simplest form of bicycle transmission in use today consists of two sprockets of different sizes connected by an endless chain. The larger sprocket, the chainwheel, is fitted to the axle of the pedal cranks, and the smaller one is fixed to the hub of the rear wheel. The rear sprocket usually incorporates a freewheel device, the first version of which was invented by White and Davies in 1881. The only bicycles now made without a freewheel are those used for track racing, which need no braking apart from that obtained by the rider resisting the motion of the pedals. The fixed sprocket also gives better control at low speeds, the lack of a braking mechanism leads to a saving in weight, and as the wheels do not have to take the strain of the action of brake shoes, very light rims and spokes can be used.

The two main forms of gearing for bicycles are the derailleur type and the enclosed hub gear or epicyclic gear. The first modern form of derailleur appeared in France in 1909, although various designs had been tried since about 1899. The derailleur is a light, positive, and reliable mechanism, using interchangeable sets of between two and six sprockets of different diameters screwed onto a common freewheel. There is a mechanism for lifting and transferring the chain from one sprocket to another to provide a gear-change, and a chain tensioning device to take up the slack chain required when making a change. Derailleur gears (named from the French word meaning 'derail') are used mainly on lightweight sports and touring cycles.

The enclosed hub gear consists of a set of gears built into an enlarged rear hub. The drive is transmitted from the chain through the rear sprocket, which drives the hub via the internal gearing. Many designs of this type of gear were produced at the beginning of this century, one of the most

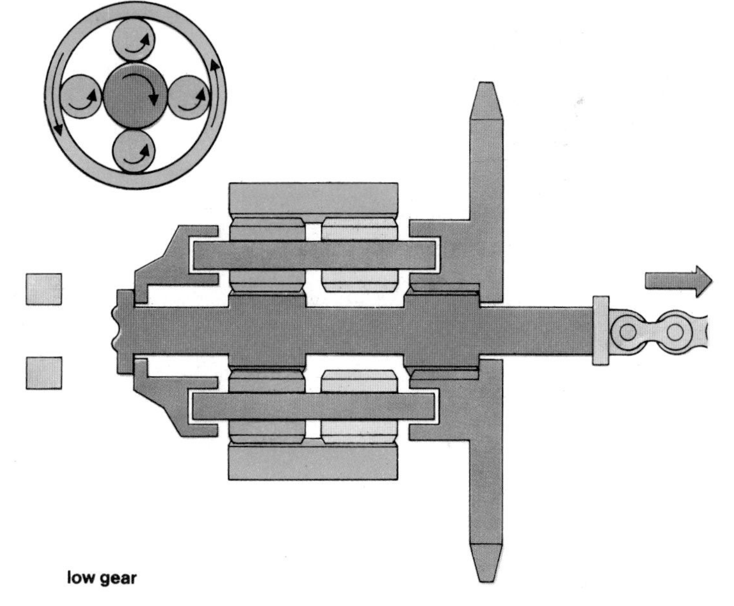

Above: a derailleur gear. The chains used with derailleurs are fully riveted, because the spring clip on the connecting link of an ordinary chain will jam in the mechanism or be forced off the chain.

Left: diagrams of a three-speed bicycle hub gear. It has a sliding shaft with two sun gears, and two sets of planet gears mounted on a single planet carrier which is integral with the chain sprocket. A single annulus gear encloses both planet sets; the annulus drives the bicycle through a freewheel (see next page). In middle gear, the whole assembly turns at the same speed as the sprocket. Another illustration of an epicyclic gear system will be found on page 147.

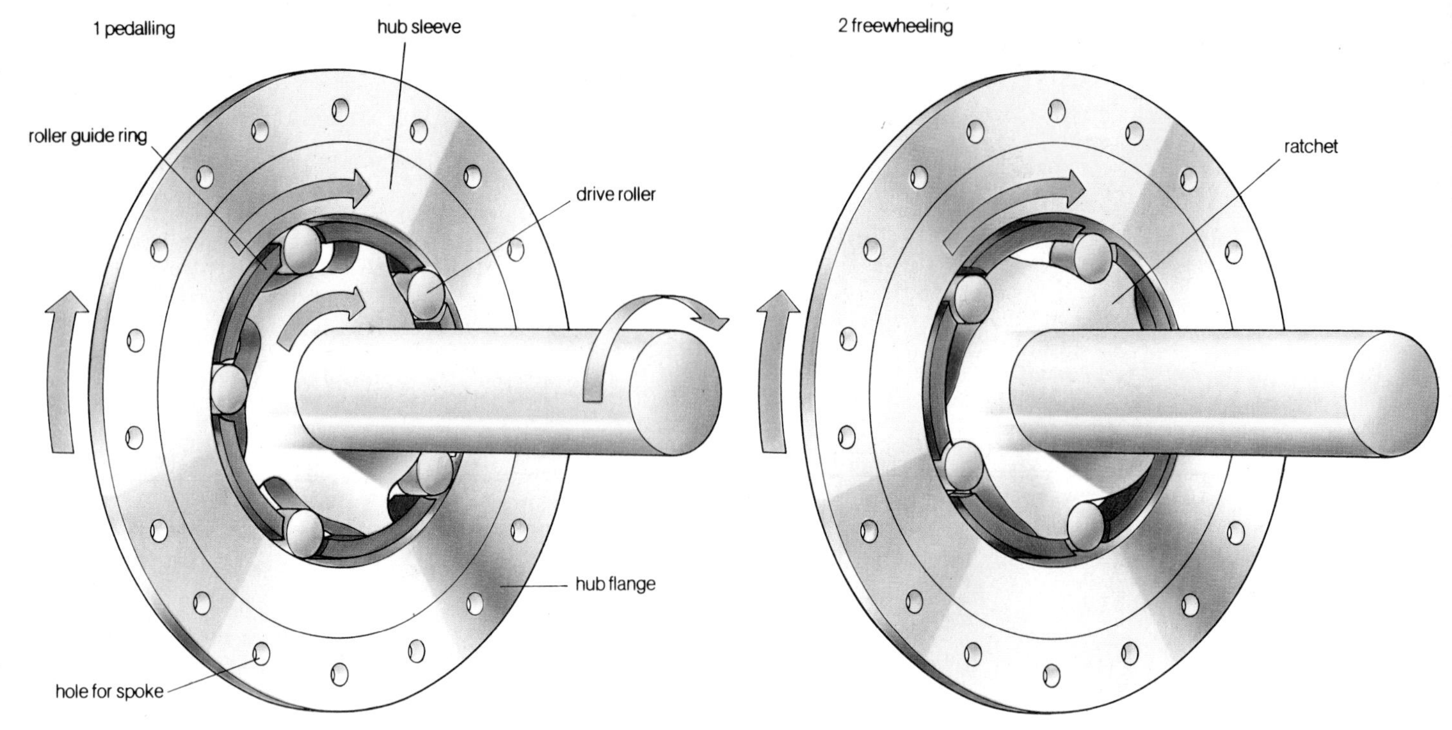

The bicycle freewheel uses rollers running up the slope of the ratchet-shaped inner race to lock the races together. Another application of the freewheel is illustrated on page 144.

successful being that of Sturmey and Archer in 1901. This was a three speed version: their four speed model did not appear until 1938, and a five speed model is now available. Many of these hub gears incorporated a back-pedal brake, and modern ones often have a dynamo built in to provide power for the lighting. The gears are selected by a trigger on the handlebars, a twistgrip control, or a lever mounted on the frame, connected to the change mechanism by a cable.

As with transmission systems, most of the advances in brake design took place after 1900. A wide range of devices was introduced, acting either on the hubs or the wheel rims. Some of the hub brakes were of the internally expanding type that is still used today, mainly on motor cycles, and were operated by cables or rods from the handlebars or by back pedalling. The coaster-hub back-pedal brake was also popular at that time, and some other forms of hub brake had a friction band which contracted around the outside of the hub.

The brakes which acted on the rim were either rod operated stirrup brakes or cable-operated caliper brakes. The caliper brakes, which have a pair of levers pivoted together at their centres to form a scissor-like mechanism, are the most popular form of brake used on modern cycles.

The first patent for pneumatic tyres for use on bicycles was issued in Britain to J B Dunlop in 1888. Except for those used on racing cycles, modern tyres have an inflatable inner tube and an outer cover of strong canvas and wire with a vulcanized rubber tread and sidewalls. Racing tyres are made in one piece, from cotton or silk bonded with rubber, and have a very thin tread.

Bicycle wheels with steel wire spokes have been in use for over a hundred years, and the rims and hubs are made from steel or aluminium alloy. Ball bearings, running on hardened steel cones mounted on the spindle, are used on most modern bicycle wheels.

The frames of mass-produced cycles are made from high quality steels, which often contain chromium and molybdenum or chromium and manganese. These alloys are strong and light with high resistance to corrosion, and the tubes are brazed together from the inside. The handlebars are made from chrome-plated steel, stainless steel, or light alloy.

The Moulton bicycle, introduced in 1962, was the first of the now well-established small-wheeled type of bicycle. These machines have wheels of between 14 and 20 inches (35 and 50 cm) diameter, compared with the 24 to 28 inches (60 to 70 cm) of the conventional models. The F-shaped frame has no high crossbar, which makes it suitable for both men and women, and its design enables the height of the saddle and handlebars to be adjusted over a wide range. Some incorporate rubber or coil-spring suspension: the Moulton, for example, uses a rubber-damped coil spring at the front, and a bonded rubber shock absorber on the rear fork. If geared transmission is used on this type of machine it is of the enclosed hub design, as the derailleur is unsuitable because of the small wheels.

The bicycle has an important place in the technology of developing countries, both as a means of transport and as a way of bringing basic mechanical and technical knowledge to people. In the industrial countries, where pollution is a major problem and the urban areas are being choked by the huge increase in motor traffic, the bicycle is gaining in popularity as a cheap, healthy and efficient form of personal transport. In many city centres it is often the quickest way to travel, and it does not make any great demands on the world's oil supplies. In 1970 more than 250 million bicycles were in use throughout the world.

The atmospheric engine

The atmospheric engine was used to pump water at the beginning of the 18th century. It used atmospheric pressure to do the work instead of direct steam power, because at that time it was not possible to build strong boilers or steam-tight machinery. The engine worked by admitting steam to a container and cooling it with cold water so that the steam condensed back to water. This caused its volume to be reduced, forming a partial vacuum; outside atmospheric pressure moved the piston.

The steam engine

The steam engine is a machine for converting the heat energy in steam into mechanical energy by means of a piston moving in a cylinder. More than any other machine it was the steam engine which made the Industrial Revolution possible, being a source of power unrestricted by site limitations such as water supply, wind, or space for draught animals. The steam engine dominated industry and transport for 150 years, and it is still found useful in certain applications because it can use any fuel or source of heat of suitable temperature, and develops its full torque at any speed. The basic steam engine was developed from the atmospheric engine in the last quarter of the 18th century by James Watt and Richard Trevithick, but the basic design soon evolved into diverse forms for different applications in the hands of many others. The steam engine is an external combustion engine, in other words the fuel is burned outside the working cylinder, and this means that atmospheric pollution by the combustion gases can be kept to a very low level.

Most rotative steam engines are double-acting, that is, there are two power strokes for each revolution of the flywheel. Steam is admitted to one side of the piston and simultaneously steam is exhausted from the other side by means of valves as the piston moves to and fro. The sequence of events at one side of the piston during one revolution of the flywheel is as follows: first, steam is admitted just before the end of the previous stroke to cushion the piston; next, the inlet port is closed (cut-off) and further movement of the piston is effected by expansion of the steam; then the exhaust valve opens just before the end of the power stroke (release); and finally the exhaust valve closes and the remaining steam is compressed into what is called the clearance volume. Similar events take place on the other side of the piston, but 180° out of phase. The sequence of events can be conveniently shown as a graph of pressure against volume for the different positions of the piston. Such graphs can be plotted mechanically while the engine is working by an indicator connected to the cylinder, and changes in their shape show the need for adjustments or repairs.

In most engines the whole sequence of events is effected by a single 'D' slide valve, invented in 1802 by Matthew Murray of Leeds. This is driven by the valve gear which, in marine and reversible stationary engines, is usually a variant of the Stephenson link motion used in early locomotives. A piston valve, also a locomotive feature, in which a piston alternately opens and closes the inlet and exhaust ports, is used where high steam pressures would cause excessive friction in a 'D' valve. Other types of valve, usually separate ones for each event, are worked by trip motions to give rapid operation when very early cut-off is needed in large stationary engines.

The to and fro motion of the piston can only be used directly for pumping or hammering, and must be converted to rotary motion for other uses. This is done by some form of crank handle driven by a connecting rod. Early rotative engines had the piston rod linked to a beam, and the connecting rod pivoted to the other end of the beam. This made a very bulky engine and was soon replaced for most purposes by the guided crosshead invented by Trevithick. A still more compact arrangement was the oscillating engine in which the cylinder rocked about a trunnion so that the piston rod could be directly attached to the flywheel crank. Steam was admitted and exhausted through the trunnion. This form of engine was often used in early screw steamers, for example Brunel's ship the *Great Britain*.

There are two positions in each revolution when the piston rod and the connecting rod are in line with the arm

A cutaway model of George Stephenson's Rocket, *the first locomotive to haul both freight and passengers. The piston, cylinder, steam valve and connecting rod can be seen.*

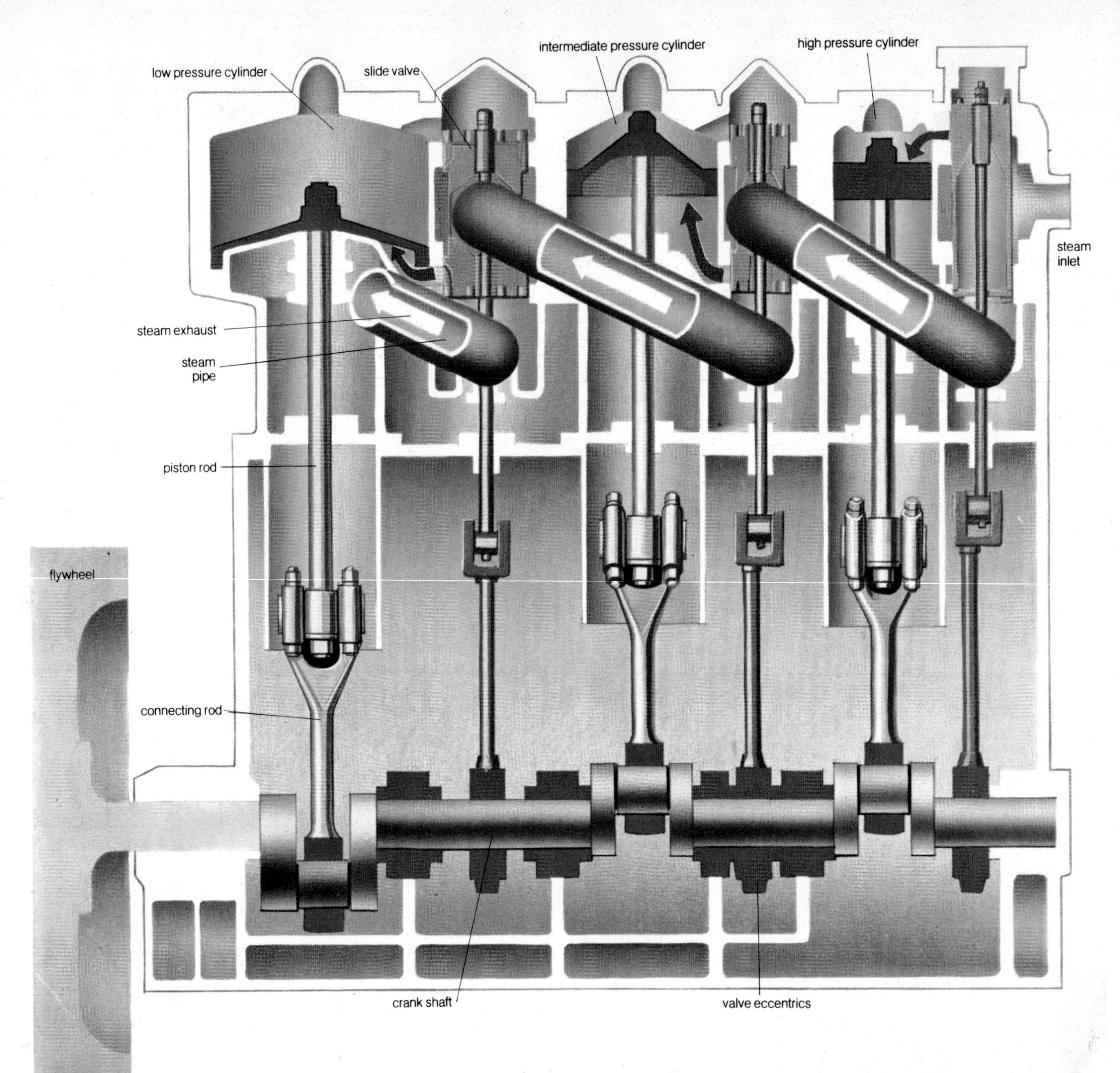

A sectional view showing the operation of a triple-expansion steam engine. Three slide valves conduct high-pressure steam from the inlet successively through high, intermediate and low-pressure cylinders to the exhaust. The valves are operated by eccentrics mounted on the crankshaft.

of the crank and have no turning effect. The inertia of the flywheel must be used to carry a single cylinder engine past these dead centres. Most engines with more than one cylinder avoid this problem by having suitable angles between the cranks.

Boilers and furnaces Early steam engines had boilers and furnaces similar to those of atmospheric engines, simply an enlarged version of the domestic *copper* seen in old houses. The limited strength of such boilers restricted the steam pressure to about 1 psi (0.07 bar) above atmospheric pressure, and this meant that a steam engine with a useful power output had to have very large cylinders. It was also difficult to provide a large enough heating surface to get the heat efficiently into the boiling water, so a great deal of the energy generated by burning the fuel went up the chimney. The first successful attempts to avoid these limitations were made by Richard Trevithick and his American correspondent, Oliver Evans of Philadelphia. They made cylindrical boilers of inherently stronger construction with a single internal furnace and flue. These were often set in brickwork containing return flues which heated the outside of the boiler shell, and they became known on both sides of the Atlantic as Cornish boilers. The Lancashire boiler, seen in many surviving steam plants, is very similar but has two furnaces and two sets of return flues.

Modern boilers have a greater heating surface, are less bulky and enable steam to be raised more quickly than early designs. They are either of the fire-tube or water-tube type. In the former, the hot gases from the furnace pass through steel tubes surrounded by water; typical of this sort of boiler is the boiler used in railway locomotives. In water-tube boilers the water flows through tubes situated above the furnace combustion chamber. Although more expensive to build, this sort of boiler can be operated at higher pressures than fire-tube boilers and is much safer. Water-tube boilers are commonly used nowadays to generate steam for marine and power station turbines.

It is not always necessary to heat a steam boiler with burning fuel, and many are heated by the waste gases from metallurgical furnaces or even the exhaust gases of diesel engines. Any combustible substance can be used as fuel, often waste materials such as sawdust, sugar cane refuse or oil refinery residues. Combustion takes place at atmospheric pressure to reduce pollution from nitrogen oxides or carbon monoxide.

Compound and multiple expansion engines Higher steam pressures allow a higher ratio of expansion from cut-off to release, but in a single cylinder engine this often demands an impracticably early cut-off and gives rise to an excessive variation in the cylinder-wall temperature. Better results were obtained by arranging for a small high-pressure cylinder to exhaust into a larger low-pressure one. Such a combination is called a compound engine and came into general use when mild steel allowed higher steam pressures in the 1860s.

Several forms of compound engine were widely used, particularly the tandem compound with two cylinders on a common piston rod and crank, the cross compound with its cylinders on either side of a rope drive pulley (often used in textile mills), and the marine type with two vertical cylinders side by side. The first two types were usually horizontal and had drop valves (reciprocating valves which move into and out of contact with annular valve seats) or Corliss valves (rotary valves which oscillate to and fro through a small angle about the axis of rotation).

Triple-expansion engines with high-pressure, intermediate pressure and low-pressure cylinders were usually designed for marine use with boiler pressures exceeding 150 psi (10.3 bar). Such engines were used in most steamers built between 1890 and 1950, and the cylinders were usually arranged vertically. Some high-power triple-expansion engines had two low-pressure cylinders. Each engine of the Atlantic passenger liner *Olympic* (1911 to 1936) delivered 15,000 hp (11,200 kW) at 75 rpm from steam at 225 psi (15.5 bar); the cylinder diameters were 54 in (137 cm), 84 in (213 cm) and 97 in (246 cm) for the high-pressure, intermediate-pressure and two low-pressure cylinders respectively. Diagonal triple-expansion engines were used in paddle steamers; the piston rods sloped upwards to the paddle shaft from cylinders set low in the ship.

Quadruple-expansion engines, with two intermediate pressure cylinders of different sizes, were also used at sea, but these were not very common because they had few advantages over a four-cylinder triple-expansion engine for the same boiler pressure.

Uniflow engines These engines were designed to allow full expansion of high-pressure steam in one cylinder by avoiding the losses which had led to the adoption of compound and multiple-expansion engines. The expanded steam was released through a ring of ports uncovered by the piston which afterwards compressed the remaining steam to boiler pressure. Steam was admitted by quick-acting drop valves loaded with powerful springs to give a cut-off of about 10% of the stroke at full load. This was further reduced for light loads.

The first successful uniflow engines were developed by Johann Stumpf of Charlottenburg, Germany, in 1908. The construction of large engines was simplified by avoiding compound working, but greater precision in manufacture was essential and serious problems in designing the valve gear had to be overcome. Many very large horizontal stationary engines were built on the uniflow principle in Germany, Switzerland, Britain and the USA during the subsequent 45 years, and these included the most powerful steam engines ever built. Some, in the USA, developed 25,000 hp (18,650 kW) from four uniflow cylinders working in parallel, and a German five-cylinder engine of the 1930s developed 30,000 hp (22,400 kW). These large engines were designed for

Fairground traction engines like this one were used both for hauling caravans from place to place and for generating electricity to power the sideshows. As can be seen in the picture, the generator (on the front of the engine) was driven by a belt from the steam engine's flywheel.

driving rolling mills. The uniflow engine was particularly successful in the USA where one manufacturer built over 2000 between 1913 and 1958.

Marine uniflow engines appeared in America in about 1930. They had from two to five vertical cylinders arranged in parallel. Such engines were a feature of American steamers built between 1930 and 1955, and they powered many of the liberty ships and small aircraft carriers of World War II.

High-speed engines Most types of steam engine had a simple drip-feed lubrication system which was unsuitable for speeds over about 150 rpm. These speeds were too low for direct electricity generation, and early power stations were equipped with speed-increasing rope drives at great cost in space and complexity. The high-speed engine overcame this difficulty by having the crossheads, connecting rods and valve gear lubricated by circulating oil under pressure inside a closed crankcase. Working was either compound or triple expansion, with piston valves and a fixed cut-off in each cylinder, the engine being governed by throttling the steam supply. The cylinders were separated from the crankcase by distance pieces longer than the stroke so that oil on the piston rods would not be drawn into the cylinder to contaminate the exhaust. Such engines ran at about 500 rpm and could be coupled directly to a dynamo [generator].

The larger condensing engines were superseded by steam turbines, but a back-pressure type, in which the steam is exhausted at above atmospheric pressure, is still made for electrical outputs up to about 500 kW. Similar engines are also made for driving the electric lighting sets in motor vessels where the exhaust from the main engines passes through waste heat boilers.

Steam cars It is generally accepted that the first automobile ever built was a three-wheeled steam-powered vehicle constructed by N J Cugnot in France in 1769. It travelled for 20 minutes at a speed of about 2 mph (3.2 kph) carrying four passengers. In the next 100 years numerous steam carriages and traction engines were built, particularly in England, but it was not until the end of the 19th century that the first steam-powered vehicles recognizable as forerunners of the modern car were produced. The most successful steam car was that built by Stanley in the USA; more than 60,000 were made between 1897 and 1927, and some remained in ordinary service until 1945. A Stanley steam car broke the world speed record in 1906 with a speed of 127 mph (204 kph). The Stanley car had a two cylinder non-condensing engine with direct drive to the rear axle and a special type of fire-tube boiler under the bonnet [hood]. Its range was very limited because it needed a full tank of pure water every 20 miles (33 km) at full speed. A condensing model was also made between 1915 and 1927.

Condensing steam cars had a much greater range but were troubled by cylinder oil contaminating the feedwater which upset the boiler and controls. The most successful of these was the car built in the USA by White between 1900 and 1911. This had a compound engine under the bonnet behind an air-cooled condenser. A spiral tube boiler was under the driving seat. The usual fuel in steam cars was paraffin [kerosene], and elaborate automatic controls were needed to ensure that the supply of steam matched the driving conditions without burning out the boiler or starting a fire from excess fuel. Steam cars went out of favour when electric starting for petrol [gasoline] engines was introduced.

In spite of their drawbacks, steam cars do have two important advantages over cars powered by internal combustion engines. First, the torque developed by a steam engine is independent of speed so a complex gearbox is not required, and second, combustion of the fuel at atmospheric pressure outside the cylinders eliminates the need for additives and substantially reduces pollution. This second consideration has led to a revival of interest in the steam car; modern research is concentrated on developing control mechanisms to replace earlier ones (now prohibitively expensive) and boilers suitable for mass production, but serious problems remain.

This Stanley steam car was built in 1899. Between 1897 and 1927, more than 60,000 Stanley 'steamers' were built; some of them were still in use as late as 1945. A Stanley broke the world's speed record in 1906 with a speed of 127mph (204kph).

The Stirling engine

Modern concern with the need to reduce chemical and noise pollution of the atmosphere has greatly revived interest in the Stirling engine. This engine was first invented in 1816 by the Rev Robert Stirling, when a junior Presbyterian minister at Kilmarnock, Scotland, but its early development was hampered by the lack of materials with sufficient strength and corrosion resistance at high temperatures and the lack of suitable materials and techniques for gas sealing. For this reason, it was unable to compete with the steam engine or internal combustion engine though it is capable of higher thermal efficiency, can be much quieter and can be designed to produce far less atmospheric pollution.

The engine cycle In contrast with the internal combustion engine, in which the fuel is burned inside the main engine cylinders, the Stirling engine receives its heat supply through the cylinder walls or a heat exchanger in contact with the heat source. The 'working gas', which usually remains permanently inside the engine, is made to undergo a continuous series of cycles of heating and cooling, which causes the cyclic expansion and contraction required for performing mechanical work.

In one popular early Stirling engine arrangement a displacer piston slides to and fro along a gas-filled cylinder which is heated at one end and cooled at the other end. The gas is therefore transferred alternately to the hot and cold spaces at the ends of the cylinder and the resultant cyclic temperature changes cause pressure changes, which are used to drive an output power piston through a connecting pipe. Many ingenious mechanisms have been devised for maintaining the correct relationship between the movements of the two pistons. In most cases, the displacer piston movements are arranged to occur in advance of those of the output power piston by about one-quarter of an engine cycle.

Stirling's original engine produced about 2hp (1.5 kW), and in common with most modern Stirling engines, the displacer and output power pistons operated in the same cylinder. This arrangement reduced the extraneous volume in the engine which is not swept by the pistons, and the consequent enhanced compression results in higher efficiency and greater power output for a given engine size. It appears that Robert Stirling had greater appreciation of this fact than most other engine designers up to the end of the 19th century.

In early engines, the working gas was air at about atmospheric pressure, but in most modern engines the pressure is raised (in some cases to several hundred times that of the atmosphere), and air is replaced by helium or hydrogen, since these gases are better conductors of heat and therefore simplify the problems of rapidly heating and cooling the gas.

Under given conditions, the power which can be produced with a given volume of working gas is proportional to the difference between the upper and lower extremes of temperature which occur in the engine cycle. It is therefore advantageous to make the temperature of the hot end of the engine as high as the properties of the construction materials will allow, and most engines now are in the 450 to 750°C (842 to 1382°F) range.

Heat regeneration An interesting feature of Robert Stirling's invention is that it embodied the earliest application of the heat regenerator principle, as a means of preventing wastage of heat by the gas as it flowed to and fro between the hot and cold spaces of the engine. The gas was made to flow through a porous material such as steel wool or a system of passages which were sufficiently narrow to ensure good thermal contact between them and the gas, so that it gave up its heat as it flowed from the hot regions to the cold and took it back again when flowing in the reverse direction.

In this type of Stirling engine, a 'Thermo-Mechanical Generator', a vibrating metal diaphragm acts as the power output piston and is fixed to the armature of an alternating-current generator.

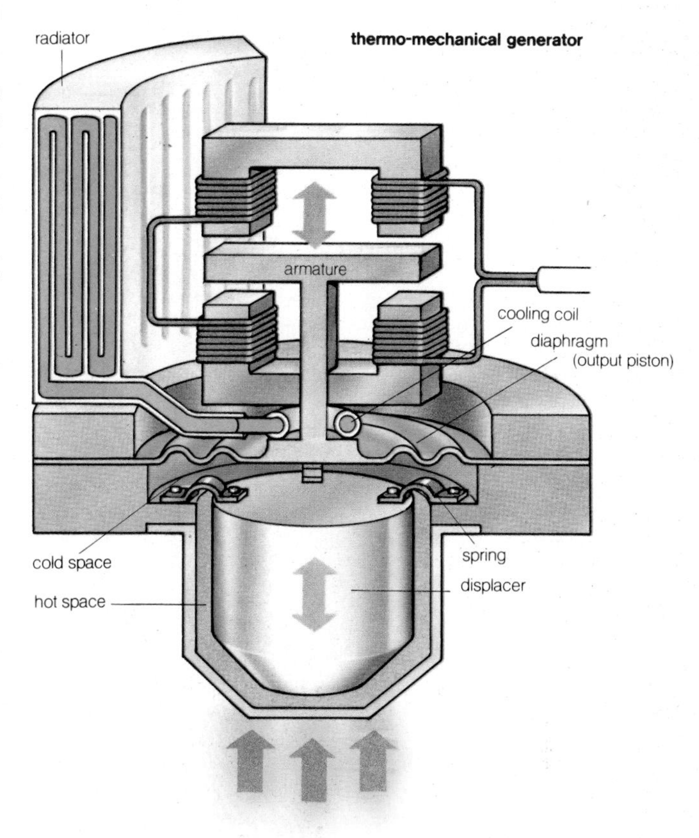

Stirling's invention of the regenerator was quite a remarkable feat of intuition. A true explanation of its function in such engines was first given as late as 1854 by W J M Rankine, and most other designers of the 19th century apart from Stirling, failing to understand its value, omitted it with consequent detriment to the performance of their engines.

Stirling engines were made in many forms during the 19th century from very low powers up to about 35 hp (26 kW), and were used mainly for driving pumps and, to a lesser extent, crushers, grinders, churns and so on. They were more efficient than the steam engines of that period but burn-out of cylinder hot-ends was a major problem.

The engine has, so far, been little used in the 20th century, but research during the last 35 years in a few organizations, such as the Philips Company in Holland, has resulted in engines with higher efficiency and similar output power for a given weight and size to petrol [gasoline] and diesel engines. The main barriers which prevent the adoption of the Stirling principle for high power engines are the well established position of petrol [gasoline] and diesel engines and the relatively high cost of the heat exchangers required for passing the large amounts of heat into and out of the engine. Perhaps the best future for the Stirling engine is at the very low power end of the spectrum.

The internal combustion engine

The development of the internal combustion engine was made possible by the earlier development of the steam engine. Both types of engines burn fuel, releasing energy from it in the form of heat which is then used to do useful work.

The steam engine, however, is an *external* combustion engine, because the fuel is burned in a separate part of the engine from the cylinder containing the piston. Anything combustible can be used as a fuel in the steam engine, such as wood, coal or petroleum products, and the liberated energy is used to heat a fluid, usually water. The hot water vapour expands in a confined place (the cylinder) to push the piston.

In the *internal* combustion engine, the burning of the fuel takes place in the combustion chamber (the top of the

Opposite page, top: how a four-stroke engine works. Left to right: intake, compression, power, exhaust. For clarity, the drawings feature twin overhead camshafts. A camshaft turns at half engine speed, opening each valve once in the cycle.
Below: cutaway view of a typical four-cylinder small-car engine, with conventional pushrod-operated overhead valves.

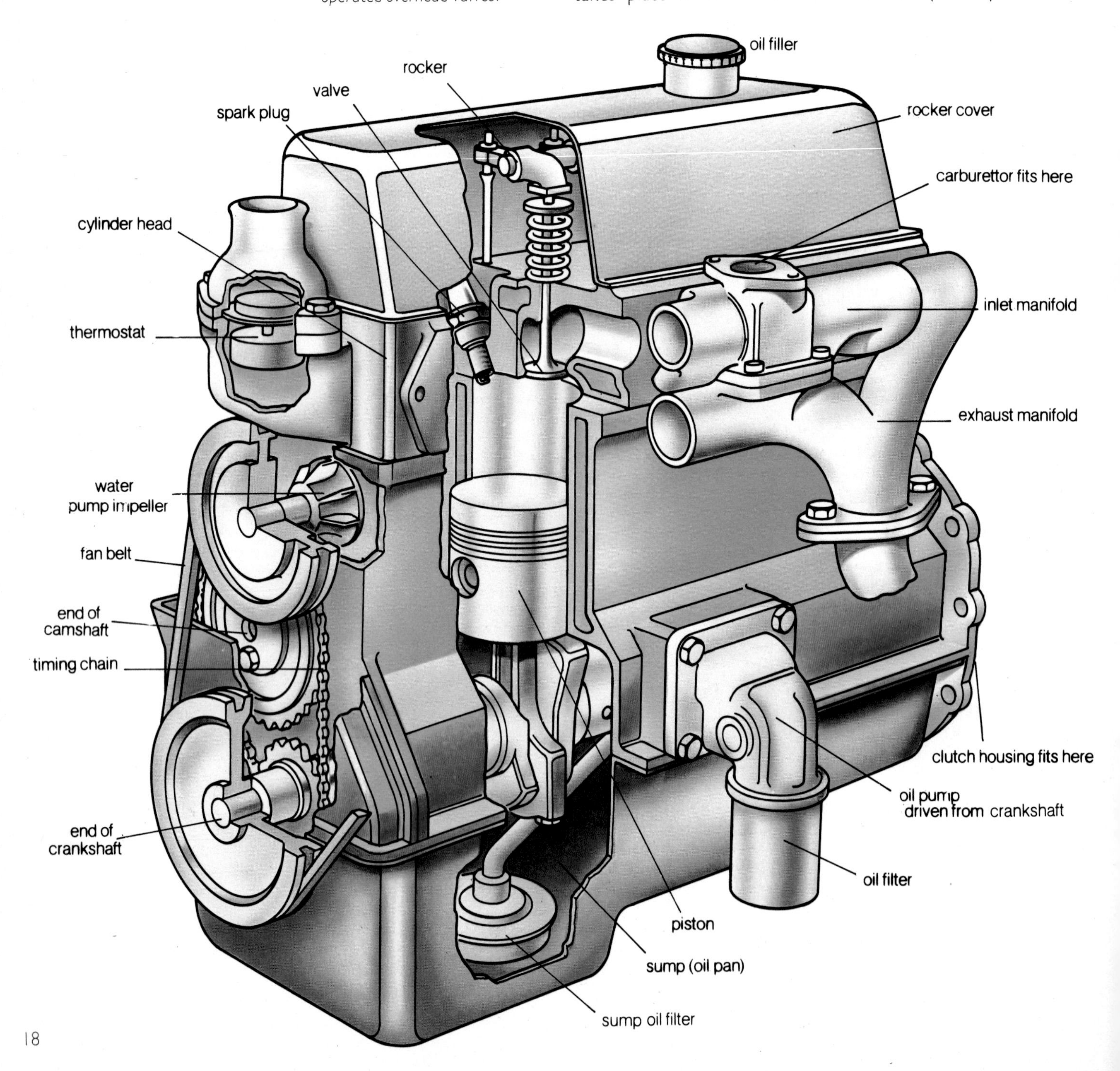

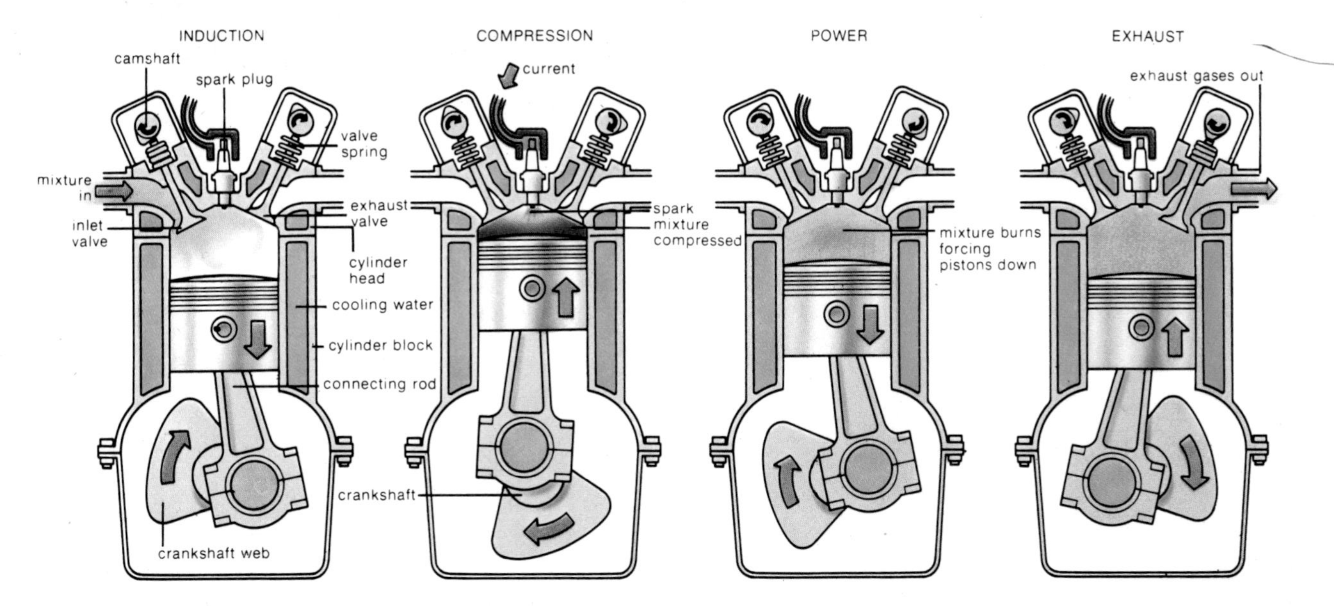

cylinder). The combustion is very sudden, amounting to an explosion which pushes the piston.

During the eighteenth and nineteenth centuries, as the steam engine was made more efficient, advancements were made in engineering and metallurgy which made possible the first successful internal combustion engines. The operation of steam engines was not fully understood at first; the French physicist Sadi Carnot published in 1824 his theories which led to the science of heat exchange (thermodynamics). Fifty years before that James Watt had already begun to develop packings and piston rings to prevent the escape of energy past the piston in his steam engines. By 1800, the British engineer Henry Maudslay was making improvements to the lathe which led to machinery capable of producing precision-made parts for engines. In the 1850s more volatile fuels were being refined from petroleum.

In 1860 J J E Lenoir, a French engineer, built a successful engine which was essentially a modified steam engine, using illuminating gas as the fuel. In 1867 the firm of Otto and Langen began producing an engine which transmitted the power of a freely moving piston to a shaft and a heavy flywheel by means of a rack-and-gear device, using a freewheeling clutch in the gear, so that it turned freely in one direction and transmitted power in the other.

Meanwhile, in 1862, Alphonse Beau de Rochas had published in Paris his theory of a four-stroke engine of the type used in the modern car. While de Rochas never built any engines, his theory included compression of the fuel mixture in order to raise its temperature, and he also realised that a four-stroke design would be more efficient at scavenging (intake of fuel mixture and exhaust of burned gases) than the two-stroke.

A two-stroke engine provides for intake of fuel, combustion and exhaust of burned gases with each back-and-forth motion of the piston (that is, with each revolution of the crankshaft). A four-stroke engine requires four strokes, that is, two complete back-and-forth movements of the piston (two revolutions of the crankshaft). The two-stroke engine delivers twice as many power impulses as the four-stroke engine to the crankshaft, but the four-stroke is much more efficient at scavenging, if all other things are equal. The two-stroke design is also wasteful because unburned fuel is exhausted with the burned gases.

In 1876, Otto and Langen began building the Otto 'silent' engine (it was a good deal quieter than their earlier model). It was the first modern internal combustion engine, a four-stroke design which compressed the fuel mixture before combustion. The design of the Otto engine was an important influence on Henry Ford, the American inventor.

Four-stroke design The four-stroke cycle operates as follows: on the first downstroke of the piston, the intake valve opens and the fuel mixture is pulled into the combustion chamber. On the following upstroke, all valves are closed and the fuel mixture is compressed. At the beginning of the second downstroke, combustion takes place; the fuel mixture is ignited by a spark from the spark plug and the expanding gases drive the piston downwards. On the second upstroke, the exhaust valve opens and the burned gases are expelled. Thus the four parts of the cycle are intake, compression, combustion and exhaust.

The fuel mixture is a mixture of fuel and air in the form of a vapour which is prepared by the carburettor. Internal combustion engines can be designed to run on anything from paraffin [kerosene] to high-test aviation fuel. The carburettor must be adjusted properly; if the mixture is too lean (does not contain enough fuel), the engine will not run properly; if it is too rich, the result will be carbon deposits fouling the spark plugs, the valves and the inside of the combustion chamber, wasting fuel and affecting the performance of the engine.

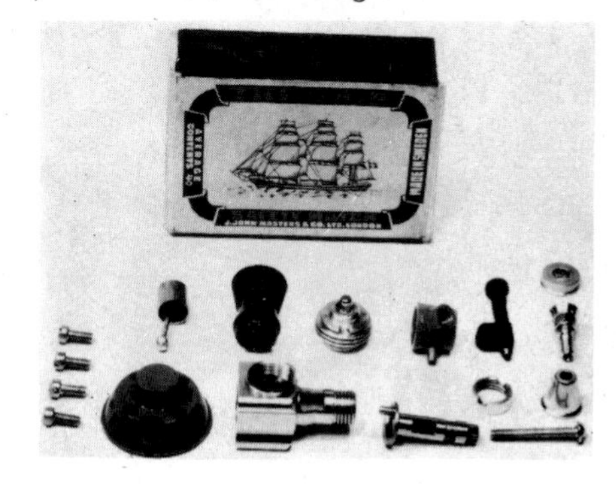

The world's smallest production internal combustion engine, the Cox .010cc motor, used in model aircraft.

Two-stroke design The two-stroke engine must accomplish intake, combustion and exhaust in one back-and-forth movement of the piston. Since scavenging is incomplete and inefficient, the proper mixture is difficult to obtain. Small two-stroke engines such as are still used in some motorcycles, lawn mowers and small cars must have oil added to the petrol, and constitute an air pollution problem; the blue smoke from the exhaust pipe of these engines is one of their familiar characteristics.

One way of improving the scavenging of the two-stroke engine is to build opposing pistons, which reciprocate in opposite directions and share a common combustion chamber. This design was chosen by Henry Ford for his first car, which was built in 1896. A big disadvantage is that each piston must drive a separate crankshaft, and the motion of the two crankshafts must then be combined through a system of gearing.

Other design aspects The block, the head and the crankshaft are all castings that require extensive machining before the engine can be assembled (some larger crankshafts are forgings, for extra strength). The cylinders must be bored and finished precisely in the block. The top of the block and underside of the head are planed or milled to fit smoothly together, and the tops of the compression chambers are also machined on the underside of the head—except in 'bowl in piston' designs, where the head is flat and the tops of the pistons are recessed. Both the block and the head must have numerous surfaces machined and holes drilled and tapped where various components will be mounted. Where the head is bolted to the block, a head gasket is included in order to prevent escape of compression in the assembled engine.

The underside of the block is open; at the bottom of each cylinder wall, bearing surfaces are machined to accept the main bearings of the crankshaft. Bearing caps are screwed down to hold the crankshaft in place. The pistons slide up and down in the cylinders and are connected to the crankshaft

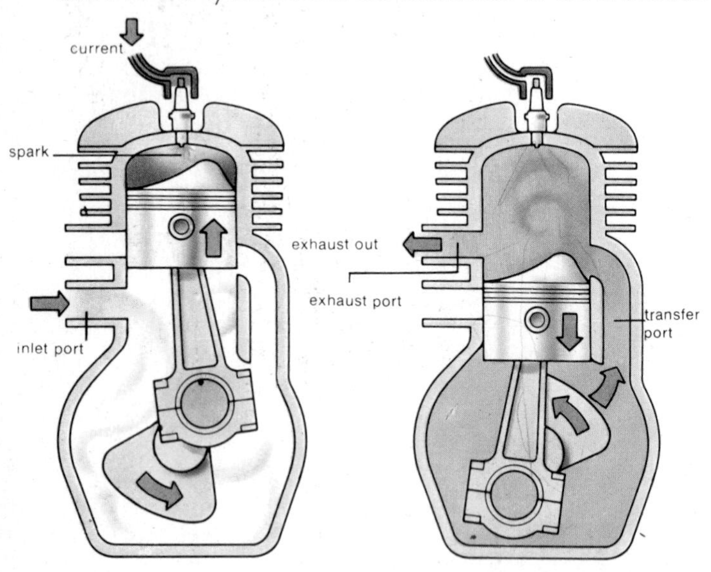

A two-stroke cycle. When the piston rises to compress the mixture above it, low pressure below it draws the mixture from the crankcase through the inlet port (there are no valves). On the piston's downstroke, the crankcase pressure goes up, driving the mixture up through the transfer port.

by means of connecting rods which pivot in the pistons and turn on the throws of the crankshaft. A sump [oil pan] made of sheet metal or cast light alloy is screwed to the bottom of the block, covering the crankshaft; a gasket is included to prevent leakage of oil, and there is also an oil seal at each end of the crankshaft where it protrudes from the block.

The crankshaft itself is a mechanical adaptation of the hand crank, used for centuries to operate simple machines such as early lathes. For each cylinder in the engine there is a separate throw (offset section forming a crank) which revolves around the axis of the crankshaft, pushed by the operation of the piston when the engine is running. Opposite each throw on the crankshaft is a web (a mass of metal) to balance it. The throws in a multi-cylinder engine are arranged equidistant around the circle described by their revolution, and the firing sequence of the combustion chambers, which depends on the crank position, is timed in such a way as to balance the engine and provide for smooth running. Internal combustion engines have been built with as many as sixteen cylinders or more, in several configurations: opposed, radial, V-formation, and in-line (all in a row). The most common type of engine today is the in-line four or six cylinder engine used in cars; V-8 engines are also common, especially in American cars.

On the front of the crankshaft, where it protrudes from the engine, is mounted a pulley wheel from which are operated, by means of a belt, the dynamo [generator] or alternator, and the water pump, if the engine is water-cooled. The crankshaft also drives the oil pump (for lubrication of the engine) by means of a skew gear.

Also mounted on the crankshaft is the timing gear. This is a pair of gearwheels, or a sprocket linked by a chain to a smaller sprocket, which turns the camshaft, which is generally located in the block, at half crankshaft speed. The camshaft in turn operates the valves, and also drives the distributor to spark the mixture at the correct moment. If the valves are located in the block, the engine is a side-valve, valve-in-head, flat-head or L-head design. In this case the valve stems ride directly on the cams. If the engine is an overhead-valve design, the valves are operated from the camshaft by means of an assembly of pushrods and rocker arms, and access to the valves for repairs is more easily obtained by removing a sheet metal cover on top of the head, instead of having to remove the head itself. In yet another variation, the overhead cam, the camshaft is also located on top.

In all cases, the valves are operated against spring pressure; there are at least two valves for each cylinder (intake and exhaust); and the adjustment of the timing gear is vital to the performance of the engine. Some high performance engines have four valves to a cylinder; some aircraft engines have sleeve valves, in which a tubular sleeve with holes in it covers and uncovers the ports.

Horsepower was originally a calculation devised by Watt to enable his customers to determine what size steam engine they needed. For an internal combustion engine, horsepower is determined by using a dynamometer to measure the amount of torque needed to restrain the rotation of the crankshaft. A horsepower rating should always be accompanied by the number of revolutions per minute of the crankshaft at which the measurement was made.

The configuration of the internal combustion engine is determined by the number of cylinders, the length of the piston stroke, the compression ratio (the ratio of the size of combustion chamber to the volume displaced by the piston) and many other factors, all of which are design decisions affecting the theoretical efficiency of the engine, and which are made on the basis of the intended use of the engine. Total weight for internal combustion engines ranges from a few pounds for lawn mower engines to more than fifteen tons for a V-16 design for diesel-electric locomotives.

The carburettor The carburettor is an important part of the internal combustion engine: it is the device used to mix the air and fuel vapour. This mixture is then fed into the engine to be burned, and provide the power which drives the engine in a variety of operating conditions from cold winter starting to fast acceleration.

The first carburettors appeared at the end of the 19th century and were called surface carburettors. They worked very simply by drawing air over the surface of the fuel, and so mixing its vapour with the air to form a combustible mixture which was fed into the engine. The wick carburettor, developed next, was similar but instead of the air being drawn over the fuel, it was drawn over wicks which had one end immersed in the fuel, which soaked the wick and vaporized into the air. To assist evaporation, hot air from the engine was used. Various versions of these designs were made until the development of the two basic types of carburettor used today: the fixed and variable jet carburettor.

Carburation The carburettor works by suction from the engine which helps to atomize (break up into tiny droplets) and vaporize the fuel. The amount of fuel drawn into the airstream in the carburettor to obtain the required air-to-fuel ratio is controlled by a narrow passage called the choke [barrel] or venturi. As the air flows through this passage its speed increases and consequently the pressure drops, which causes fuel to be sucked into the airstream from a hole or jet at this point.

The fuel atomizes and is mixed with air, usually in the ratio of about 15 parts (by weight) of air to 1 part (by weight) of fuel. In cold starting, however, the mixture may need to be much richer, say 2 parts of air to 1 of fuel.

The amount of fuel-air mixture allowed into the engine is controlled by a butterfly valve or throttle, which is positioned after the venturi. The valve is a simple device which if it is opened (when the accelerator pedal is depressed) allows large amounts of the mixture through, and if it is closed cuts off the supply. The throttle therefore controls

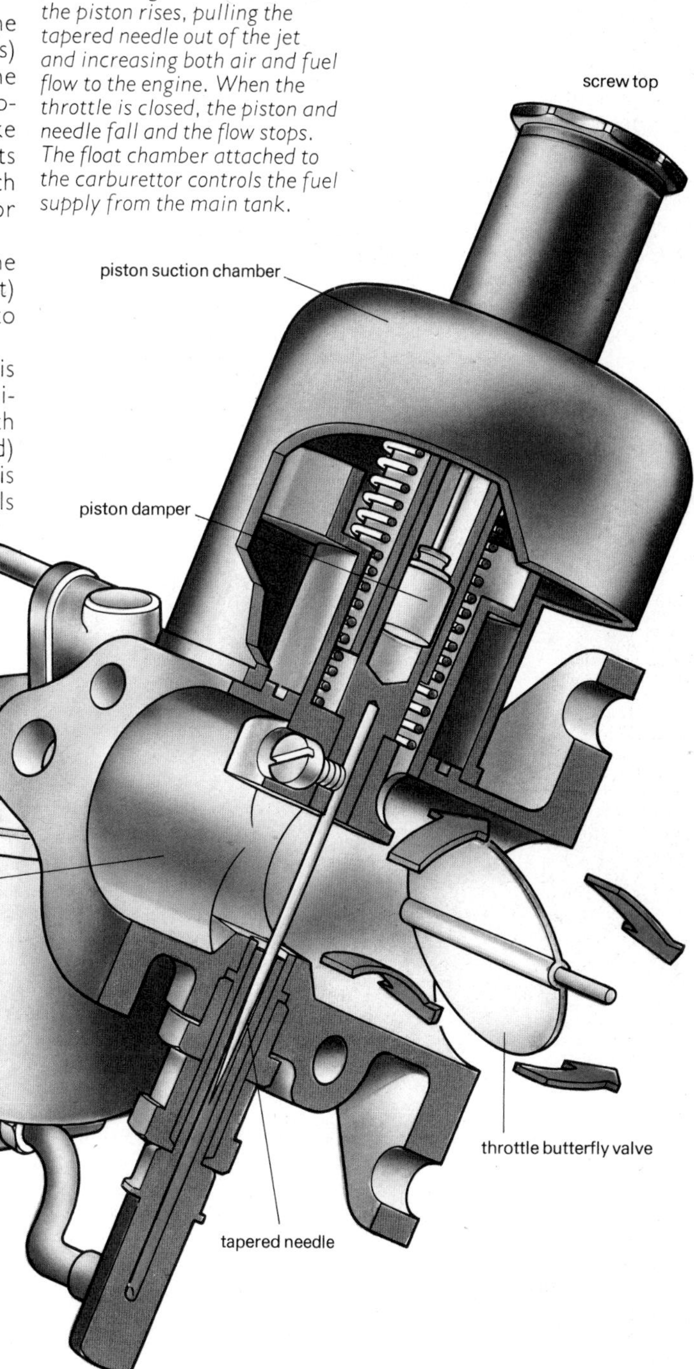

Cutaway diagram of a variable-jet SU carburettor. When the throttle is opened, the suction from the engine increases and the piston rises, pulling the tapered needle out of the jet and increasing both air and fuel flow to the engine. When the throttle is closed, the piston and needle fall and the flow stops. The float chamber attached to the carburettor controls the fuel supply from the main tank.

the speed at which the engine runs. The fuel-air mixture is sucked into one cylinder of the engine, where a valve closes to seal it in. The piston rises to compress the mixture before it is fired by the spark plug. The force of the burning mixture pushes the piston down, the valve opens, and the cycle starts again.

The fixed-jet carburettor This design has several jets of a fixed size, and an accelerator pump which is used to boost the fuel supply when necessary, as in sudden acceleration. Each jet has a function; the idling jet bypasses all of the other jets and allows a constant small flow of fuel to reach the air flow. It is used to keep the engine turning over at low speeds. The other jets are designed to mix some air with the fuel before it reaches the venturi to prevent the mixture from becoming too 'rich'. They include the main jet which operates once the butterfly valve is opened, supplying fuel for constant high speed running, and the compensating jet. This jet functions when the butterfly valve is opened to supply extra fuel to the engine and enables it to accelerate to a high speed at which the main jet takes over. There may be more jets found in this type of carburettor but basically they just assist the three jets which have been mentioned above.

The variable-jet carburettor This type works on the suction of the engine and also depends on the butterfly valve to control engine speed. Basically it consists of a main jet and a tapered needle which is mounted on a piston. In this type of carburettor the air is always drawn in from the side of the carburettor, as the piston and needle have to be vertical to operate smoothly. They are mounted through the air tube or the venturi. The tapered needle sits in the main jet and as the butterfly valve is opened the suction from the engine is increased; this suction acts on the top of the piston which is sucked upwards. As the piston rises it pulls the tapered needle out of the main jet and so more fuel is allowed to flow through and mix with the air. A damper on top of the piston slows its rise when a richer mixture is demanded for sudden acceleration.

Attached to both types of carburettor is a small reservoir tank of fuel called the float chamber because it has a float which rises with the fuel level until it reaches a certain level. At this stage it cuts off the fuel supply from the main tank. This means that it also stops too much fuel being passed into the carburettor and so acts as a control valve, preventing 'flooding' of the engine.

The two common types of supercharger are the Roots and the turbocharger. The rotors in the Roots turn very closely together but without touching, so they do not need lubrication. They compress the air and deliver it to two sets of ports, one for each bank of a V-8 engine. In the turbocharger, exhaust gases turn the shaft, driving air through a volute, similar to that of a centrifugal pump. The volute is shaped like a snail's shell and is graduated in size, so that the flow of the gases is changed from a high-velocity, low-pressure stream to a low-velocity, high-pressure stream.

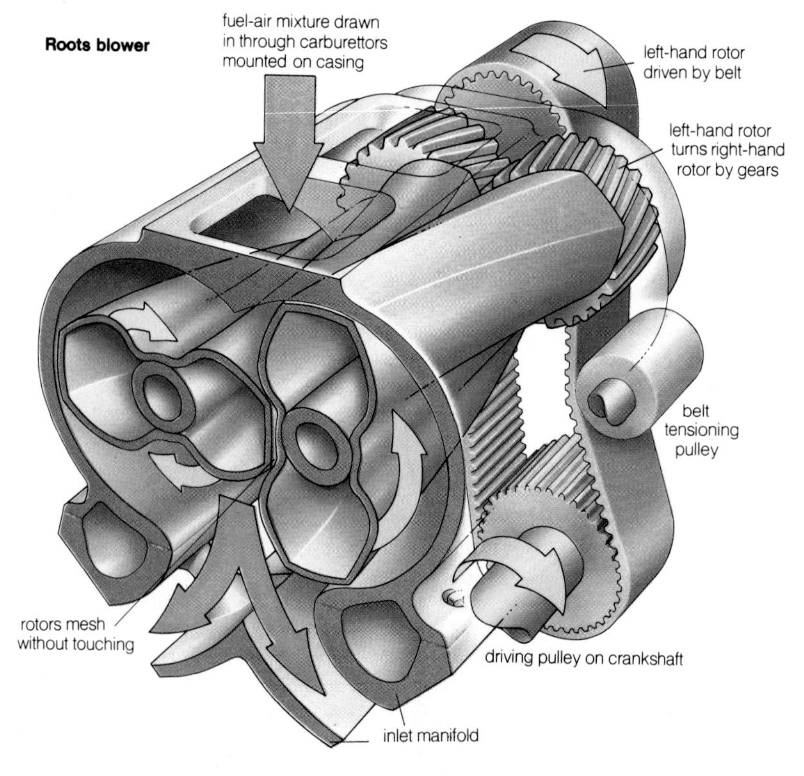

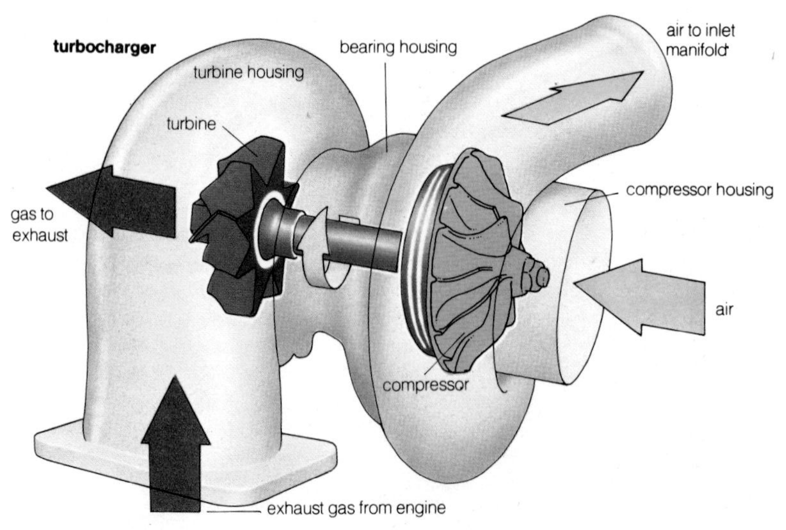

If better performance is wanted from an engine, as in racing cars, one of the first things to be modified is the carburettor. Engines may have two carburettors instead of one, or a carburettor with two or four chokes side by side. These are said to give a better distribution of the mixture to the engine. Short open pipes may be fitted to the air intakes of the carburettors; these are called ram pipes and are said to improve the air flow to the venturi to give a better mixture. Sometimes larger main jets are fitted to allow more fuel to flow. Other high performance engines have fuel injection systems, where fuel and air are precisely metered and mixed in the engine itself.

Fuel injection Fuel injection is a system of supplying the fuel to the combustion chamber under pressure by means of a pump, which eliminates the need for a carburettor. Development of fuel injection was accelerated during the 1930s because of the increasing requirements of aircraft engines. In á fuel injection system used in a car, an electric pump supplies fuel to an engine driven metering distributor, which in turn delivers fuel to an injector device at each combustion chamber. Electronic metering is also available.

Supercharging The only way to extract more power from an internal combustion engine is to burn more fuel. Since the amount of fuel burnt requires a given amount of oxygen, the only way to burn more fuel is to supply more air to the combustion chambers. This is what a supercharger does, acting as an air pump and compressing the air at the engine intake. Superchargers are mechanically driven by the engine, or they are turbochargers, which are operated by the excess energy in the exhaust gases. They are not common on cars, but are used on racing cars, engines for heavy duty transport, and aircraft engines.

The ignition system The ignition distributor is operated by the camshaft of the engine and distributes a spark to each combustion chamber at just the right time to ignite the fuel mixture. A high voltage spark is produced by the induction coil from the low voltage supplied by the car's battery. As the spindle of the distributor turns, a rotor arm at its top end points toward a contact for each cylinder in turn. At the same time, a raised edge on the cam on the spindle separates a pair of contacts in the low tension circuit, so producing the spark. There are as many raised edges on the spindle as there are cylinders. The contact 'points' become worn and have to be replaced periodically.

The spark is 'distributed' to the spark plug, a ceramic-clad

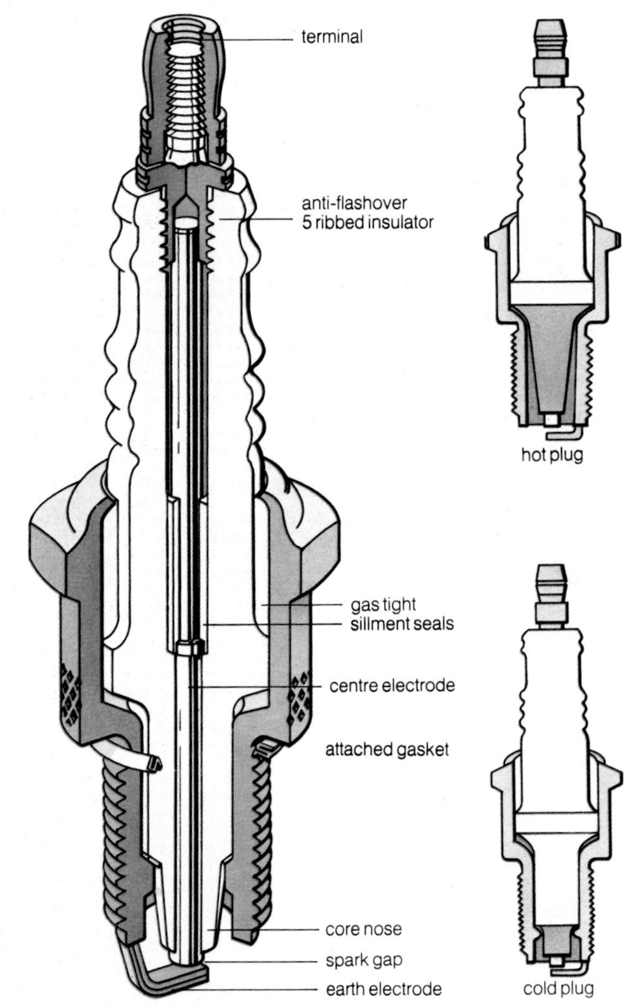

Above: the component parts of a modern spark plug. The high-voltage pulse of electricity passes down through the centre electrode and jumps across the gap to the earth [ground] electrode, igniting the fuel mixture in the combustion chamber. The smaller drawings show the difference between two types of plugs.

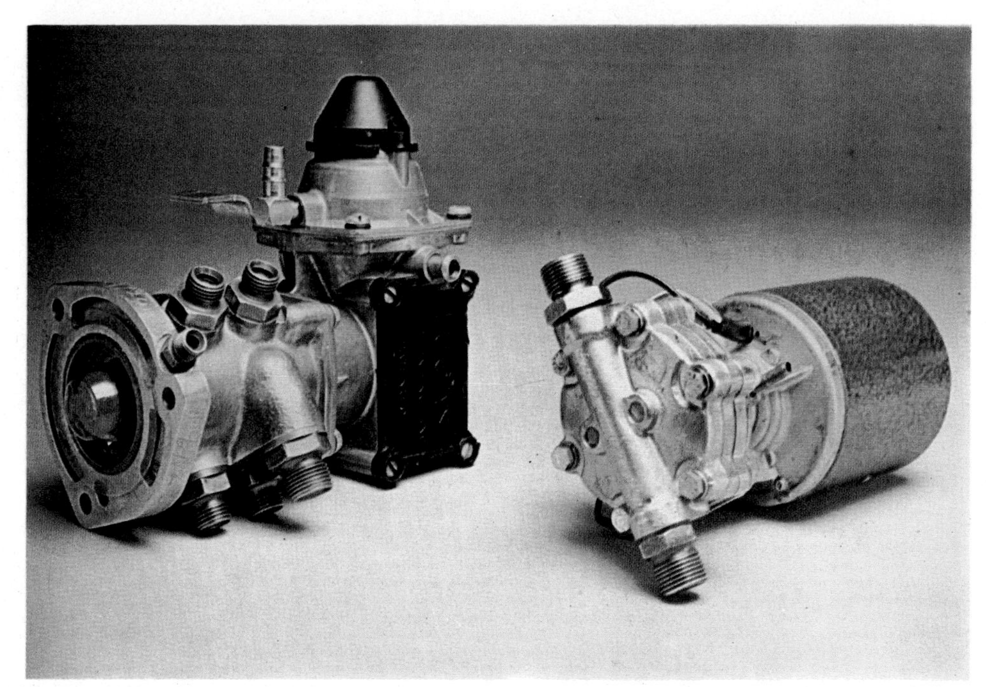

Left: (left to right) a metering distribution pump and a fuel pump, as in a fuel injection system. The fuel pump supplies the fuel to the metering distributor, which feeds a measured amount to an injector at each combustion chamber.

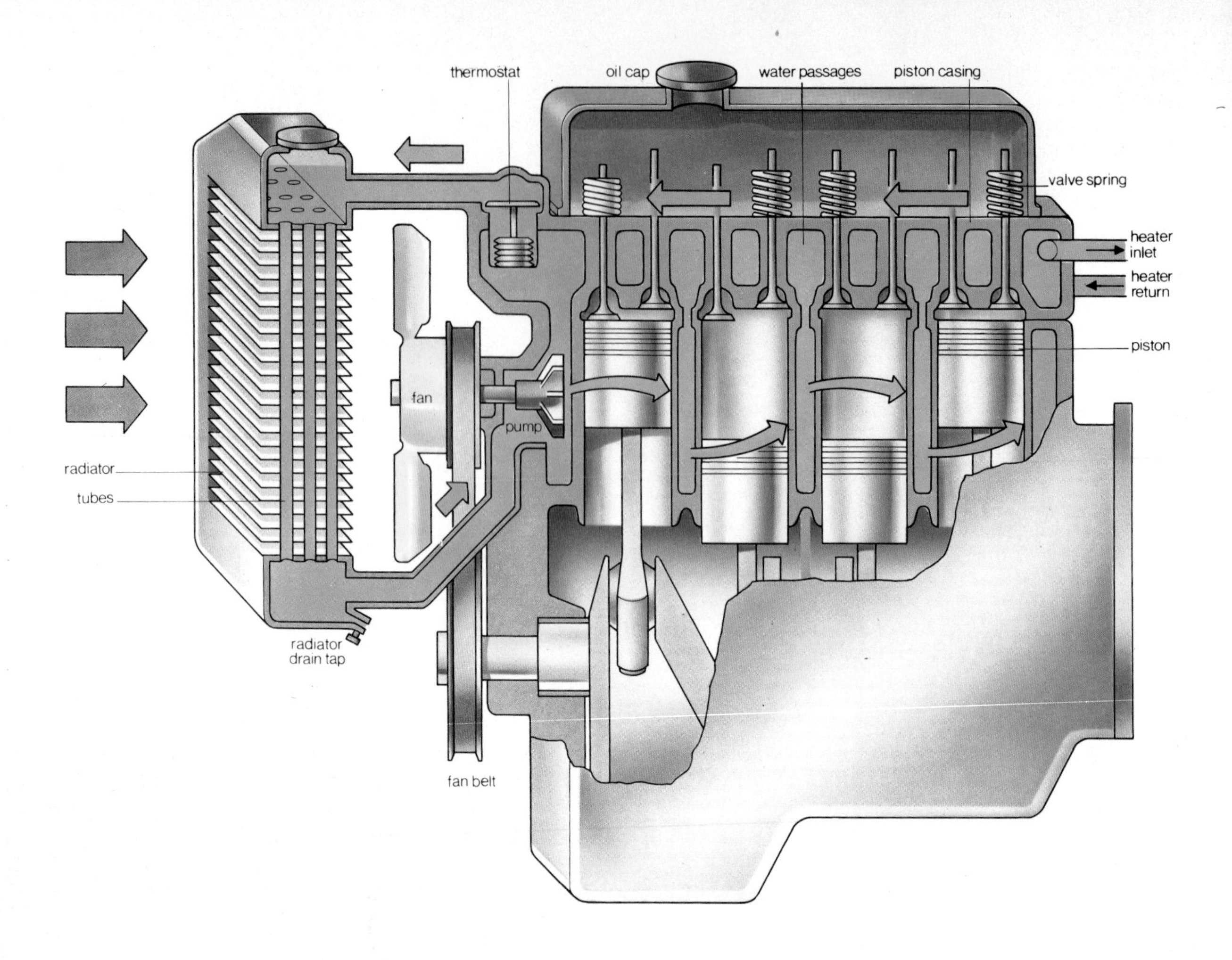

Top of this page: a cutaway view of a car's water-cooling system. The water is cooled by the fan as it passes down through the radiator; it is then pumped through the water jacket in the engine block. The heated water returns to the radiator via a thermostat which controls the cooling capacity by regulating circulation.
Right: the high performance Porsche 2.7 litre engine. It has overhead valves and camshaft, six cylinders, eight main crankshaft bearings for stability and an air-cooling system. The exhaust valves are hollow and filled with sodium to facilitate cooling of the valve head; the sodium melts and circulates. The engine is fitted with an oil filter which does not restrict flow; the belt-driven fan and its ducting are seen in the foreground.

device which is screwed into the cylinder head so that its point protrudes into the combustion chamber. The tip of the spark plug comprises a gap which must be set according to the manufacturer's specification and the spark jumps across this gap, igniting the fuel mixture. There are a variety of different types of spark plug available, and the right one must be chosen according to the design of the engine, because the operating temperature of the firing end of the spark plug must be kept within the correct range.

Electronic ignition systems are also widely used; these replace the contact points and distributor cam with electronic devices such as photo-electric cells, giving more accurate timing, longer life and higher operating speeds.

The cooling system Because of very high operating temperatures within an internal combustion engine, the system must be cooled in order to prevent seizing of moving parts, that is, jamming of the pistons in the cylinders. There are two main types of cooling system: water cooled and air cooled.

Water cooled systems employ a radiator which is made of copper and brass. Such a radiator is expensive but does not rust. A brass tank at the top of the radiator is connected to a brass tank at the bottom via a mesh of copper tubing. The other major component is the water pump, which is usually mounted on the front of the engine and driven by the fan belt from the crankshaft.

The engine cylinder block and the cylinder head are castings which have waterways cast into them. The block and head are connected to the radiator at the top and bottom by rubber hoses. When the engine is running, the water is pumped out of the bottom of the radiator into the block, where it circulates around the cylinders and through the head gasket into the head past the combustion chambers (the hottest part of the engine, where the fuel mixture is ignited). Then it reaches the thermostat, a valve which is operated by heat. The thermostat stays closed if the engine is cold until the water is heated to the correct operating temperature. Then it opens and the water flows through the top hose back into the radiator, where it is cooled as it passes down through the copper tubing.

The fan is mounted on the front of the engine just behind the radiator. When the vehicle is moving, the air rushing through the radiator may be sufficient to cool the water, but the fan is necessary to cool the system when the engine is idling. The fan is normally operated by the fan belt from the crankshaft, but there are some types of fans with blades whose pitch is variable, controlled by a thermostat, or by the speed of the fan by means of centrifugal force. Other types of fan are driven by electric motors, turned on and off as necessary by a thermostatic switch.

Water expands when frozen, and if allowed to freeze in the engine block will crack it, so in cold weather anti-freeze must be added to the system. Anti-freeze is essentially alcohol, which has a much lower freezing temperature than water. Nowadays many cars are equipped with 'permanent' anti-freeze at the factory and are designed to operate on it all year round.

Water-cooled systems are usually pressurized to raise the boiling point of the system, thus improving the efficiency of the engine and the thermal coefficient of the radiator. The radiator cap on such a system will be designed to cope with the pressure, and if the system overheats great care must be taken if the cap is removed while the fluid is still hot.

Air-cooled engines have fins on the cylinder bores to give the largest surface area for disposal of heat. The air is drawn in by a large fan. A thermostat, which is operated by the heat of the engine, directs the air toward or away from the fins. The main advantages with this system are that the engine casting is simpler, and the radiator, water pump, hoses and antifreeze are all unnecessary.

The larger the engine, however, the less efficient the air cooling system becomes, because much larger fans would be needed. Water is a far more efficient coolant than air. Air cooled engines also tend to be more noisy in operation, because fans are more noisy than pumps and because the water jacket in a water cooled engine block deadens much of the mechanical noise from the engine.

Lubrication The lubrication system of the engine comprises an oil filter and a pump driven by the engine. Oil is circulated to all the bearing surfaces in the engine, to reduce wear of moving parts due to friction but also to disperse heat, reduce corrosion and help the sealing action of the piston rings.

The exhaust system The engine exhaust system includes a manifold, a silencer [muffler] and a tailpipe or exhaust pipe. The operation of the engine amounts to a continuous series of explosions and the sound would be unbearable without some sort of muffling device; in addition the exhaust gases leave the engine at a temperature of about 1700°C (more than 3000°F) and are poisonous. They must be ducted to the rear of the vehicle and released without harm to the occupants or to passers-by.

'Manifold' means 'many-branched'; the manifold bolted to the engine is a set of pipes, one for each cylinder exhaust port, which converge into one or two pipes. The silencer is a set of baffles which slows down the exhaust gas, incidentally supplying back-pressure to the engine which is calculated when the engine is designed. The correct silencer must be fitted.

This 1928 Bugatti 'straight 8' engine has eight exhaust ports converging into two exhaust pipes.

The diesel engine

The diesel engine is named after its inventor, Rudolf Diesel, whose first working prototype ran in 1897 after many years of research work. Until the late 1920s most of the development of diesel engines took place in Germany, and a great deal of experience was gained from the production of engines for submarines during World War I. The first successful diesel engines for road transport appeared in 1922, although unsuccessful attempts had been made to produce such engines since as early as 1898. The use of diesels for marine applications dates from about 1910, but it was not until 1929, with the introduction of designs by Cummins in the USA and Gardner in Britain, that they became a practical proposition for powering small boats. These engines were subsequently adapted for road use. Power outputs increased steadily during the 1930s, and by the beginning of World War II diesel engines were in widespread use in road transport, rail locomotives, tractors and construction plant, ships and boats, and as industrial power sources (including electricity generating sets).

The diesel engine, like the petrol [gasoline] engine, is a form of internal combustion engine, and although the two have much in common there are important differences in their respective operating principles. In a petrol engine, fuel and air are mixed in the caburettor, and the mixture is drawn into the combustion chamber at the top of the cylinder during the downward stroke of the piston. The next upward stroke of the piston compresses the mixture to between a sixth and a tenth of its original volume, and as the piston reaches its upper limit of travel the mixture is ignited by an electric spark created by the spark plug. The resulting expansion of the burning mixture forces the piston back

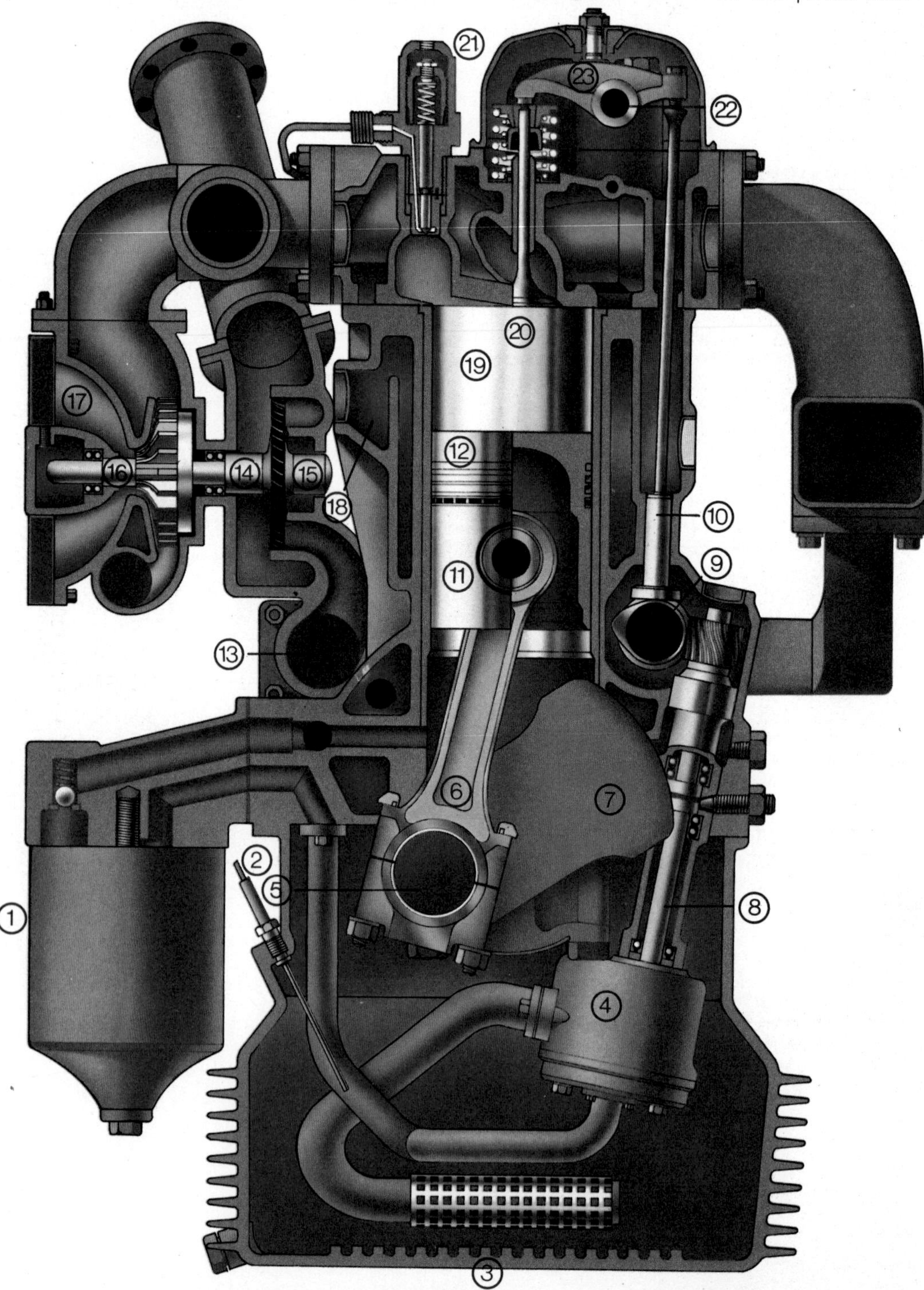

A cutaway diagram of an indirect injection diesel engine, showing the following features: 1) oil filter; 2) dipstick; 3) oil scavenger pipe; 4) oil pump; 5) crankshaft; 6) connecting rod; 7) crankshaft web; 8) oil pump drive; 9) camshaft; 10) cam follower; 11) piston; 12) oil scraper and compression rings; 13) turbocharger exhaust duct; 14) turbocharger spindle; 15) turbocharger turbine; 16) compressor; 17) air inlet duct; 18) water galleries; 19) cylinder bore; 20) inlet valve; 21) injector; 22) rocker shaft; 23) rocker.

down the cylinder (the power stroke).

In a diesel engine, however, as the piston moves down, only pure air is drawn into the cylinder and compressed as it moves up again, but it is compressed to a much higher degree than in a petrol engine (with compression ratios of between 12 : 1 and 25 : 1) with the result that its temperature is raised considerably, to well over 1000°F (538°C). As the piston nears the top of its travel a fine spray of fuel is injected into the cylinder by an injector nozzle near the top. The fuel mixes with the air, which has become so hot due to compression that the fuel/air mixture ignites spontaneously without the need for a spark.

As the volume of air drawn into the cylinder is always the same in a diesel engine, its speed is controlled by the amount of fuel that is injected.

A diesel engine can be adapted to run on almost any fuel from vegetable oils to natural gas and high octane petrol, but the most suitable and widely used diesel fuel is distilled from crude oil and closely related to kerosene. It is much less volatile than petrol, with a flash point (temperature at which a heated petroleum product gives off enough vapour to flash momentarily when a small flame is placed nearby) of around 168°F (75°C) whereas the flash point of petrol is between 70°F (21°C) and 100°F (38°C).

Fuel injection The fuel is delivered to each injector by a fuel pump, and there is either one pump for each injector or else one main pump supplying all the injectors in turn by means of a distributor valve. Where there is a pump for each injector the pumps may be grouped together in a single unit, supplying the injectors through feed pipes, or else the pumps and injectors may be combined into individual units with a separate unit mounted on each cylinder. The pumps are of the reciprocating type, with spring loaded plungers actuated by a camshaft driven by the engine. The accelerator control is connected to the pump mechanism, and alters the engine speed by varying the amount of fuel delivered to the injectors. The injectors have spring loaded needle valves that are opened by the pressure of the shots of fuel delivered by the pumps at the correct instant in the firing cycle. The fuel is sprayed out through holes in the end of the injector, which break it up into a fine mist and distribute it correctly around the combustion chamber.

Combustion chambers The fuel-air mixture should burn evenly and progressively, as a violent detonation of the mixture causes an uneven running condition known as 'diesel knock'. To achieve correct combustion the fuel and air must be thoroughly mixed. On engines which have the fuel injected directly into the combustion chamber, more effective mixing may be achieved by creating turbulence in the air in the cylinder as it is compressed. This is often done by contouring the crown of the piston so that the air is moved around within the cylinder during compression.

Other designs of engine use swirl chambers or pre-combustion chambers to improve combustion. A swirl chamber is a small spherical chamber above or at the side of the main combustion chamber, and connected to it by a passage. When the air in the cylinder is compressed some of it is forced into the swirl chamber, where a turbulent effect is created due to the shape of the chamber. The fuel is injected into the swirl chamber, and preliminary combustion occurs forcing the mixture into the main combustion chamber where complete combustion takes place.

The pre-combustion chamber is connected to the main combustion chamber by a number of fine passages, and the fuel is injected into it. Part of the mixture in the chamber ignites and expands, forcing the remaining unburnt fuel through the connecting passages, from which it emerges into the main chamber as a fine spray and ignites smoothly.

Many diesel engines work on the two-stroke principle, and as they need to draw in only pure air instead of the usual air-fuel-oil mixture needed by the two-stroke petrol engine they are more efficient. The intake of fresh air and the expulsion of the exhaust gases is known as 'scavenging', and the two most common methods used on diesels are loop scavenging and uniflow scavenging, both of which employ a blower unit to blow the air into the inlet ports.

In the loop-scavenging system, as the piston nears the bottom of its stroke it uncovers the inlet and exhaust ports; the inlet port directs air from the blower into the cylinder in an upward direction, and this forces the exhaust gases downwards and out of the exhaust port on the opposite side of the cylinder. As the piston moves back upwards it covers the ports which effectively seals the cylinder, and the clean air is compressed before the fuel is injected into the top of the cylinder.

The uniflow system also has an inlet port in the side of the cylinder, near the bottom, but the exhaust gases are expelled through one or more valves in the top of the combustion chamber. The valves open just before the inlet port is uncovered, and at this point the gases are still under some pressure which starts them flowing out of the cylinder, the remaining gases being expelled by the upward flow of clean air from the blower.

Rudolf Diesel's first successful engine was built in Augsburg, Germany, at the Maschinenfabrik works, in 1897.

Some two-stroke diesels work on the opposed-piston principle, with two pistons in the same cylinder acting in opposition to each other, moving towards the centre of the cylinder from opposite ends. The pistons may be connected by a crank arrangement to the same crankshaft, or may have separate crankshafts coupled by a gear train. The inlet and exhaust ports are near the opposite ends of the cylinder, and the fuel injector is at the centre. At the point of ignition the two pistons are very close together, crown to crown, and the force of the combustion forces them in opposite directions down the cylinder. One piston uncovers the exhaust ports slightly before the other uncovers the inlet ports, and most of the exhaust gas rushes out under pressure, the remainder being expelled by the incoming air when the inlet port is uncovered.

The power output of internal combustion engines can be increased significantly by supercharging, and the diesel is well suited to this as only air has to be blown in as opposed to a fuel-air mixture.

Diesel engines which have the fuel injected directly into the combustion chamber do not present any special difficulties when starting from cold, other than the need for very powerful starter motors on the larger versions. Engines fitted with pre-combustion or swirl chambers, however, can be difficult to start and usually employ some form of heater plugs or coils, electrically powered and usually mounted next to the injectors, which pre-heat the air in the combustion chambers and help the fuel to vaporize until the engine has warmed up.

Hand starting is quite easy with smaller engines, and can be used on larger industrial models by means of some form of energy storage. This can be done by spinning a large flywheel and coupling it to the engine when it is spinning fast enough, or by building up pressure in a hydraulic cylinder by means of a hand pump, then releasing the energy to a toothed rack which engages a pinion on the engine crankshaft. In some cases a small, easily started engine may be used as a starter motor for a large engine.

After World War II the economic reasons for use of diesel power on the world's railways became overwhelming. Diesel engines can be started instantly, even in cold weather, and do not need to be kept running while on standby, unlike steam engines. Their thermal efficiency is also higher.

Diesel-electric propulsion

Railway locomotives and railcars are usually powered by a diesel-electric system, in which the mechanical power of the diesel engine is converted to electric power to drive the wheels.

Locomotives with internal combustion engines mechanically coupled directly to the drive wheels have been built, but have not been satisfactory, because they cannot pull the tremendous weight from a standing start and are not easily adapted to changing load requirements. Using a diesel engine to generate electricity, which can be delivered to the drive wheels in a variable amount as required, has been the practical solution.

The first successful diesel-electric vehicle was a 75 hp (56 kW) railcar on the Mellersta-Sodermanlands Railway in Sweden in 1913. Development was retarded in most countries by World War I, but by 1925 the 1250 hp (933 kW) Lomonosov locomotive had appeared in the Soviet Union. Power rose steadily until the outbreak of World War II, particularly in North America, though the higher powers were achieved by double units, for example the 4000 hp (2980 kW) locomotive of the Romanian State Railway. After the war, diesel traction really began to get under way. During the 1950s large numbers of 350 hp (260 kW) and 400 hp (295 kW) electric shunters were followed by locomotives with up to 3300 hp (2460 kW), the latter being the famous Deltic locomotive, though even this was fitted with two engines.

The power outputs in use today range from about 150 hp (112 kW) to 7000 hp (5220 kW), the smaller outputs being used for shunters and the largest coming from two engines. Often the higher powers are used for pulling low speed heavy freight trains in hilly districts, and not necessarily for high speed running, although the 3300 hp (2450 kW) British Rail locomotive can haul passenger trains at speeds in excess of 100 mph (160 km/hr).

Construction In a typical design, the diesel engine is mechanically coupled to a DC generator which feeds a series-wound DC traction motor (for some small vehicles there may be an alternator which feeds an AC traction motor). The motor is normally mounted resiliently on the bogie truck (the undercarriage assembly of wheels and axles) by a nose suspension, which consists basically of compression springs or rubber pads. Sometimes, however, the motors are rigidly mounted on the bogie and a resilient drive is used. A pinion gear on the motor transmits power to a gearwheel mounted on the axle next to the roadwheel. The gearwheel is usually about four times as big in diameter as the pinion; a typical gear ratio is 19 : 92. The road wheel is 25% to 50% bigger than the gearwheel.

Depending on the power output required, there may be up to six traction motors. The speed of the locomotive depends on four factors: the speed of the diesel engine (hence of the generator); the excitation of the generator (controlling the voltage supplied to the traction motors); full-field or weak-field operation of the traction motor; and the gear ratio of the pinion and gearwheel.

The engine is usually started by connecting the generator as a series-wound motor by means of the battery, an additional starting field winding being provided: the engine turns over, fires and runs to idling speed. The starting circuits then revert to normal.

There is also an auxiliary generator, driven by belts or gears from the main generator or mounted on an extension of the main generator shaft. This generator, normally DC, is used to supply the control circuits, the auxiliary circuits and to charge the battery. The output voltage of the auxiliary generator, hence the battery-charging voltage, is maintained substantially constant by a voltage regulator.

Batteries range from two 12 V batteries on the smallest models, to enable 12 V and 24 V commercial components to be used, to 110 V, which is standard in many parts of the world; some American and Australian vehicles use 74 V.

The controls Control of the vehicle is by means of a master controller, which is essentially a series of cam-operated switches arranged in groups and interlocked to prevent incorrect operation. The control group controls the speed of the vehicle and the reverse group controls the direction. The reverse group operates a power switch in the field circuit of each traction motor; it switches the direction of field current, the direction of armature current staying the same. This causes the armature, and hence the vehicle, to reverse.

The function of the control group and circuits is to adjust the excitation of the main generator so that its output voltage applied to the traction motors is also adjusted. This is often accomplished by varying the amount of resistance in series with the generator shunt-field. The variable resistor is part of the load regulator which is under the control of the diesel engine governor.

With the engine at maximum speed and the generator at maximum speed and excitation, the vehicle is at maximum speed under full-field conditions. If a resistor is now connected in parallel with each traction-motor field, some of the field-current is diverted, and this causes an increase in traction-motor speed. This is known as traction-motor field diversion, or weak-field, and makes full use of available engine power. The various circuits are protected by overload relays, circuit breakers or fuses, so that if a fault occurs power is suspended or cut off.

The Wankel engine

A piston engine is one which has a working chamber whose volume can be altered by the movement of the piston. The reciprocating internal combustion engine achieves this by means of a sliding piston in a cylindrical chamber; in the rotary engine the volume is varied by two or more elements revolving with respect to each other. Thus the fundamental difference is that the centre of gravity of the moving power output member of the reciprocating engine oscillates to and fro in a straight line, while that of the rotary engine moves in one continuous circular motion. Since there is no to-and-fro movement of the working parts, the rotary piston engine is, or can be, in perfect balance, and so for vehicular applications has the advantage of inherently smoother operation.

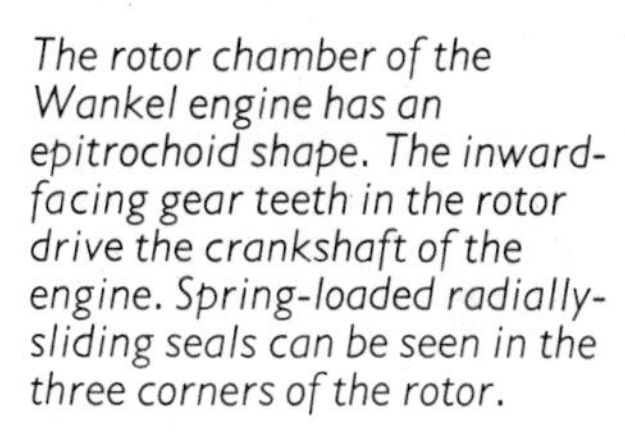

The rotor chamber of the Wankel engine has an epitrochoid shape. The inward-facing gear teeth in the rotor drive the crankshaft of the engine. Spring-loaded radially-sliding seals can be seen in the three corners of the rotor.

The Wankel has a rotor shaped like a slightly rounded triangle which rotates inside a chamber shaped like a fat figure 8. The geometrical shapes of the rotor and housing are derived from a group of curves generically called trochoids, which are found by revolving one circle around another and plotting the path of a point either on the circumference or on an extension of the radius of the revolving circle. The housing of the Wankel, for example, is shaped like an epitrochoid.

The rotor has a hole in it with inward-facing gear teeth. It rotates about a smaller gear rotating about a fixed centre; these represent respectively the larger rolling circle and the smaller fixed circle. The shape of the housing can be visualized by imagining an arm of length R attached to the rotor, and following the path traced out by the end of the arm as the rotor revolves around the fixed gear. The centre of the rotor itself describes a circle around the fixed gear, and the radius of this circle is called the eccentricity of the rotor, e. The ratio of R to e determines the basic geometry of the engine. When this ratio is large, the swept volume is comparatively small and the fixed gear must be small, which limits the size of the crankshaft that has to pass through it.

Opposite page: the Wankel has the same four-stroke cycle as a conventional piston engine, but the geometry is different. The diagrams show the cycle for one side of the triangular rotor, but the other two sides are going through the same cycle.
Below: the engine and gearbox unit of the Suzuki RE-5 motorcycle. The liquid-cooled engine has a capacity of 497cc (measured as the swept volume of one chamber) and develops 62 hp (SAE) at 6500rpm, giving the 507-pound (203kg) machine a top speed of 115mph (185kph).

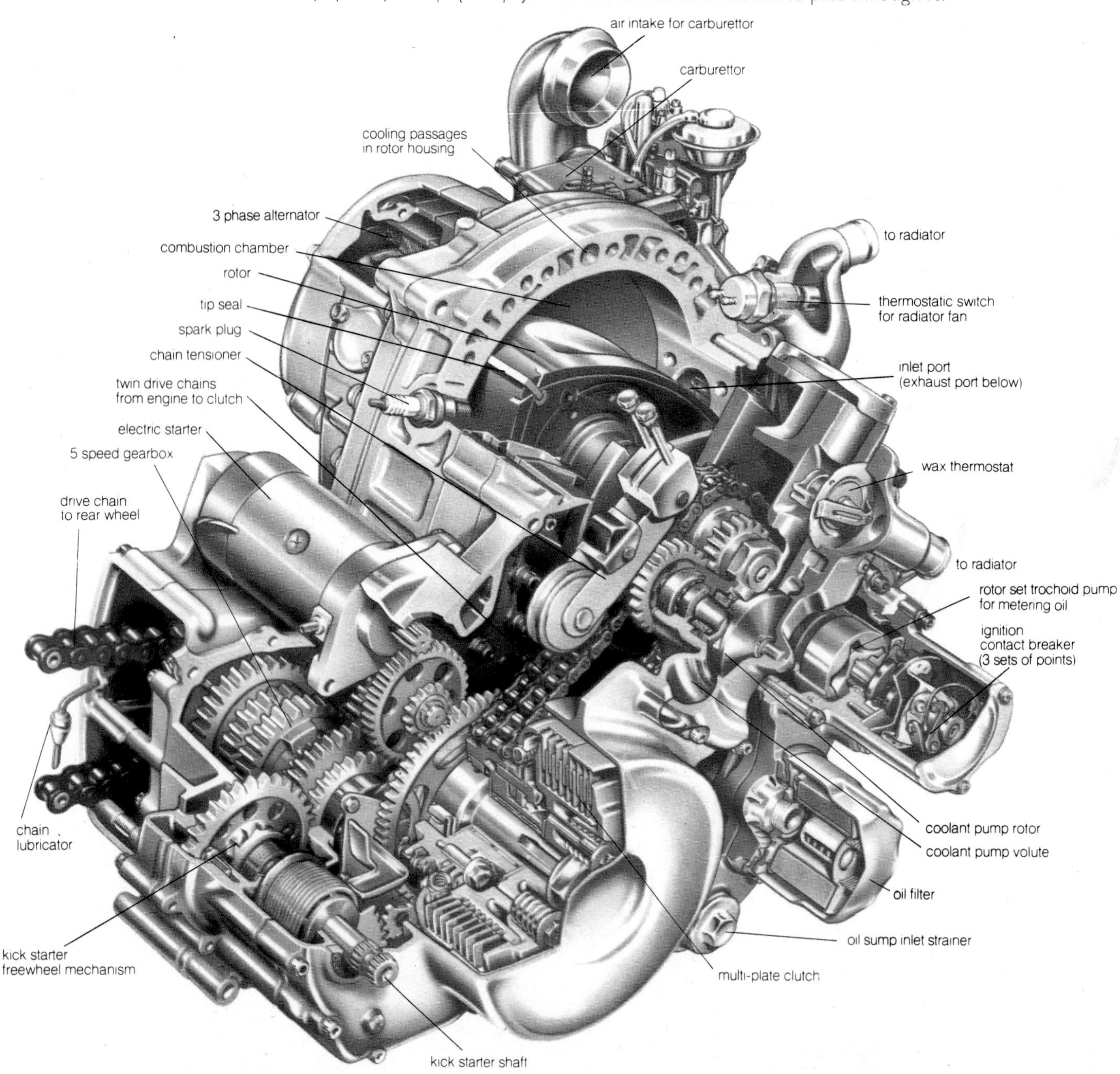

As the rotor rotates, the gap between the flanks of the rotor and the walls of the housing fluctuates cyclically, expanding and contracting to give the four strokes—induction, compression, expansion and exhaust—of the reciprocating engine. Note that there are three firing strokes per revolution of the rotor (there are three sides to the rotor, so a four-lobed rotor would have four, and so on) but only one per revolution of the crankshaft, since the rotor rotates at one-third the angular velocity of the crankshaft.

Since each flank of the rotor acts effectively as a piston in a reciprocating engine, it is only necessary to consider one flank to follow the sequence of operations. Consider first that the leading seal of one face has just passed the intake port in the housing. As the rotor moves on, the gap between the side of the rotor and the housing increases, and mixture is drawn in. When the trailing seal passes the port the mixture is trapped in the space, and as the rotor continues this gap becomes progressively smaller, compressing the mixture. When the gap is at or near its minimum, corresponding to top dead centre, the spark plug fires and the mixture ignites and expands. Since the centre of rotation of the rotor is eccentric to the centre of the casing, the side of the rotor in question is pushed around, and in turn turns the small gear on the crankshaft. Finally the leading seal passes the exhaust port and the gases escape.

Theoretically the Wankel engine has a number of important advantages. One has already been mentioned: its inherent smoothness. Shaft speed in the Wankel engine must be high for optimum performance, but this is possible because of the absence of reciprocating parts. The only out-of-balance loads come from the orbital motion of the rotor, and these are easily counterbalanced, for example by having two rotors, each with its own combustion chamber, set at opposite points of their orbit on a common crankshaft. Another advantage is that it has fewer moving parts. Unlike the ordinary reciprocating engine with pistons, connecting rods, crankshaft, camshaft and valve gear, the Wankel has effectively only two moving parts: the rotor and the crankshaft. In addition it is basically an extremely compact unit for its output. It generated considerable interest among engineers almost from the outset.

In 1954, in co-operation with NSU, Felix Wankel presented his ideas on a rotary engine. Initial studies took place on a compressor rather than an engine, but by 1957 a prototype engine was running, in a version wherein both the rotor and housing rotated at different speeds. This meant a fairly complex structure, and in 1958 the basic construction was changed to the planetary rotation type with a fixed housing. Shortly thereafter it appeared in production, buried away in the tail of a small sports car, the NSU Spider.

Production problems The first serious problem to appear was seal wear. (Wankel himself had begun as a specialist in sealing devices.) The three tips or lobes of the triangular rotor must be fitted with seals to serve the function of piston rings in a conventional engine: to maintain compression by preventing the escape of gases. Seals are not a problem in rotary pumps and other devices, but the high pressure and operating temperature in an internal combustion engine caused seal wear, manifested by 'chatter' marks on the housing. An intensive investigation by NSU and other licensees resulted in special materials which solved the problem, but at some cost. NSU use Ferrotic (a ceramic/metal material) for the apex seals and a coating of Elnisil (a composite nickel-silicone material) on the housing, which is first sprayed on and then ground. Other materials used include special cast iron or tool steel for the seals and hard chrome or tungsten carbide coatings for the housing.

Another problem was the result of two inter-related factors. Like a conventional two-stroke engine, the Wankel runs on a mixture of petrol [gasoline] and oil so that the apex seals can be properly lubricated, which leads to a

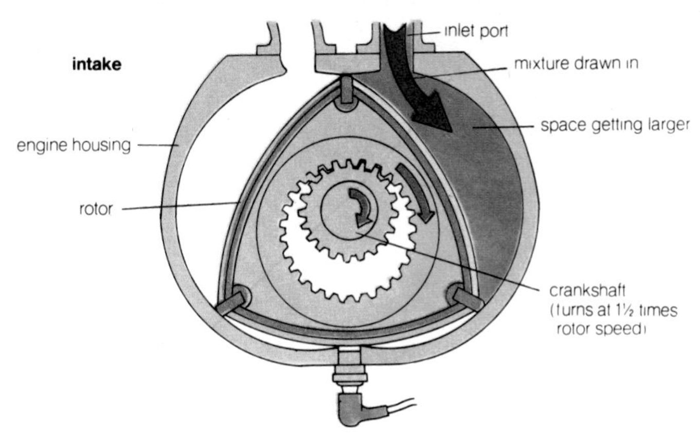

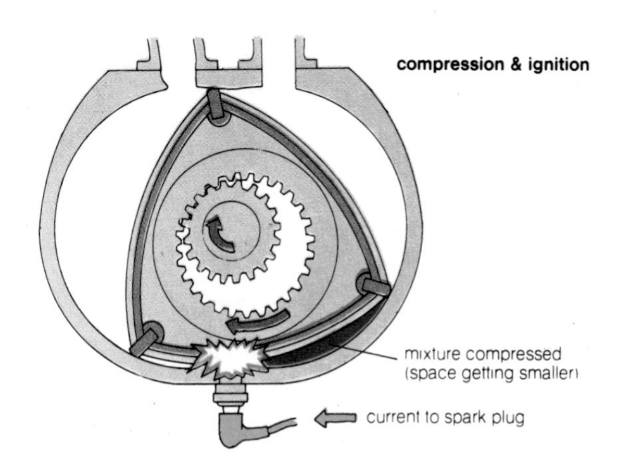

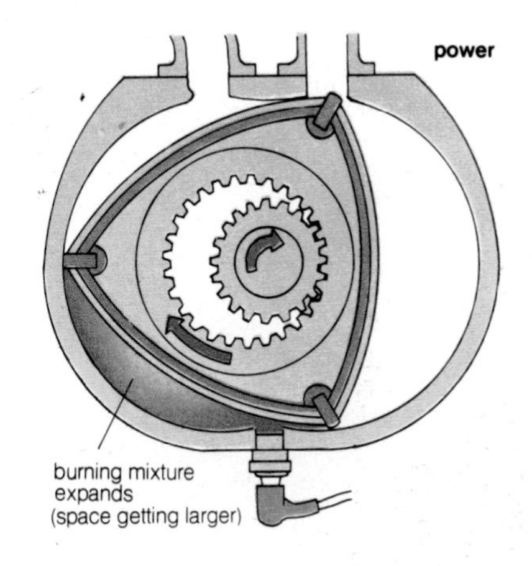

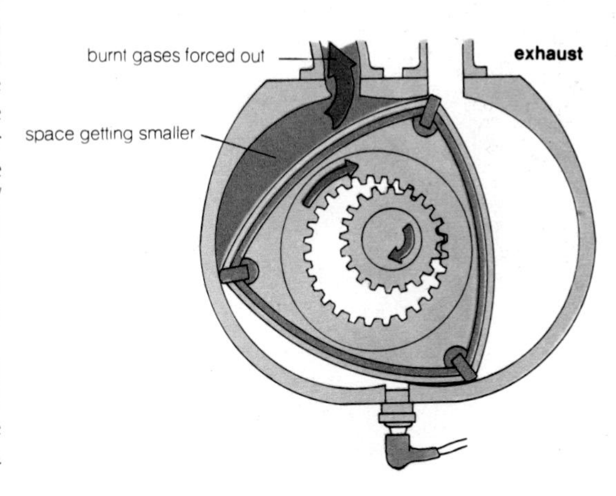

smoky exhaust. In addition the combustion chamber is long and narrow, having a high quench action, which cools the mixture and 'puts the flame out' at an early stage in the cycle, as it were: the result is a very high HC (hydrocarbon) content in the exhaust, and the Wankel would have had trouble meeting the American emission laws. In fact this problem was easily solved, for the exhaust was rich enough to maintain combustion in the exhaust pipe; all that was required was the addition of a thermal reactor and an air injector to get one of the cleanest engines in production, as used by the Japanese firm of Mazda for its REAPS (Rotary Engine Anti-Pollution system) equipment.

Fuel consumption The Wankel has thus been developed to a stage where the mechanical problems are no longer major problems; but the high fuel consumption is. The Wankel is a relatively new engine compared to the reciprocating engine, and the latter has had years of development behind it, particularly in relation to the shape of the combustion chamber, the design of which (after taking into account compression ratios, ignition timing and intake and exhaust systems) probably has more effect on engine efficiency than any other single factor. The Wankel combustion chamber will require research break-throughs before the specific fuel consumptions (a measure of the efficiency of the engine in converting fuel energy into useful mechanical work) approach those of the reciprocating engine: the simple bathtub-shaped cut-outs in the rotor sides may well have to be modified in some way to improve matters. Other ideas which may prove useful include stratified charging by injecting the fuel directly into the combustion chamber as close to the spark plug as possible, reducing the R/e ratio to give a more compact chamber (although this may create mechanical problems), and further investigation into ignition timing and the number and position of the spark plugs.

This close-up photograph of the Mazda's Wankel engine shows the curve of the housing, which is coated with a composite nickel-silicone material by means of a spray torch. The heat from the torch causes it to be bonded to the housing material, and it is then finish-ground on a machine specially built for the purpose. Felix Wankel began his career as a specialist in sealing devices; the special problems with seal wear and maintaining compression were the first to appear in development of the rotary engine.

The aircraft engine

When the Wright brothers came to search for an engine to put into their glider in 1903, they thought they could manage to fly with one of only 8 horsepower, provided it was not too heavy. They approached, without success, half-a-dozen makers of car engines. Eventually, they built their own engine and got 12 hp from it, but it was still relatively heavy at 15 lb (7 kg) to the horsepower. Thirty years later, engine designers were aiming at a ratio of 1 hp per pound of engine weight (2.2 hp per kg). In the years leading up to World War I, the French led the field in aircraft engine design, producing several 50 hp and two 100 hp engines by 1908. But the best of these still only had a power to weight ratio of 3.7 lb to the hp.

Early engines were water cooled, with the cylinders arranged in line or in a V formation as in a car. But in 1907 a new and highly successful type was introduced: the rotary engine. In this, the crankcase and cylinders revolved in one piece around a stationary crankshaft. The pistons were connected to a single pivot mounted off-centre, so that they moved in and out as they revolved with their cylinders. The propeller was connected directly to the front of the crankcase and turned with it. The rotary engine had fewer parts than a conventional engine, and since the cylinders moved rapidly around, they could be air cooled by fins mounted so as to take advantage of the draught. Both factors contributed to make it light. Rotary engines always had an odd number of cylinders. This reduced vibration, since there were never two pistons moving in exactly the same direction at the same time. The original 1907 Gnome engine had seven, and later types nine.

Other types of engine produced at the time included the Spanish Hispano-Suiza, a design well ahead of its time with

The Wright brothers' first aircraft had a tail-first layout; the rudder was an afterthought, added when they discovered that wing-warping would not steer an aircraft, but only tilt it. They built their own engine for it, a four-cylinder in-line design which produced 12 horsepower and weighed 179 pounds (81kg). This was a poor power-to-weight ratio even for the time; an earlier five-cylinder radial produced 52 horsepower and weighed only 151 pounds (69.5kg).

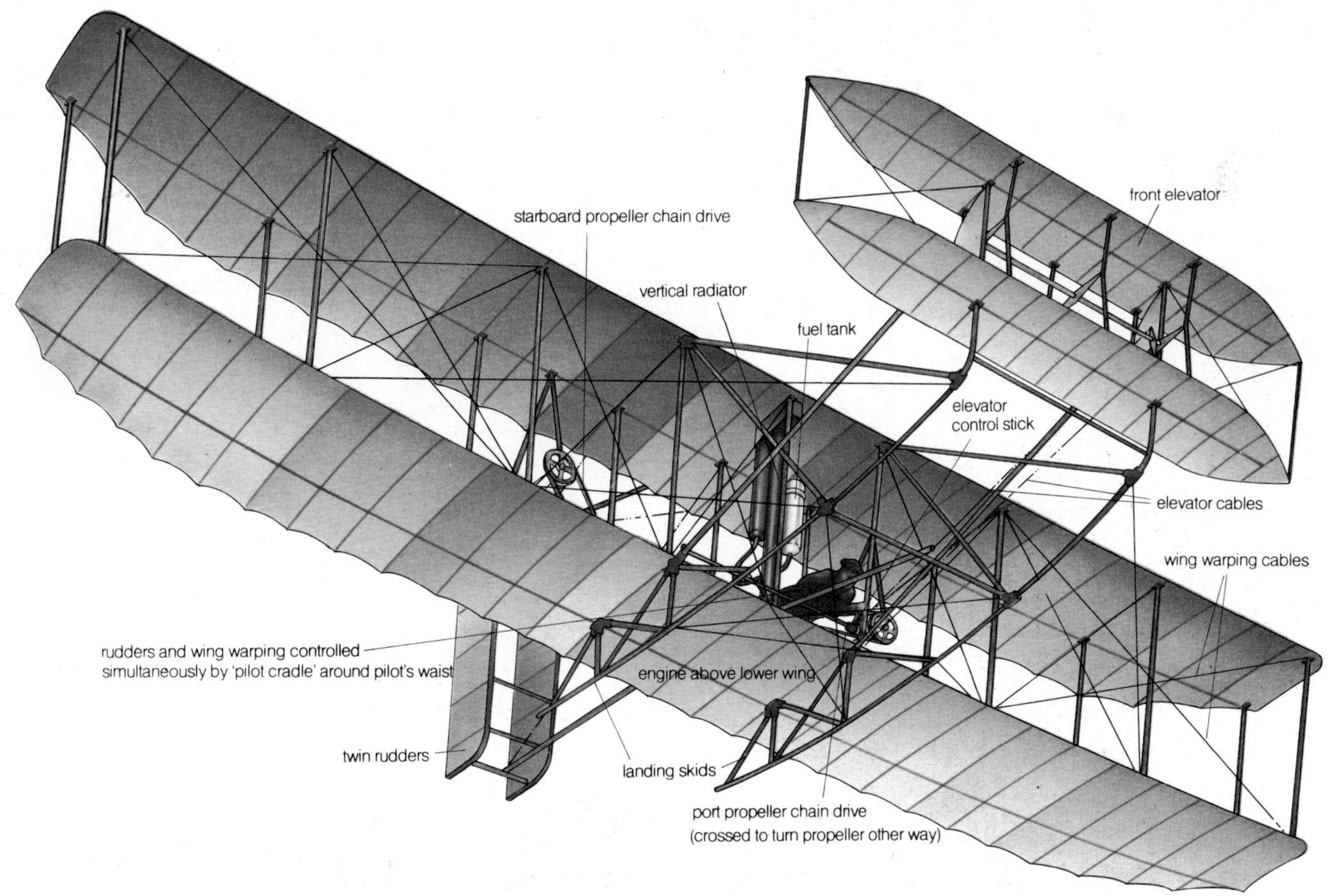

This Wright TC18 (Turbo Compound) R3350EA Series engine, as fitted in the DC7 aircraft, produces 3400 hp at take-off for a dry weight of 3645 pounds (1653kg). Many are still in service today. It has conventional valve gear, not sleeve valves as fitted to Bristol engines. The impeller of the two-speed gear-driven supercharger can be seen at the rear. The energy of the exhaust gases is fed back to the engine by passing them through three 'blow down' power recovery turbines, as shown in the small drawing below.

eight steel cylinders arranged in a V and screwed into an aluminium block. In the later years of the war this engine yielded, in successive versions, 150, 220 and 300 hp. The Rolls-Royce Eagle, a V12 engine with a broadly similar layout adapted from an original Mercedes design, produced 360 hp in its Mark 8 version of 1917. This was the engine that carried Alcock and Brown across the Atlantic in 1919.

Both the Rolls and the Hispano engines had conventional water cooling. This often gave trouble, since vibration and the shock of landing caused the plumbing to break. It was to overcome this problem that a third type of engine was introduced: the air-cooled radial engine, in which static cylinders were arranged in a circle and cooled by the backwash of the propeller.

Proper cooling is one of the most critical points of aircraft engine design. Such engines have always produced far more power for their size than automobile engines of the same date, and consequently have generated much higher temperatures. These problems led to great rivalry between the designers of air and water-cooled engines. Their object was to produce engines that were adequately cooled with the lightest possible system—thus improving the vital power to weight ratio—and at the same time were utterly reliable.

As far as reliability went, the water-cooled engine seemed to have all the advantages. Any capacity of radiator could be used to produce the desired temperature. The temperature of the engine was kept within safe limits by the boiling point of the cooling water, since it could rise no higher than this until the water boiled away completely. Some engines used this feature in evaporative cooling systems, where the water was allowed to boil at the engine. The steam was ducted off, re-condensed into water and returned to the engine. The system had been used as early as 1907 in the French Antoinette engine. In other engines, ethylene glycol (anti-freeze) was used as a coolant, raising the boiling point to 140°C (284°F) to provide an additional safety margin.

The principal trouble with this type of engine was the weight and complexity of the cooling system—it was one more thing to go wrong. Air-cooled engines did not suffer from this problem, since their system had no moving parts. Their cylinders were always arranged radially in one or more circular rows. This placed them just behind the propeller, an ideal position for cooling. They were also spaced quite wide apart, so that the outside could be covered with large fins to increase the surface area and thus improve heat dissipation.

Early radial engines had their cylinders completely exposed to the air, but in the early 1930s a shaped ring cowling was added around the engine to improve air flow around the

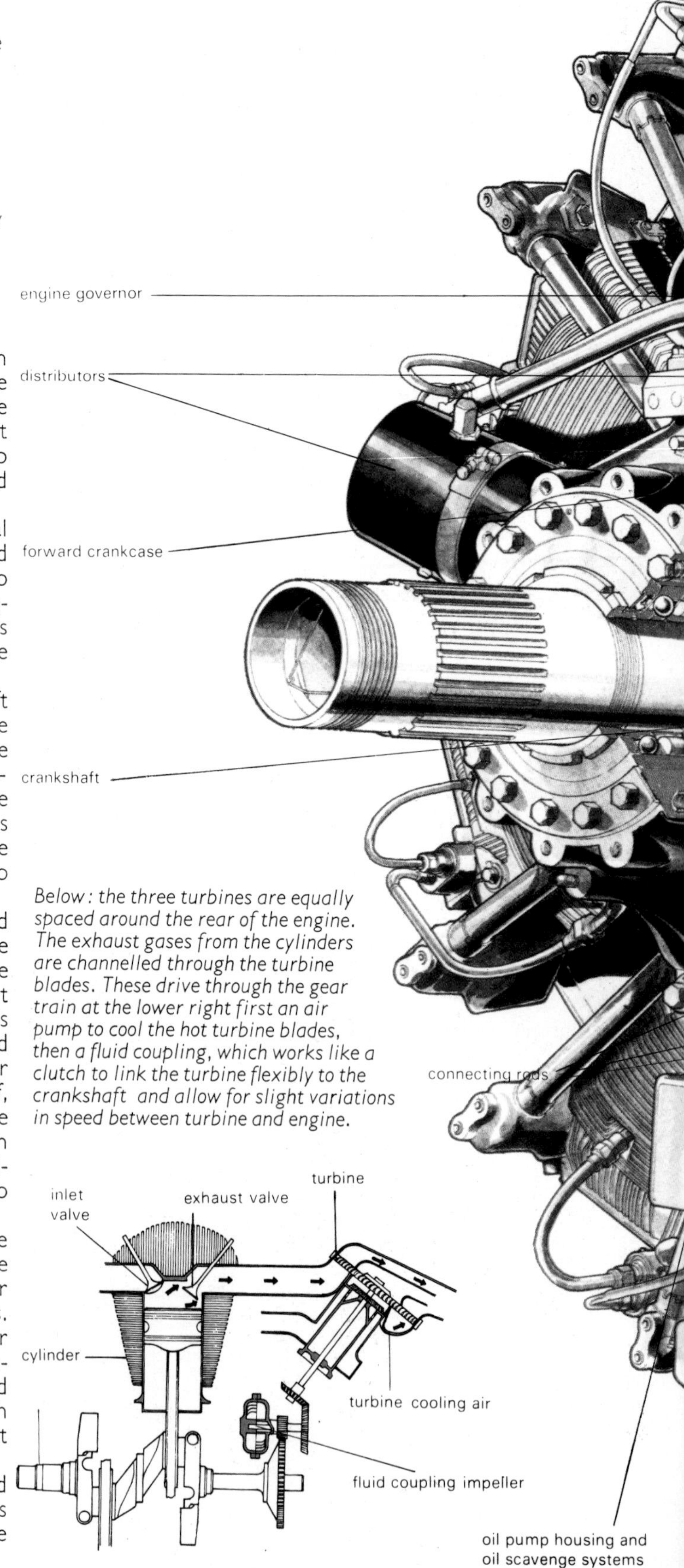

Below: the three turbines are equally spaced around the rear of the engine. The exhaust gases from the cylinders are channelled through the turbine blades. These drive through the gear train at the lower right first an air pump to cool the hot turbine blades, then a fluid coupling, which works like a clutch to link the turbine flexibly to the crankshaft and allow for slight variations in speed between turbine and engine.

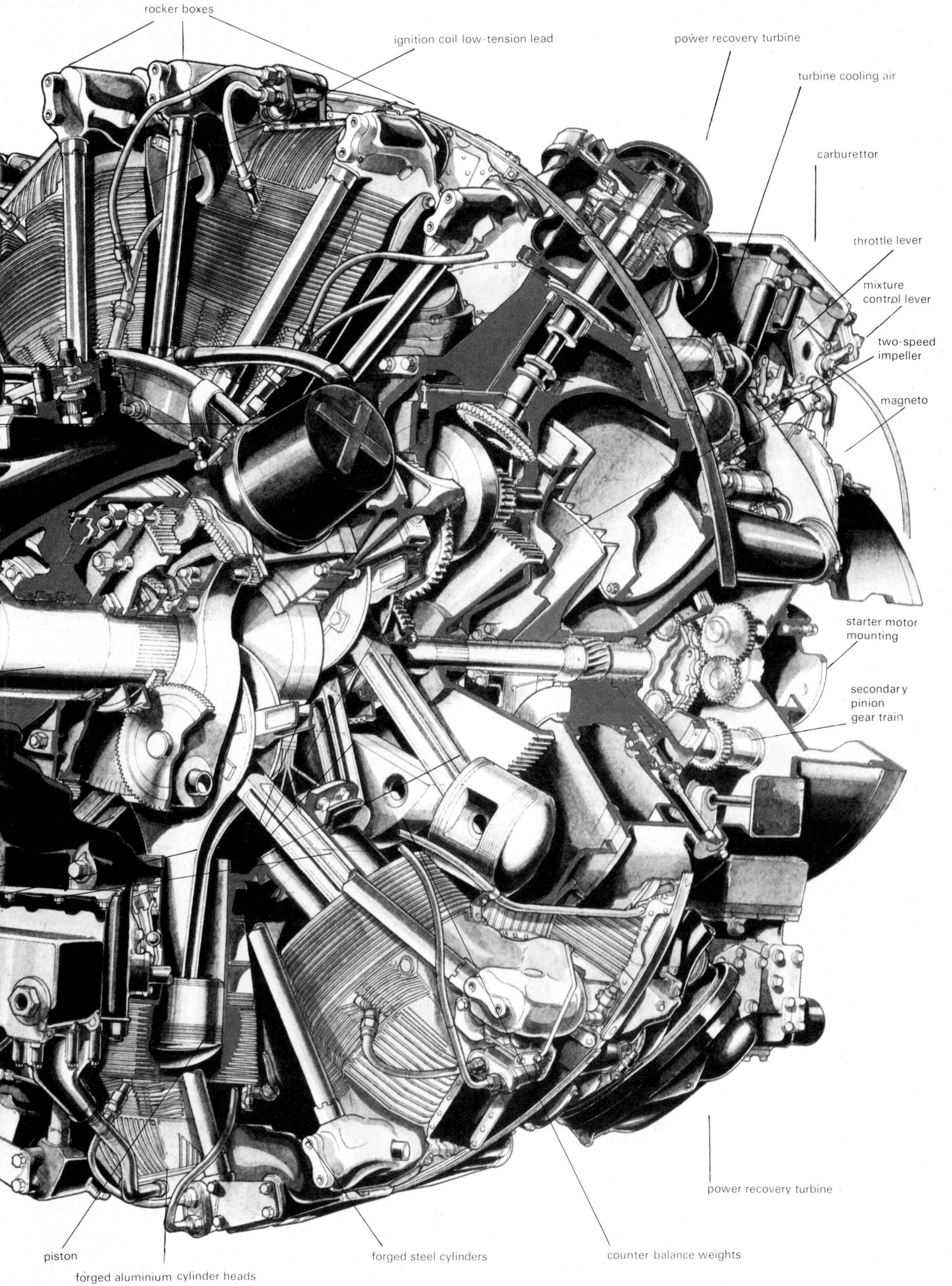
rocker boxes
ignition coil low-tension lead
power recovery turbine
turbine cooling air
carburettor
throttle lever
mixture control lever
two-speed impeller
magneto
starter motor mounting
secondary pinion gear train
power recovery turbine
piston
forged aluminium cylinder heads
forged steel cylinders
counter-balance weights

cylinders and reduce the drag caused by the wide, flat-fronted engine. The main trouble with the air cooled radial was that there was no fixed upper limit on its temperature, so it would overheat very quickly if over-extended. This problem, however, led to the production of high quality heat resistant alloys which made the development of the jet engine possible later on.

The engine designers of the 1920s and 1930s managed to produce reliable engines with ingenious new features by sheer good design and workmanship. One of the best of these improvements was the sleeve valve, which replaces the complex valve gear of a conventional engine with a single tube sliding up and between the piston and the cylinder—it completely encircles the piston. It has ports, or holes, in its upper end. These slide past matching ports in the cylinder head which are connected to the fuel-supply and exhaust systems, thus opening and closing them at the correct time. This greatly reduces the number of moving parts in the engine, particularly as the sleeve can be moved by quite simple machinery set around the inner edge of the ring of cylinders instead of the conventional long train of rods and levers reaching to the outside.

The alloy of which the sleeve is made is vital, because of its expansion as it heats up. If it expands too much it jams against the cylinder; too little and it jams against the piston. Fedden had to consult 60 firms before he found the right alloy.

Many engines had superchargers—compressors to force extra fuel and air into the cylinders and thus improve the engine's performance. These had been used as early as 1910, but were never entirely satisfactory because the compressor needed power to drive it, thus wasting some of the extra power it gave. Several attempts were made to build a turbocharger powered by a turbine driven by the exhaust gases, but there was no alloy that would withstand the high temperature. This was found later.

By the mid-1930s, engines were producing so much power that the propeller was being driven at an excessive speed. The tips of the blades broke the sound barrier and created shock waves that reduced the propeller's efficiency. The difficulty was overcome by gearing the propeller down. The more advanced American engines had variable gearing. By the end of the 1930s most propellers also had variable pitch (blade angle) so that they could run efficiently at different speeds.

There was always an incentive for designers to produce more and more powerful engines. During the 1920s and 1930s it was the glamorous (and lucrative) Schneider Trophy; later it was the desperate need to build fast aircraft in the Second World War. Rolls-Royce produced a V-12 water-cooled engine for the Schneider Trophy which gave 2600 hp, though it could only maintain this for one hour. This was all that was needed for two successive races, but the basic design of the short-lived engine was used for the famous Rolls-Royce Merlin, which powered the Spitfire, Hurricane and Mustang in the Second World War. The original 1934 Merlin produced only 790 hp, but by the end of the war this had been increased to well over 2000 by successive modifications.

By this time, the piston aircraft engine had reached the end of its possibilities. It was used for years afterwards, and still is, but leadership in design had passed to the jet engine.

The Rolls-Royce Merlin is one of the most successful aircraft engines ever built. Developed from a design which produced 2600 hp in short bursts for the Schneider Trophy race, the Merlin continued in production beyond World War II, and was even built under license in the USA. The 1934 production model developed 790 hp; the Rolls-Royce design philosophy of improving an existing device by successive modifications raised this to well over 2000 by the end of the war. The P-51 Mustang, one of the most celebrated aircraft of the war, did not achieve its potential until fitted with a Merlin. The engine in the photograph is a Mark 60 of 1943, which produced 1250 hp. The large domed casing at the rear contains a two-speed two-staged supercharger.

The jet engine

The gas turbine and its variant, the jet engine, are the latest developments of old ideas. The word 'turbine' comes from the Latin 'turbo', which means a whirl or eddy. Water wheels, which have been in use for centuries, are turbines.

In 1791 the gas turbine as we know it was described in a patent by John Barber. His drawing showed the essential features of the gas turbine engine: a compressor, from which air is passed to a continuous-flow combustion chamber, in which the fuel is burnt; and a turbine, through which pass the resultant hot gases. The hot gases turn the blades of the turbine and this motion is used to drive the compressor. There is a surplus of power over that needed to drive the compressor, because it is a feature of the behaviour of gases that the work available from the expansion of a hot medium is greater than that required to compress the same medium when it is cold.

Other internal combustion engines work on an intermittent cycle. Any single part of the engine is exposed to combustion gases for only a short time and is then washed by the relatively cool charge of the next cycle. Practically the whole of the oxygen in the air can be utilized, giving temperatures of over 2500 K (2227°C, 4040°F) in the combustion products, while few parts of the engine exceed 500 K (227°C, 440°F).

The Rolls-Royce RB 211 is one of the best known jet engines in service. Problems with carbon fibre fan blades on the original engine were solved with new titanium blades. The modular design means low maintenance costs, and the engine has been developed as a stationary pump for pipelines. The three-shaft layout results in a rigid, stable engine with relatively low noise and low fuel consumption for the power obtained.

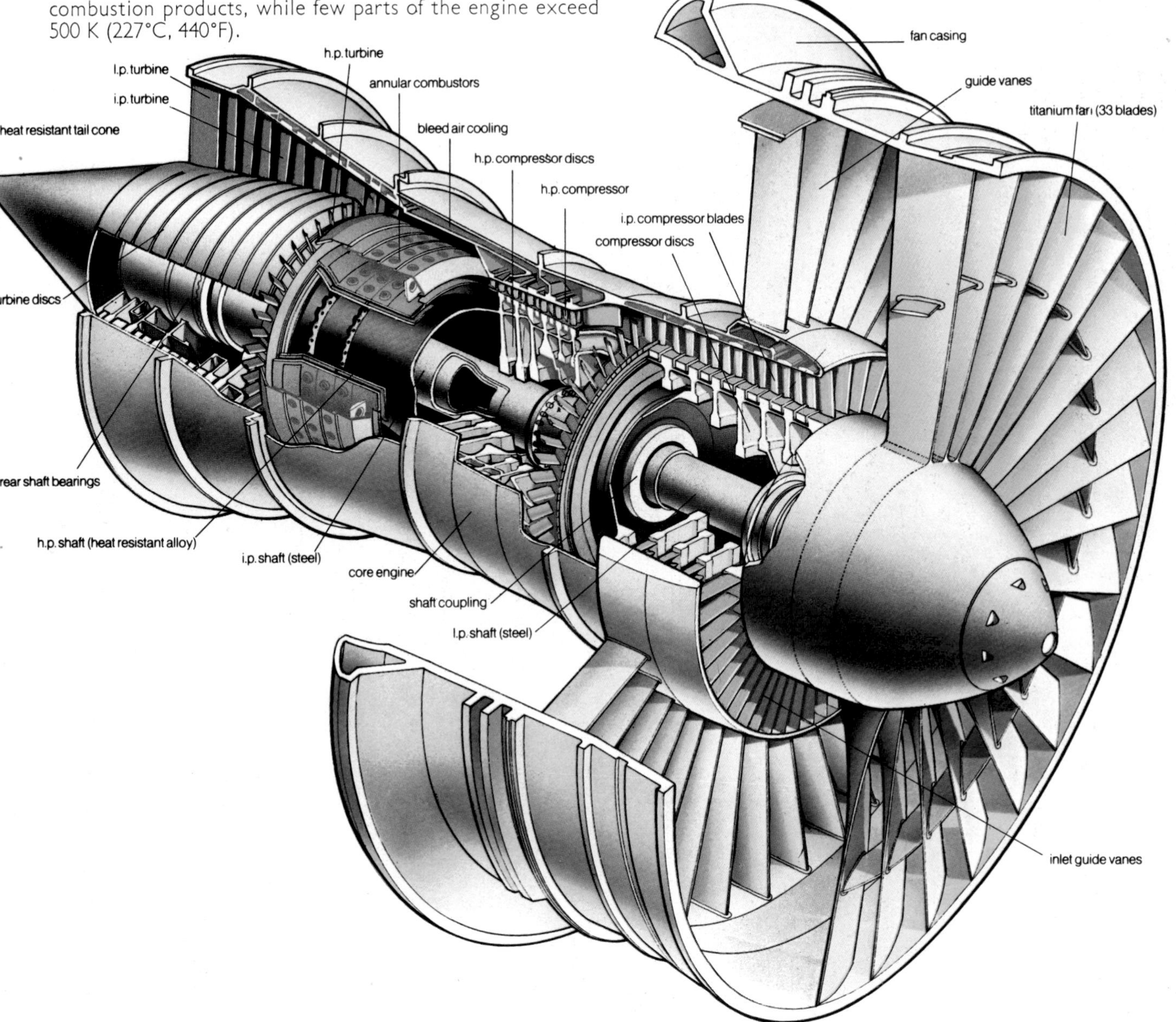

(K stands for Kelvins, the unit of measure of heat in thermodynamics, in which temperature is measured from absolute zero. Temperatures quoted in Kelvins may be converted to Celsius by subtracting 273.)

By contrast, the blades of a turbine, which experience a continuous flow of combustion through them, can only withstand a temperature of about 1100 K (827°C, 1520°F) if they are to be reliable over a useful period of time, even when they are made of complex modern alloys of nickel and cobalt. Only part of the oxygen in the air can be used if the temperatures of critical parts are not to exceed this figure. This fact seriously limits the output and efficiency of the gas turbine and provided a stumbling block for many years. There were no practical developments until the beginning of this century.

The turbocharger was the first successful use of the principles of the gas turbine. It is a variant of the gas turbine which produces no power in itself but increases the power output of the engine to which it is attached. A turbine is operated by the flow of hot exhaust gases from the engine; the turbine operates a compressor which raises the density, and hence the mass flow-rate, of the air charge to the combustion chamber of the engine. The first turbocharger was designed by Alfred Büchi and built by the Swiss firm of Brown-Boveri in 1911. In 1916 turbochargers were first used in aircraft, and were subsequently used in the most successful aircraft of World War II.

The General Motors Allison GT-404 gas turbine engine is being built in both America and Britain. With a heat exchanger and other design features, it is said to be the first gas turbine engine suitable for road vehicles.

The centrifugal compressor was the only machine available when the turbocharger was first developed, and was capable only of low pressure ratios and modest efficiency, insufficient for a power-producing engine.

Just before World War II gas turbine engines were built by Brown-Boveri and installed for electrical generation. These had multi-stage axial-flow compressors, which have rotating blades and stationary blades arranged so that pressure is built up in stages. More efficiency and higher pressure ratios are available from this type of compressor.

The specific power output (the power obtained divided by the air-mass flow rate, a size criterion) and the thermal efficiency (the power obtained divided by the fuel energy rate, a running cost criterion) both increase with turbine inlet temperature, compressor pressure ratio (up to a point) and the efficiencies of compressor and turbine. Efficiency is determined by comparing the power of the actual machine with that of a theoretical machine having no frictional or other heat losses. Efficiencies of both compressors and turbines can be as high as 90% nowadays, but were much lower at the beginning of development.

The first man to envisage the use of gas turbines for aircraft propulsion was Frank Whittle, a Pilot Officer in the Royal Air Force, who was later knighted for his achievement. He patented in 1930 the combination of a simple gas turbine and a nozzle to provide a jet-propulsion device, or turbo-jet. In 1936 he formed a company, Power Jets Limited, to develop it, and in 1939, the company received a contract for a flight engine. The same year a German design, the Heinkel He 178, made a short flight; the Gloster-Whittle E 28/39 was more successful in 1941. In that year a Whittle W1X engine was shipped to the USA and the General Electric Company began development; the first American jet aircraft, the Bell XP59A, flew in October 1942.

These engines all had centrifugal compressors; it was not until well after the war that axial-flow compressors were reliable enough for aircraft. The promise of the turbo-jet idea was to provide greater thrust for less weight and with less vibration than the conventional combination of internal combustion aircraft engine and propeller, as well as overcoming the forward-speed limit of the propeller itself. In war-time the advantage of jets would have been their ability to climb quickly to intercept bombers, as well as being able to fly faster than bombers or enemy fighters. The Germans had begun development of axial-flow compressors, but at the end of the war they were only just beginning to develop new alloys which would have made possible longer flying time without overhaul of the engine.

After the war, many successful jet engines were sold with centrifugal compressors, but ultimately they were a dead-end development for aircraft, because not only was efficiency low but the frontal area of such engines was high, leading to engine-nacelle drag. Axial-flow designs made possible better aerodynamic design of aircraft.

In modern jet aircraft which fly at below the speed of sound, modifications to the engine are made. The efficiency of the engine is high only if the pressure ratio and turbine inlet temperature are high, and this leads to a high jet velocity which is wasteful at lower forward speeds. This is aggravated by the use of air-cooling for the turbine blades, which allows inlet temperature to go still higher. Accordingly, the turbine is extended to extract more energy from the gases, and this energy powers a fan which is placed in front of the compressor. This fan blows air back between the body of the turbine and the outer concentric casing, and a greater flow emerges at lower velocity, giving a greater thrust with the same amount of fuel. An advantage of this type of engine, called a turbofan or fanjet, is that it is quieter because the jet velocity is lower. The ratio of the airflow through the outer duct to that going through the turbine is the bypass ratio. Fanjets with high bypass ratios propel the large airliners of today. This system has been developed with multiple swivelling nozzles for the Harrier 'jump-jet' aircraft.

For supersonic flight the simple jet engine is entirely suitable, as the high jet speeds match more closely the high forward speed of the aircraft. Jet speeds are sometimes increased by burning extra fuel between the turbine and the propelling nozzle, using surplus oxygen left in the turbine gases. This is called afterburning and increases thrust considerably. During take-off, fuel consumption increases faster than thrust, but at cruising speeds afterburning can be quite economical. Afterburning is incorporated in the Rolls-Royce/SNECMA Olympus 593 engines for the Concorde, which cruises at 2.2 times the speed of sound. A characteristic of afterburning is that it greatly increases the noise level of aircraft taking off.

The turboprop engine, like the fanjet, makes use of the extra turbine energy to save fuel, but uses it to turn a conventional propeller rather than a fan inside the engine cowling. During the 1950s turboprops were used on large jetliners, but cruising speeds have increased since then with development of the jet engine, and turboprops are used today only on smaller aircraft. The best turboprop designs are lighter, less noisy and more free from vibration than comparable piston engines; the Rolls-Royce Dart Mk 525, which had a two-stage centrifugal compressor and was used in the Vickers Viscount, was well known for its smooth operation.

An engine similar to a turboprop, but used to drive a transmission shaft rather than a propeller shaft, is called a turboshaft engine. Such engines have been used to drive helicopter rotors, notably by the Bell Aircraft company of the United States, on several models beginning with the XH-13F of 1955. The engine is usually mounted on top of the fuselage directly adjacent to the rotor.

Ramjets and pulse-jets are also called athodyds, an acronym from Aero THermODYnamic Duct. They have no rotating parts.

In the ramjet, incoming air is compressed by a specially shaped inlet nozzle which slows its velocity and raises its temperature. After combustion the hot gases are allowed to expand and leave the rear nozzle at a velocity greater than that of the aircraft, resulting in thrust. The ramjet operates only at forward speed, so a take-off assist is necessary. The ramjet has been used to power guided missiles which do not fly outside the atmosphere. They have been used to power target drones, and have also been fitted to the tips of helicopter rotor blades, with inlet passage on the front edge of the blade. Ramjets are most efficient in the 1500 to 2500 mph range.

The pulse-jet, as its name implies, operates intermittently. The air enters through a valve which then closes; combustion then takes place and thrust is produced; when the pressure on the combustion chamber drops, the valve opens and the cycle begins again. Pulse-jets were used during World War II in the German V-1 flying bomb. They were unreliable, extremely noisy and had high fuel consumption, but were simple and inexpensive to build and were a good choice for this application, because they only had to operate for a few minutes to get the bomb to the target. The pulse-jet tends to lose thrust with speed, and is most efficient at subsonic speeds. Since the war, they have been used for target drones, and were also used on the tips of helicopter rotor blades.

The turboramjet is a hybrid jet engine in which the turbojet section is closed off from the airflow when speed is attained, and the bypassed air is used in a large afterburner.

Gas turbines for powering road vehicles such as trucks and buses have been talked about for years, but have not been available until recently. Since the engine of a road vehicle would not be running at a continuous optimum speed, as would the engine of a jet plane or a power plant generator, fuel consumption would be unacceptably high, because turbine inlet temperature drops as turbine rpm drop. When the engine is running at 60% of load, for example, 80% of full load fuel may be necessary to obtain 60% load power.

In 1971 the Allison division of General Motors began building a gas turbine engine which was said to be competitive with diesel engines for lorries, buses, marine propulsion and small stationary duties. It is also clean enough to meet the tough exhaust emission control laws of the state of California. The latest model of this engine is now being built in both the USA and Britain. It uses a heat exchanger so that the hot turbine gases can heat the cold air coming from the compressor in order to maintain efficient operating temperature of the turbine inlet. The engine also has an oil-cooled power-transfer clutch between the turbine shaft and the power shaft. The clutch assists economy by automatically slowing the turbine shaft when the power shaft slows down, thus slowing the compressor and thereby raising the temperature to the turbine inlet. For safety, the clutch also prevents the power shaft from running faster than a preset level, and provides braking power to the vehicle equal to the rated power of the engine.

This is a cutaway view of a Rolls-Royce jet engine. The hot gases from the continuous combustion are ducted through the engine in such a way as to do the most possible work. The engine castings and turbine blades are made of special alloys to withstand high temperatures.

The rocket

The generic term 'rocket' loosely applies both to any non-air-breathing jet engine and to the vehicle it propels. Small rockets, which carry scientific instruments on short parabolic flights to the edge of the atmosphere, are called sounding rockets. Large multi-stage devices designed to carry spacecraft into orbit are more correctly called boosters or carrier vehicles, while their propulsion units are usually described as rocket engines (if liquid-fuelled) or rocket motors (if solid-fuelled). The terms are flexibly applied and, particularly in the USA, 'engine' tends to mean larger propulsion units of both types and 'motor' refers to smaller ones.

Newton's third law of motion—to every action there is an equal and opposite reaction—is the basic principle of rocket propulsion. Turn up the water in a garden hose and watch the nozzle jump back. Cover half the nozzle with your thumb and feel the extra strain on the nozzle as the water suddenly streams farther and faster. Likewise, throw anything off the back of a rocket and the rocket will move forward. Speed of movement will naturally increase if you throw more at once or throw it faster—or, best of all, do both. So the escape of burning expanding gases, accelerated by passing through a constricted nozzle, propels a rocket engine in the opposite direction.

The first known form of rocket appeared in the early 13th century in China, following close on the invention of gunpowder, and was simply an arrow with a tube of powder lashed to it. By the mid-13th century, first the Mongols and then the Arabs were using rockets in battle. The French crusaders brought them to Europe, and French troops under Joan of Arc defended Orleans with rockets in 1429. But, by this time, cannon and small arms were proving more accurate and effective, so that the rocket faded from the military scene, if not the festive one, for the next 350 years.

Then, in 1792, British troops fighting in India were heavily assailed by small metal-cased rockets with an effectiveness that revived British respect for the military potential of such devices. It was the director of Woolwich Arsenal, Colonel (later Sir William) Congreve, who, by 1804, developed this still fairly crude device into an efficient and destructive naval weapon with an incendiary of explosive warhead.

The accuracy, however, remained far from satisfactory until the mid-19th century when the Englishman William Hale applied spin-stabilization by means of curved vanes in the exhaust nozzle. Range was still limited in relation to size until 1855 when an idea (actually conceived by a Frenchman named Frezier) for stacking two rockets together was first applied by Colonel Boxer. The result was a two-stage, line-carrying, life-saving rocket in which the exhausted first stage ignited the second stage via an explosive separation charge.

Meanwhile, the propellant potential of combustible liquids and the broader applications of rocket power were becoming increasingly appreciated and numerous people are credited with devising rocket machinery of amazing foresight or absurdity. The first realistic theoretician was Russia's Konstantin E Tsiolkovsky who, as early as 1883, recognized the potential of the rocket in space. He first understood the significance of increasing exhaust velocity, tne importance of mass ratio (launch weight to engine burn-out weight) and their relationship in increasing vehicle velocity. This led him into extensive study of multi-staging techniques, both in parallel and tandem arrangement, efforts which subsequently earned him the title of the 'father of spaceflight'.

His work also inspired the pioneers of modern rocketry, notably in Russia, Germany and the USA. In Russia, rocketry enjoyed official status from the start in 1929 when research began at the Gas Dynamics Laboratory in Leningrad. In 1933 GDL linked up with Moscow's enthusiast 'Group for Jet Propulsion Research' (GIRD) and, with military finance, built liquid-fuelled rockets which flew to a record 3.5 miles (5.63 km) altitude in 1936.

Germany's pioneer was Hermann Oberth, a theoretician whose concepts in liquid-fuel rocketry prompted a group of young engineers to form the VFR (Society for Space Travel) in 1927. The VFR's practical experiments laid the foundation for Germany's wartime lead in rocketry. After it dispersed in 1934 under Nazi opposition, military research absorbed some of its members, notably Wernher von Braun, later to lead development of the V-2 rocket and then to become the guiding light of the USA's work up to the Apollo programme. In the USA, however, the pioneer was physicist Robert Goddard whose own small group struggled on limited private financing from the early 1920s, launching the world's first liquid-fuelled rocket in 1926 and continuing with distinctive success until Goddard's death in 1945.

But by 1945 the USA, if still uninterested in spaceflight, had become very aware of the new vogue of war rockets—the ballistic missile. It was the fall of Germany and the seizure of its world-beating rocket technology by both Russia and the USA that lent the final impetus to full-scale rocket develop-

ment, unchecked until man stepped on to the Moon.

The increasingly diverse methods used for generating rocket exhaust mean that the basic Newtonian law often remains the only common factor. The comparative criterion, however, of any propellant (or propulsion system) is its specific impulse (I_{sp}). This indicates the number of pounds (or other unit) of thrust available per second from every pound of propellant consumed. I_{sp} is measured in 'seconds' and is directly related to the speed of the exhaust (an I_{sp} of about 102 seconds provides an exhaust velocity of 1 km/s). The higher the I_{sp} the lower the fuel mass needed for any specific thrust level.

High thrust-systems, where thrust greatly exceeds the engine weight, are necessary where gravity must be opposed, as in planetary lift-off and soft landing. The main contenders are chemical and nuclear rockets. Chemical types take two forms: solid fuel and liquid fuel. Both produce exhaust through combustion and must therefore carry their own oxygen supply. Solids are basically powder-packed squibs, the charge being a mix of dry fuel and a dry, oxygen-rich chemical (such as a mix of polyisobutene and ammonium perchlorate). They offer simplicity and therefore reliability,

After an air raid in August 1943, when over 800 people in the Peenemuende area were killed, V2 production was moved to several locations such as this one at Nordhausen. On 10 April 1945 the town was taken by the US First Army, who took over the underground factory and liberated the inmates of the nearby slave labour camp. The V2 was the world's first guided missile, and more than 5000 were built.

but a lower I_{sp} and, being all combustion chamber, a heavy structure. Nor can they be controlled except in burn rate, determined by the shape of the propellant grain or of the core cavity.

Modern liquid-fuel engines, in which fuel, such as hydrazine or liquid hydrogen and an oxidizer, such as liquid oxygen, are pumped separately by turbopumps into a small combustion chamber, can be stopped, restarted and throttled at will. But the price is complexity and reduced reliability.

Some form of nuclear propulsion is seen as the high-thrust system of the future. Solid-core devices which release heat from a fission reaction have already been ground-tested in Russia and the USA. Fuel pumped through perforations in the core is superheated and expelled through a nozzle. The fusion rocket, which at the moment is purely speculative, would exploit a hydrogen conversion process similar to that in the Sun itself—and with comparably high energy output. Fission (as in an atomic bomb) of 1 lb of uranium, or fusion (as in the hydrogen bomb) of 1 lb of heavy hydrogen, produces two million times the energy obtained from burning 1 lb of kerosene.

But greater attention is now being given to the idea of streaming nuclear 'bomblets' behind a thrust shield where they are positioned by a magnetic field and detonated by laser beam. This could provide the hard acceleration for quick orbital escape or entry, or fast interplanetary flight. Contamination would prevent its use for Earth launch, which means that inefficient chemical rockets are likely to be in use for a long time to come.

Scaled-down chemical rockets are used today to power orbit change, escape or capture, interplanetary course-corrections and, in miniature form, for attitude control of probes and satellites and for rocket stage and payload separation. Solids are used if only one burn is needed (as in stage separation). Spacecraft control thrusters now tend to use monopropellants (fuel and oxidizer in a single fluid) or hypergolic chemicals which ignite on mutual contact. Tank pressurization forces the propellants into the chamber under the control of simple valves; I_{sp} is low but reliability is greatly improved. Simple venting of compressed cold gas is also exploited. But the low I_{sp} of all of these means a heavy or limited fuel supply.

Unfortunately, under current technology, very high I_{sp} systems offer only a minute thrust in relation to engine weight, so that they are practical only where inertia is the sole opposition. The most promising of these is the electrostatic or ion rocket already test-flown in space, which works by isolating ions (electrically charged atoms) and accelerating them to produce exhaust. The thrust is tiny (about 1/100 oz, 0.2835 g), but with its compact fuel load and a nuclear generator, an ion-powered spacecraft could keep accelerating gently for months and years, eventually developing phenomenal speeds. Deep-space missions are the obvious application but, within ten years, Earth satellites will also be carrying ion engines for orbital positioning and attitude control.

Ion rockets have specific impulses some ten times higher than chemical rockets—that is, the fuel for a given thrust weighs ten times less. Ion propulsion motors commonly use caesium or mercury as the propellant. The mercury is vaporized by passing it through a heated porous plug, and the vapour is fed through a narrow tube which is the cathode, or negative terminal, of a high-voltage supply. An arc strikes between this and a positive anode, called a keeper, creating positive mercury ions and electrons from the mercury vapour. More vapour is fed separately up to the chamber, where it in turn is ionized by collisions with the free electrons. The electrons are attracted towards the main anode, but their path is deliberately lengthened by means of a weak magnetic field, ensuring a greater number of collisions with neutral atoms.

The positive ions are accelerated toward and through

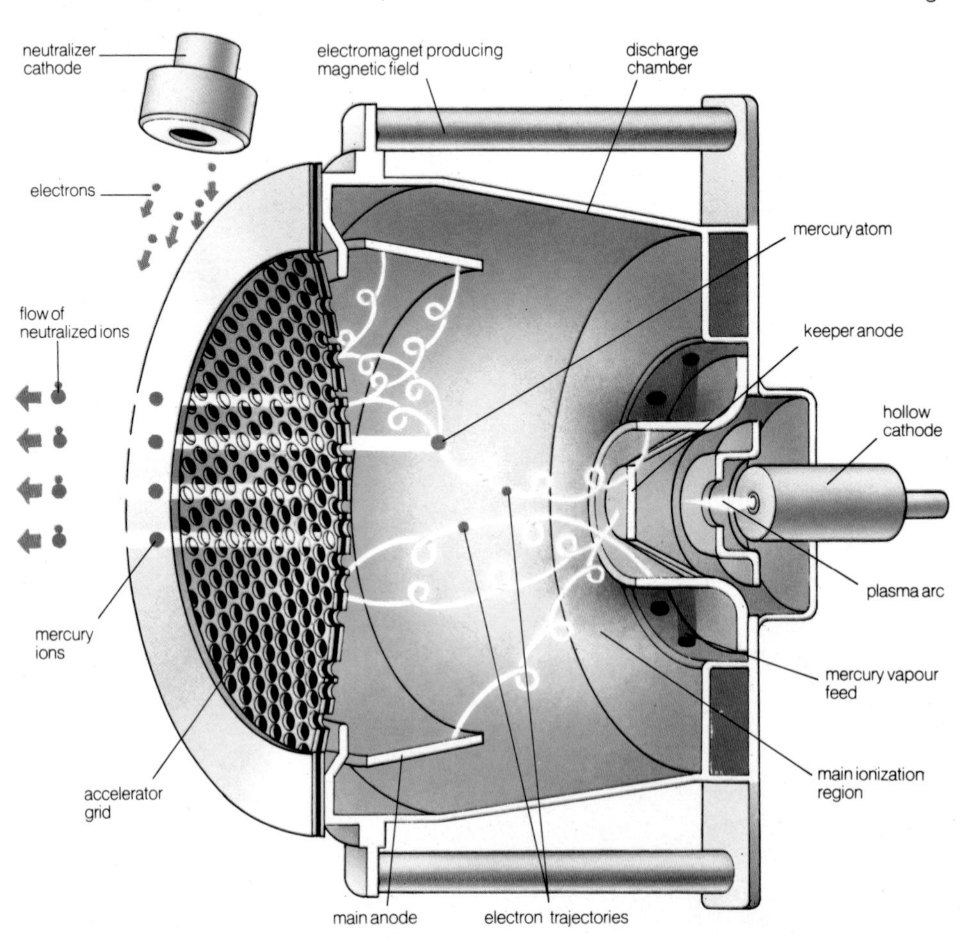

the negatively charged acceleration grid, applying a reactive thrust to the motor. To prevent the ions being attracted back to the motor, they are neutralized by a stream of electrons.

In a similar vein is the still theoretical electromagnetic or plasma rocket, in which a fuel such as hydrogen is converted to a neutral plasma or conductive-gas state by an electric arc and then accelerated out by a magnetic field. Nuclear fission is also under study for this category in the form of a gaseous core rocket where fuel is passed through a gaseous reactor, suspended in the chamber by magnetic fields. Like other concepts using magnetic fields for containment and positioning, this faces the major technological hurdle of accuracy in field control.

But drawing boards abound with even more exotic ideas. To achieve extreme velocities such as a realistic percentage of the speed of light, emissions of near-light-speed itself would be necessary, hence the theoretical photon rocket—in principle just a concentrated beam of light shone rearwards. Its application, however, would be more for the humbler role of minimizing fuel loads. Then there is the space ramjet, fed from the sparse hydrogen content in space collected by a gigantic bow scoop. Or the solar rocket in which hydrogen is superheated by solar energy captured in a huge parabolic dish.

Rocket research is still largely focused on the chemical engine—finding better propellants and more efficient ways of exploiting them. The main challenge is in the materials required to cope with ever increasing extremes of operating temperature and pressure or to produce the right flexibility, rigidity, lightness, purity or thermal conductivity. In general copper alloys tend to be used for lining combustion chambers and aluminium alloys for structural elements. Glass fibre is also finding a place in small engines. The engine that will power the 'space shuttle' (NASA's reusable rocket plane due to enter service in 1981) will operate at a record chamber pressure of 3000 psi (206.8 bar).

In the USA, engines are usually designed specially for a particular booster stage, each stage being a self-contained rocket in its own right. Specialist firms are paid by NASA to produce competitive preliminary designs so that the best can be selected. Usually one of the major aircraft companies builds the structures, tanks and controls and installs the engines, different stages of a booster often being built in different parts of the country. The stages are brought separately to the launch site by air, road or (for the giant Saturn 5 lower stages) by barge, and assembled vertically on the launch pad. The Saturn 1B and 5 Apollo launchers, however, are stacked in Cape Kennedy's Vertical Assembly Building, the world's largest building, and then carried erect by the huge crawler transporter to the launch pad 1.5 miles (2.4 km) away.

Countdown can begin days before launch to schedule in complex fuelling, checkout and contingency operations. Final ignition sequence is usually automatic from about zero minus three minutes. Ignition is achieved either electrically, explosively or from hypergolic chemicals. Restraining arms hold the rocket for 2 to 4 seconds or until full thrust has built up. As the rocket ascends, the thinning air allows the exhaust flame to broaden and shorten. (Upper-stage engines designed to operate purely in space tend to have longer, narrower engine 'bells' to contain the exhaust flow. Directional control is often achieved, especially on solids, by swivelling this bell). During ascent, power is sometimes reduced to ease acceleration loads, either by early cut-out of one engine in a cluster or from burn-out of small, strap-on, 'half stage' rockets. First stage burn-out and jettison usually occur at 30 to 50 miles (48 to 80 km) altitude and one or more upper stages takes the payload on to the orbital height of 200 miles (322 km) or more and a speed of 17,500 mph (28,160 km/h). Orbit is achieved 12 to 13 minutes after lift-off.

Above: this 20 inch (½m) ion engine, shown in a vacuum test chamber, was designed to help find the optimum size for such engines.
Opposite page: a diagram of a T4 ion engine. The arc between the hollow cathode and the keeper starts the process off, but the main ionization region is in the centre of the chamber. The overall trajectory of each electron is a gentle curve, but the effect of the magnetic field is to make it spiral. When an electron ionizes a mercury vapour atom, a spare electron is produced; this too spirals away and the process continues, though it has reduced energy and the spirals are tighter. When the electrons strike the walls of the chamber, they bounce off until they reach the anode. This thruster has an exhaust velocity of 30km/sec, and a thrust of 10 millinewtons.

WATER TRANSPORT

In 1947, a party of Norwegian scientists sailed the Pacific in a raft proving that long sea voyages could have been made by ancient peoples. But the modern story of transport begins in the Age of Discovery, when in the fifteenth and sixteenth centuries brave men sailed into the unknown to discover the world. The world's navies have long since advanced beyond the technology of sails, but the traditional lore of the sea is still highly valued: in this picture, a training ship for officer cadets of the Chilean Navy calls at Sydney Harbour.

Boatbuilding

Boats developed from a number of primitive water craft in different places and at different times. The earliest such craft may have been a log used as a float to cross a river. The log became a dug-out canoe when fire and axes were used to hollow it out. Two such canoes were lashed together for stability, or an outrigger was added. In pre-historic Europe, stability seems to have been achieved by placing weights in the bottom of the canoe.

In Egypt, where timber was not available, boats were built by lashing together bundles of reeds, and pulling the two ends of the boat up from the water by means of ropes. In China, flat bamboo rafts were in use for river haulage, probably from about 4000 BC. Later, perhaps about 1000 BC, they began to adapt these rafts by laying the bamboo along the curved sides of semi-circular wooden planks, thus creating a vessel which had a number of solid bulkheads down its length. This continued to be developed until it resulted in today's Chinese 'junk'. Another ancient type of boat, the wooden frame covered with skin or bark, survives today in the form of the Irish curragh, the Eskimo kayak and the American Indian canoe.

For thousands of years, wood was the most common material used in boatbuilding, but in the 1950s glass reinforced plastic brought rapid changes, allowing mass production techniques and greater flexibility in hull design. Today about 70% of boats built in Europe and the USA are made of GRP.

Plank construction Boatbuilding in wood is a skilled craft involving the use of a great many wooden components to build a watertight structure. This has to combine stability in the water with the ability to withstand stresses often comparable with those experienced by jet aircraft.

Wood boatbuilding follows two principal styles, clincher (sometimes spelled clinker) with overlapping planking, and carvel or caravel where the planking is smooth. The clincher or lapstrake technique gives a 'monocoque' or stressed skin construction. The shape of the craft is developed by the addition of successively fastened planks, sometimes using only a single mould or guidance shape amidships. After the planking is completed, light frames are steamed or sawn to shape and added to the interior to strap the planking together as a safeguard against a plank splitting along the grain. This construction makes the boat very light but vulnerable to hard use. The overlapping edges or lands of the planking are liable to wear and it is difficult to keep the

planking watertight once it has been disturbed. It is therefore normally used only for small light craft such as beach boats.

In carvel construction wooden planks are fitted edge to edge over a completed framework which determines the shape and forms the structural support of the finished craft. The framework consists of a centreline piece called the keel with a companion piece called a keelson, a stem at the front which is joined to the keel with a wooden knee (angle piece) called a foregripe. At the other end the upright part of framing consists of a sternpost with a supporting knee. The framing across the boat is normally built with timber sawn from branches whose grain lies roughly in the required curves. These frames are in turn strapped together inside with full length wooden planks called stringers, and at the deck edge with an inwale or interior plank which is sometimes called a beam shelf if it has to carry the deck beams. Originally a number of heavy planks called wales were arranged longitudinally and fastened to the outside of the frame before the skin planking was fitted. Their function was to stop the planking from spreading when the caulking compound was hammered between the planks to make the hull watertight. With modern improvements in building techniques, wales have largely disappeared.

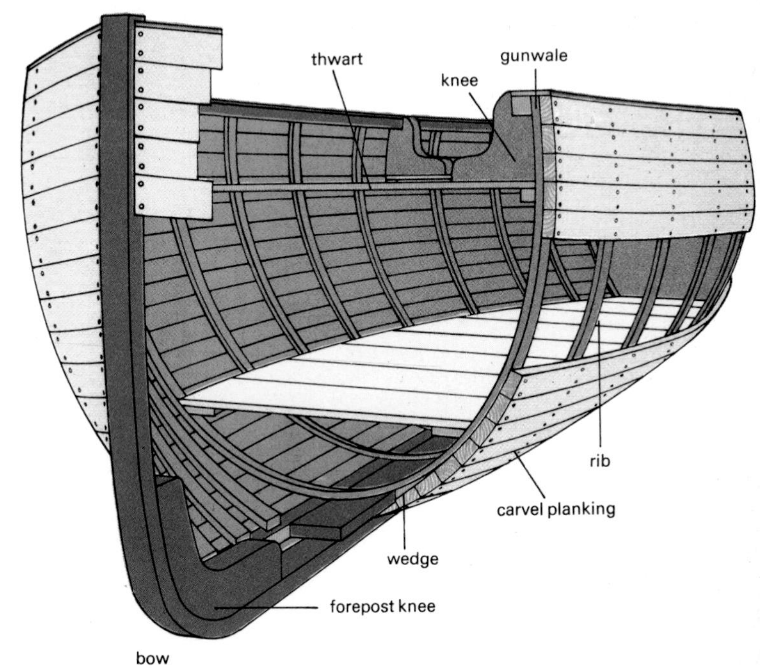

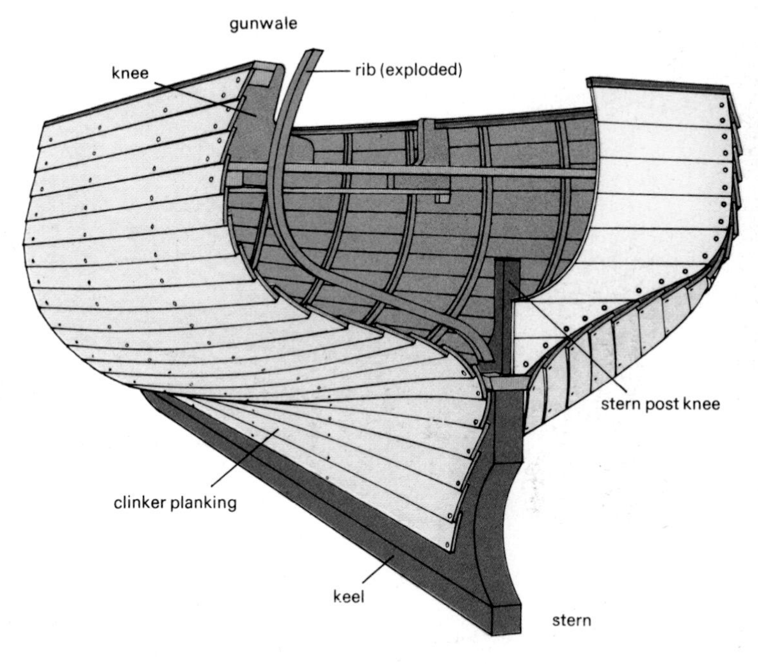

Above: diagrams of carvel (top) and clinker construction. In the carvel type, the planks are laid on the framework with edges smoothly together, usually from the top down; in the clinker, planking starts at the keel and goes up, each plank overlapping. The ribs are bent into place afterwards.
Left: Lake Titicaca, in Peru, is the world's highest large freshwater lake, and one of the deepest. The traditional boat of the local fisherman is made of bundles of papyrus reeds lashed together, and is remarkably similar to ancient Egyptian boats.

There are many variants of the two techniques and light, steam-bent strap frames are often used to augment and lighten the sawn frame structure. Other common variations include composite constructions where steel frames are used with a wooden keel and planking; multiple planking where two or more layers are placed diagonally to make a strong skin which is more watertight, though difficult to repair; and strip planking. Very narrow planks are used in the normal way for strip planking but are nail fastened through their thickness to the previous plank as well as being fastened to the frames.

To get the best performance from a fast power boat a knuckle or chine (projecting corner) must be built at the division between the bottom and sides of the hull. This chine is built like the centreline keel and the technique is known as a chine construction.

Plywood During the 1920s boatbuilding techniques using plywood were developed, largely in the USA. The invention of waterproof resin adhesives led to the development of water resistant 'marine grade' plywood. This quickly became popular for building bulkheads or vertical wall divisions of the hull and later it became a very common material for the planking skins of inexpensive small craft. This technique, still widely used in 'do it yourself' built craft was the first real attempt to use glue to stop water from entering between the components of the hull.

Despite great ingenuity, the shapes which can be constructed from bending flat sheets of plywood are very limited and were invariably angular in appearance. The next step therefore was to make the actual plywood sheets over a curved mould of the required shape. This type of building is now used for high-quality racing yachts where the high costs involved can be balanced against the strength and light weight. This moulded plywood construction involves the planking of the mould with very thin wood sheets of veneer thickness. These are added in successive glued diagonal layers until a finished thickness approximately half that of a carvel planked hull is achieved. If cold setting adhesives are used each layer is held down with staples until the glue sets. For hot moulded construction the layers are held in place by air bags or a vacuum press and moved into an oven to give a quick 'cure' to the hot setting glue.

A variation of this construction is the use of laminated structural members. Here the keel, stem, beam, and ribs are built up over a mould to the required dimensions and curves.

Glass reinforced plastics The most popular form of construction today uses polyester resin reinforced with glass fibre, generally spoken of as fibreglass or glass reinforced plastics (GRP). The normal building process starts with the construction of a full size solid model of the final boat, called the plug. Over this a hollow mould is formed by a similar process to the hull construction. The hull and deck mouldings, and even mouldings for the interior accommodation, are formed in separate moulds.

The materials used in GRP construction consist basically of polyester resin reinforced with finely spun glass fibres either in cut pieces or made up into a woven cloth.

The mould is first coated with wax or some other release agent to prevent the resin from adhering to the surface. Next a gel or surface coat of resin, usually impregnated with pigment to suit the final colour scheme, is sprayed or painted onto the surface of the mould. A very light supporting mat of glass fibres is then placed on top of the resin and pressed in with a roller until completely saturated with wet resin. Subsequent layers of moulding resin with heavier glass reinforcement carefully rolled into it are added until the required hull thickness and strength are attained. The resin hardens in three stages: first to a soft gel, then to a point where the moulding can be removed from the mould, and then over a period of weeks it matures to full strength. Bulkheads and reinforcements are added to the moulding either in or just out of the mould. Considerable skill and care

are necessary in glass fibre construction in order to make certain that the resin is properly supported by glass at all corners and edges, that no air is trapped between the layers and that the glass is thoroughly saturated with resin.

Another common method of fibreglass construction uses a hand held spray gun to deposit both the resin and chopped glass strands which leaves a mat of partly saturated fibre that must then be rolled to complete saturation as before. This method involves skilled spraying and accurate control of quantities to ensure uniformity. Glass fibre construction is also mechanized in other ways such as the use of glass mats previously saturated with resin, and vacuum or pressure resin saturation of previously laid glass reinforcements.

The undoubted advantage of the glass reinforced plastics method of construction is the monocoque or stressed skin nature of the hull, without any joins where water might enter. Another benefit is the reduction of maintenance, which can be one tenth of that for a conventional wooden hull. Further benefits lie in being able to use the moulds for production building, and the ease of achieving a high finish. The material, however, is fairly heavy (flotation chambers are often included) and inconveniently flexible. One way of correcting these faults is a 'sandwich' construction where another material is placed between the GRP layers to improve stiffness and reduce weight. End grain balsa wood slabs or foam plastics are commonly used, laid over areas or in patterns as required during the course of the hull moulding.

The cost of building a single hull in glass fibre is very high owing to the cost of the plug and mould. To overcome this some craft are built with a PVC or polyurethane foam core planked over moulds like a traditional wooden hull and then covered with fibreglass inside and out. The outside surface of the hull is ground, sanded smooth and paint finished.

Other plastics boatbuilding methods include simple foam plastic castings and vacuum forming, where a plastic sheet material such as ABS or polyethylene is heated until soft and then sucked down over a hullshaped former. This is either thick enough and the right shape to be rigid in its own right, or else it may be formed of two or more mouldings which are filled with injected foam to give buoyancy.

Other materials Boats are also built in metal. Steel, for instance, has been a common material for building boats as well as ships. Its strength and weight characteristics limit its use to larger craft but with welded construction and the new anti-rust coatings, steel is becoming more popular again for one-off larger craft.

Aluminium is also a popular material for high performance one-off yachts, increasingly since new alloys have reduced the original serious corrosion problems of aluminium in salt water. The metal's lightness also makes it suitable for small craft which have to be manhandled. Some small aluminium hulls are made by stretch forming, where a sheet of material is stretched bodily into shape over a hull-shaped former.

Above: 'plating' a 40 foot aluminium racing yacht. The frame is made first, then the contoured plates of the skin are welded or bonded on to it.
Left: concrete boats are cheap, strong and easy to build. They make good utility craft and are popular in underdeveloped countries.
Opposite page: the bottom hull section of a moulded plywood racing yacht is being lifted from the mould after curing in the large cylindrical oven at the rear. The bottom mould is put on to another mould where the upper hull is built up from layers of thin plywood.

Sailing

The history of man's ability to keep himself and goods afloat on boats must go back many tens of thousands of years, and these earliest logs, bundles of reeds or branches, could travel only where water currents or primitive paddles allowed. After the discovery of paddles it would soon have become obvious that the wind could be as powerful an opponent to muscular propulsion as could be the currents and waves, thus frequently forcing an undesirable drift to leeward. That man fairly early learned to use this drift to advantage is clear from vase paintings and clay models of Egyptian origin, variously dated by archaeologists to be 7000 to 11,000 years old. Certainly by 3000 BC this controlled drifting, or downwind sailing, using large square-shaped sails was a firmly established seagoing technique for the transport of men and goods.

From the details of reliefs depicting these vessels we can deduce that the sails were hung from a horizontal spar (yard) which could be set at different angles to the wind by means of ropes attached to its ends. The square shape is obviously consistent with the earliest technologies of woven fibre or reed, it can be easily hung, raised or furled on to a simple spar, and when controlled by ties from its corners it is naturally blown by the wind into a near optimum curvature

for downwind sailing, also allowing some angular variation from this course.

As in the much quicker development of modern fuel-driven transport devices, the technological development of sailing vessels has always been a compromise between demands for speed, cargo-carrying capacity and manoeuvrability, qualities which due to the complexity of their interaction and to local traditions and available materials throughout the world have led to the building of a tremendous variety of boats—by no means all successful.

The next important ability of a sailboat, after that of sailing fast downwind, is the ability to sail at angles departing from the downwind direction, and the greater this angle the greater will be the period of the voyage when wind can be used and oars put away. Thus an exceedingly important stage of this development is when a boat becomes capable of sailing reliably at 90° to the wind direction and so *holding station*. This achieved, the crew is completely independent of oars whatever the wind direction may be, as it is always possible to sail to any destination by holding station during periods of unfavourable winds, oars only being required for extreme wave-conditions and inshore navigation. Longer journeys would also become more acceptable as man's knowledge increased of the large scale circulations of wind over ocean surfaces, enabling routes to be picked which involved holding station for even less time. Easily established wind patterns, such as the monsoon in the Indian Ocean for example, might even eliminate the need for development of anything better than vessels that would just hold station, such as the *dhows* of the area.

Windward sailing Gradually boats achieved the capability of sailing slightly into wind, a feature so contrary to intuition that it is not surprising that the associated technology seems to have been learned, lost and rediscovered many times and by many apparently well-separated societies, even though the ability itself might be considered to be the best method for its distribution. This only reflects the heavy overlay of tradition in sailing vessel construction. A success in some detail, once achieved, was held firmly in local collective memory to such an extent that it often inhibited, or made impossible, the incorporation of further improvements. The advantages of sailing closer into the eye of the wind are considerable; to be able to sail closer than a trading adversary often ensures the quickest delivery of cargo, or a wartime adversary can often be outmanoeuvred or eluded irrespective of his possible weapon superiority.

The way a sailing boat works is almost intuitive for downwind sailing. In common experience, any toy boat can be blown along irrespective of its shape of simple sail. It is sailing against the wind which needs explanation.

It is impossible for a boat with normal sails to sail directly into the wind. It is possible, however, to sail along a line at right angles to the wind direction (*reaching*) by setting the sail closer to the centreline of the boat, rather than across it; it is easier for the boat to move through the water in the direction it is pointing, rather than sideways, so the result is that it moves at right angles to the wind. If the rudder is set so as to continually point the boat further into the wind, it is then possible to sail even closer to the wind direction. To progress into the wind, a series of paths at an angle to the wind are followed first one way then the other, always getting a little further upwind. This procedure is called *beating*.

The special design features required for good windward sailing were only being realized in the 19th century and the racing yachts of today have windward performances well exceeding the fast clippers of the 1850 period, although their maximum speed when reaching at 90° to the wind is only slightly superior. These special features are those associated with sideforces at least as much as with the dragforces most important in downwind sailing.

Above: the felucca, seen on the Nile, has a standard lateen rig. The billow of the lateen sail is distorted by the mast when the wind is on the wrong side, as in this picture, unless the sail is rearranged.
Opposite page, top: prahus, seen on Sanur Beach on the Indonesian island of Bali, are double outrigger canoes carrying light sails.
Opposite page, below: the appearance of the Chinese junk has changed little over the centuries, since the design of the hull and the sails are very efficient. The battens in the sails maintain their shape even when sailing close to the wind, as in this photograph taken near Hong Kong.

Opposite page: the Polish ship Dar Pomarza *sails with the wind in the English Channel. 'Ship' here has a special meaning: ships are fully square rigged, with topmasts and topgallant masts.*
Below: the US Coastguard barque Eagle *is 'reaching' (sailing at an angle to the wind) during the 1972 Tall Ships race. A barque has a fore-and-aft, rather than a square sail, on her after mast, requiring a smaller crew.*

Imagine a boat sailing close to the wind—that is, pointing to within about 45° of the wind direction. There are various forces on the sail and on the hull which can be split up into components in each direction. In each case, the sideforce is that force acting perpendicular to the flow of either air or water, and the dragforce is that acting along the flow. In each case these forces have a combined *resultant*. The sideforce on the sails and the sideforce on the hull act in roughly opposite directions because the wind is blowing on the sail, but the water is resisting the sideways movement of the hull.

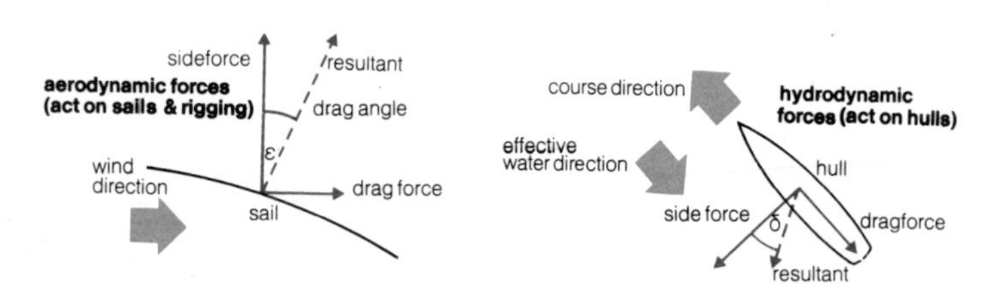

For successful sailing into the wind, the ratio of sideforces to dragforces must be a maximum, for both the aerodynamic forces acting on the sail rigging, and also for the hydrodynamic forces acting on the hull. That is, the sideforces should be as large as possible compared with their respective dragforces.

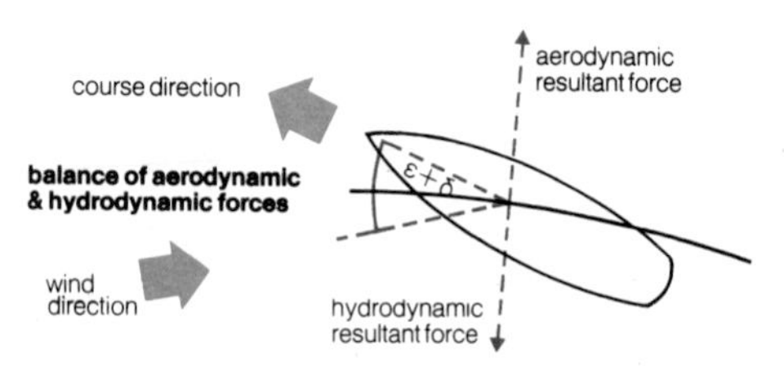

When a sailboat is in a steady sailing condition, the resultant aerodynamic and hydrodynamic forces must balance each other out completely. Therefore another way of stating the special conditions required for good windward sailing is that the drag angle of the sail (ε) and the drag angle of hull (δ) must each be as small as possible because, as the diagram shows, the angle of sailing into the wind is equal to the sum of ε and δ. As this sum decreases, so the boat will be capable of sailing closer into the wind.

The dragforce results from the fluid particles being slowed by the object itself, the drag being less as the shape is streamlined and its surface is smoothed. The sideforce (or lift in the case of aircraft) depends subtly on the curvature and thickness of the shape, which is similar to an aerofoil. The high values of sideforce coupled with low dragforce required by a sail for good windward performance demands thin spars and rigging, a stiff forward edge (luff) and a small curvature (camber) so that the shape is not destroyed when used at the low angles of wind incidence demanded for windward working. For hulls, good performance demands general smoothness, well-rounded bilges, deep keels, gentle stern exits from the water level, well-designed rudders and a low and smooth silhouette above water.

The origin of sideforce, at its simplest, is the Newtonian reaction to the deflection of fluid particles with greater momentum toward one side of the hull than to the other side due to the hull being aligned asymmetrically to the water flow. Bernoulli's equation, used in hydro- and aerodynamics, shows that the static pressure is lower on the side which has the higher water velocity past it. Thus a sail (or hull) is essentially sucked sideways by the leeside static pressure being lower than the mean value due to its higher streaming speed, assisted to a lesser degree by the excess pressure or slower streaming speed on the windward side.

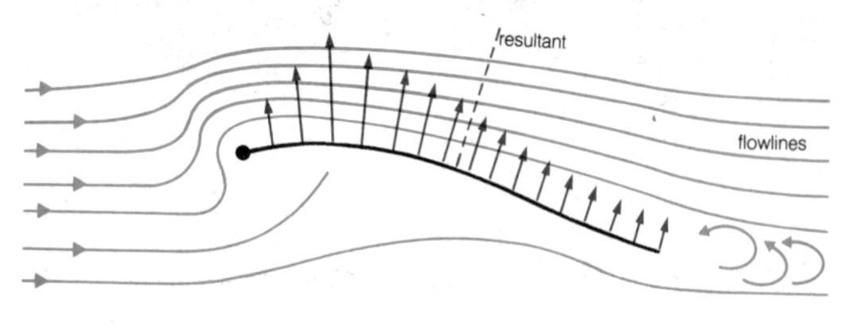

1 jigger topsail
2 spanker
3 mizzen royal
4 mizzen topgallant
5 mizzen upper topsail
6 mizzen lower topsail
7 mizzen crojack
8 mizzen topmast staysail
9 mizzen topgallant staysail
10 mainsail
11 main lower topsail
12 main upper topsail
13 main topgallant
14 main royal
15 main skysail
16 main royal staysail
17 main topgallant staysail
18 main topmast staysail
19 main staysail
20 foresail
21 fore lower topsail
22 fore upper topsail
23 fore topgallant
24 fore royal

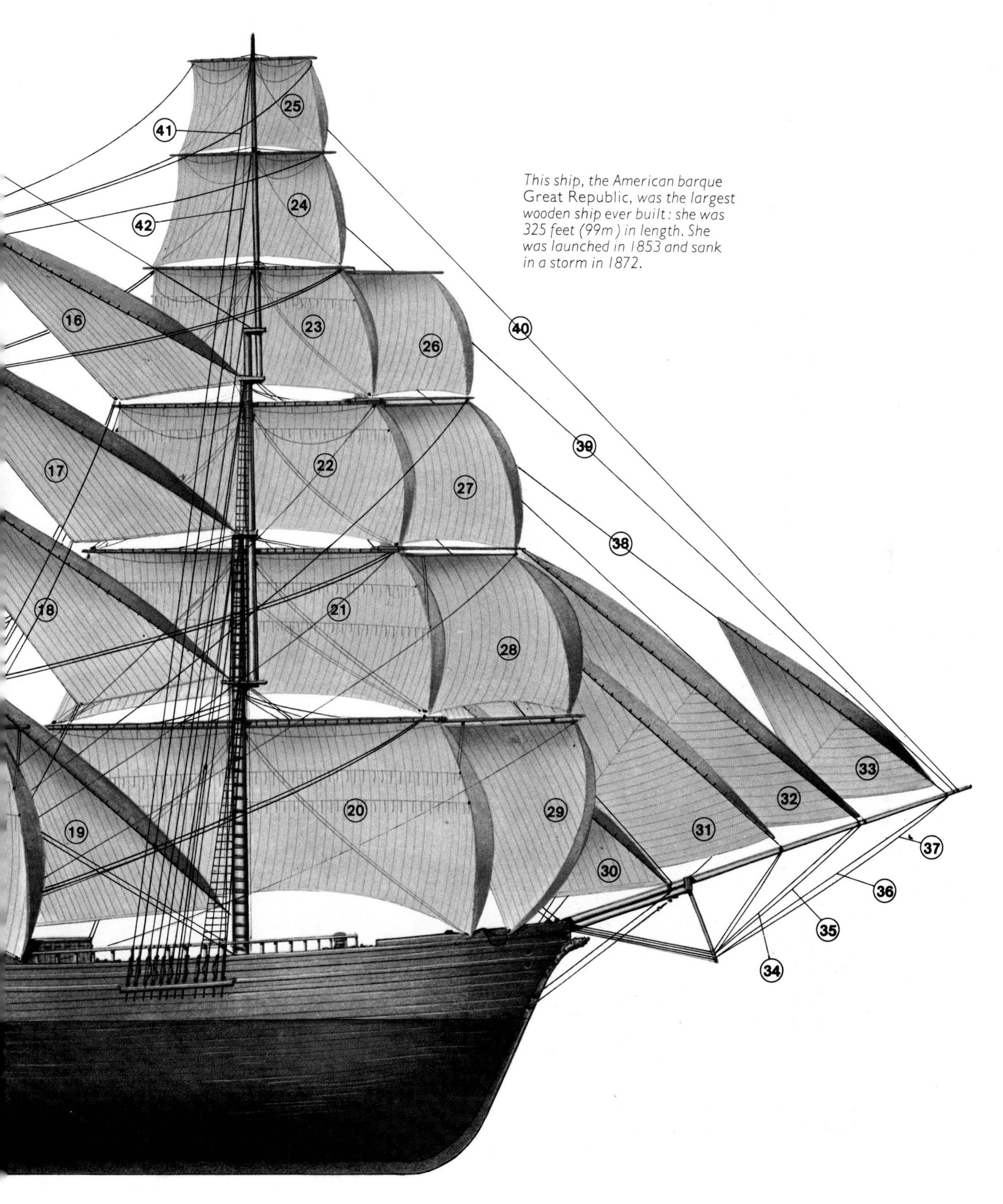

This ship, the American barque Great Republic, *was the largest wooden ship ever built: she was 325 feet (99m) in length. She was launched in 1853 and sank in a storm in 1872.*

25 fore skysail
26 upper studding sail
27 upper topsail studding sail
28 lower topsail studding sail
29 lower studding sail
30 fore staysail
31 fore topmast staysail
32 jib
33 flying jib
34 jib martingale
35 jib stay
36 flying jib martingale
37 flying jib stay
38 flying jib stay
39 fore royal stay
40 fore skysail stay
41 fore skysail backstay
42 fore royal backstay
43 skysail stay
44 foot ropes
45 skysail braces
46 royal braces

Finally, it should be noted that however small the drag angles can be made for sails and hulls, there will be no point if the opposing aerodynamic and hydrodynamic sideforces in practice cause such heeling of the vessel that the sideforce is seriously reduced. Therefore additional qualities required for windward performance are those associated with righting stability: low centre of gravity or deep heavy keels, lightweight spars and topsides coupled with high centre of buoyancy.

The particularly successful sailing vessels of the past can now be assessed in terms of these criteria for good windward performance. There is much evidence to suggest that the ability to hold station with the early square sail was first exhibited by the phenomenal North Atlantic voyages of the Norse longships of the 7th to 10th centuries. These ships, ranging in size from 70 to 270 feet (21 to 82 m) had hulls which even today are considered to have an excellent low drag profile, while the sail rigging exhibited for the first time special cordage (bowlines) and spars (beitass) so arranged as to tighten the luff and to ensure the sail remains correctly shaped at low angles of wind incidence.

Square sails can also be kept in shape at low incidence by incorporating stiff horizontal battens of bamboo, as in the early Chinese junks, which also had low silhouettes and good underwater characteristics, having developed the more efficient central rudder centuries before European vessels. These junks were so well developed by the 13th century, and bear such strong resemblance to the large ocean-going junks of today, that it is probable they were sailing well into the wind centuries earlier.

Sails in which the yard was used at such an angle to the mast that it acted as a stiff leading edge, the so-called lateen (after 'latin') sail, could be set close to the wind and so contributed much to the technique of windward sailing. Its earliest appearance in the ocean-going Polynesian sailing canoes of the 5th century, and especially in the Mediterranean from the 9th to 13th centuries, led to classes of boats whose high efficiency contributed much to the well-documented exploration and merchanting exploits of the Venetians and Portuguese caravels, continuing a chain of development which includes the Scandinavian 17th century jachts (present word yacht) and the famous Massachusetts schooners, which became the direct forerunners of the modern ocean-racer. In this development process, it seems probable that the lateen sail was split into two more manageable sails: the foresail whose leading edge was sharpened by replacing the oblique yard by a permanently set sail luff, and the aft or mainsail suspended on the aft portion of the yard which then became the gaff. This latter spar eventually disappeared when it became necessary to reduce topside weight and in the process left the modern Bermudan sail which exhibited windward advantages now ascribed to its height compared with its length, or aspect ratio.

Developments of sailing vessels in this century are based

Above: on this Thames barge, the use of the bowsprit, a spar extending from the bow of the ship, enables an extra foresail to be carried.
Right: a fore-and-aft rig can sail particularly close to the wind.

on long experience, together with all the current scientific measuring and deductive techniques. Models of sails and rigging are tested in wind tunnels in the same way as aircraft models are tested, the major measurements being those of dragforce, sideforce and centre of effort (the point on which the combined effect of the sails and rigging can be said to act) as the speed and angle of incidence of the wind, and also sometimes the heel angle of the sail are varied. The forces measured on the models by dynamometers are then scaled to predict the actual forces to be expected on a full-size rig. Similarly, models of hulls are towed to different speeds and attitudes in towing tanks and the dragforce, sideforce and centre of effort (in this case usually called centre of lateral resistance) acting on the model are measured and scaled to predict the actual forces expected on the full-size hull. Such measurements are most often used as comparisons to check possible advantages of small variations. Occasionally, model hull and sail measurements can be used together to predict the performance of designed but unbuilt boats, a process needing considerable engineering skill and large computer facilities.

It can be recognized, perhaps, that the many measurements required to completely predict an unbuilt yacht's performance requires the skills and instrumental resources of both aerodynamic and hydrodynamic laboratories and demand more actual measurements than required for an aircraft and ship together. It is not surprising, therefore, that such a full programme of measurement is rarely performed for an end product which is essentially recreational. More usually, yacht development arises from a mixture of limited wind tunnel and towing tank tests, together with experience of successful yachts and also trial and error with full-size new boats.

To improve the performance of a given boat, modern shipboard electronic instruments will provide the crew with instantaneous readings of windspeed, wind direction, waterspeed, course direction and even complicated combinations of these readings, the most useful for beating being the effective yachtspeed as if travelling directly into the eye of the wind (speed made good). Most sailors today, however, as throughout history, use their own experience or quite simple devices to estimate the same measurements in order to improve performance.

Sailing boats, as for all other technologies, have always advanced as a result of improvements in materials of construction. Sails need strong, lightweight, fairly stiff, smooth surfaced, bacterial and sunshine resistant fabrics with little mechanical fatigue or creep and which will not let the wind through. Vast improvements in most of these properties have been made by replacing natural fibres by woven and hot calendered (passed through hot rollers) synthetic fibres, particularly of the Terylene [Dacron] family, but there is still room for improvement. Rigging needs materials of high intrinsic tensile strength (strength for weight) and density, while spars need materials of high intrinsic stiffness together with high corrosion resistance. Hemp, then iron, then galvanized steel has been replaced by stainless steel rigging, while anodized aluminium alloy tubes, fibre or honeycomb composites now replace wooden spars. Hulls need corrosion resistant materials suitable for forming into smooth, lightweight skins of great strength and shock resistance and that most excellent material, wood, is being replaced particularly by aluminium alloys and fibre-reinforced plastics. Thus each new material is rapidly incorporated into new designs as its cost allows.

Above: a fishing schooner champion of the North American fishing fleet.
Left: modern lightweight materials have taken some of the hard work out of sailing. This crank hoists the mainsail, while the mainsail boom can be turned to shorten sail, or 'reef', quickly.

AMY HEWES

Ships

In 2500 BC the Egyptians were building fairly sophisticated sailing vessels; from then until the 19th century, ships were built mostly of wood and powered by sails.

At the end of the 16th century, the French inventor Denis Papin outlined plans for a boat with revolving paddles powered by a simple steam engine. Early steam engines, however, were far too big and heavy to be installed in boats. In the middle of the 18th century, the trickle of ideas, drawings and patents began to turn into a flood, and finally in 1783 a paddle steamer called *Pyroscaphe* steamed against the current of the River Saône in France for fifteen minutes.

Three years later a steamboat built by John Fitch was tested on the River Delaware in America, and there were other early experiments. The *Charlotte Dundas* is known as the first successful steamboat, because she towed two loaded vessels along the Forth and Clyde canal in March 1803. This steamboat was inspected by Robert Fulton, an American inventor and painter, whose *Clermont* began carrying fare-paying passengers on the Hudson River in New York State in September of the same year, making the steamboat a commercial success.

Today nearly all merchant ships and warships are built of metal and powered by Diesel engines or steam turbines. Merchant ships can be divided into the following categories: dry–cargo vessels, bulk carriers, container ships, LASH (lighter aboard ship) vessels, passenger vessels, oil tankers and LPG (liquified petroleum gas) vessels.

Dry–cargo vessels The basic orthodox design for a dry cargo vessel consists of a double bottom, several holds, a midship engine room and a forward and after peak tank. Usually they have one or two decks and three main super-structures. These superstructures are a forecastle, bridge and a poop located at the bow, the middle and the stern of the ship respectively, and they extend to the sides of the vessel. The ship is sub-divided with steel divisions called bulkheads, which are watertight from the bottom of the vessel up to the main strength deck. Their main function is to restrict flooding if the hull is damaged, but they also support the deck and prevent the hull from distorting owing to cargo or sea pressure.

The double bottom is a safety device in case the bottom shell is damaged, and it also provides a space for storage of fuel oil, water ballast or fresh water. The double-bottom structure gives great strength to the bottom of the ship, which is essential for dry-docking operations. The forward and after peak tanks are normally exclusively used for water ballast to give adequate draught when the vessel is unloaded and to adjust the trim if necessary.

The forecastle tween decks (short for 'between') are used for bosun's stores, the storage of wire ropes and rigging equipment and for paint and lamps. On the forecastle deck, each anchor cable passes from the windlass down through a spurling pipe into the chain locker, where the ends of the cables are connected to the fore peak bulkhead by a cable clench.

Above: this photograph of Brunel's steamship Great Eastern *was taken in 1857, just before she was launched. She was driven by both paddles and screw, and was the largest ship afloat, designed to carry 4000 passengers as well as 12,000 tons of coal for fuel. Opposite page: the steamboat* Amy Hewes *puffs through the Bayou Teche in Louisiana.*

At the after end there is a steering gear compartment where a hydraulic mechanism is used to move the rudder. The control for the steering gear is transmitted from the wheelhouse by a telemotor system. Directly below the steering-gear compartment is the rudder trunk which houses the upper rudder stock that is used to turn the rudder. The poop and bridge are used for accommodation and for provision stores, some of which may be refrigerated.

As diesel machinery is thermally more efficient than other types it is often used in dry-cargo vessels. The propeller is driven directly from a slow-speed in-line engine. The propeller shaft passes to the after end through a shaft tunnel; this tunnel protects the shaft from the cargo in the holds and it provides access for maintenance of the shaft and its bearings.

In addition to the main engine, the engine room contains auxiliary machinery such as diesel generators, oil purifiers, air compressors, ballast and bilge pumps, cooling water pumps and many other essential items of equipment. Just forward of the engine room are the settling tanks, oil fuel bunkers and a deep tank port and starboard, which may be used to carry liquid cargoes or a dry cargo such as grain or sugar. The accommodation is practically all amidships with the officers berthed on the bridge deck or boat deck. The wheelhouse, chartroom and radio room are usually together and the captain may have his dayroom, bedroom, toilet and office on the same deck. Galleys, pantries, lavatories and recreation rooms are carefully positioned to control the noise level and prevent annoyance to the off-duty crew.

The latest dry-cargo vessels have the engine room nearer to the stern. This shortens the shaft length and leaves a clear deck space forward of the bridge to work the cargo. Many vessels now have deck cranes for cargo handling instead of derricks operated by winches, and some vessels are fitted with special heavy lifting equipment.

Bulk carriers These are single-deck, single-screw vessels which carry large quantities of bulk cargo such as grain,

sugar, bauxite and iron ore. The engines are installed at the after end to leave the better spaces in the hull for cargo, and the accommodation is all aft above the engine room, so that services and sanitation are concentrated in one region of the vessel. Upper and lower wing tanks extend over the whole length of the cargo holds and they are used for water ballast when the ship is in the unloaded or light condition to give sufficient draught to immerse the propeller and give a better control over the empty vessel in heavy seas. The slope of the upper wing tank is designed to restrict the movement of a grain cargo, which may otherwise cause the vessel to become unstable. The double-bottom tank is used for oil fuel or for water ballast, and these tanks can be used to make adjustments to the trim of the ship. Some bulk carriers have their own derricks or deck cranes, but many rely entirely on the dockside amenities for loading and discharging cargo. The hull construction for these vessels is a combination of two framing systems in order to obtain the best strength characteristics from each. The deck, wing tanks and double bottom are longitudinally framed and the side shell is transversely framed.

Container ships These vessels are a relatively new concept in cargo handling which reduces the time that the vessel stays in port. The containers also form a complete load for road vehicles without further handling. British built vessels are normally designed for 20 ft (6.1 m) long containers, but they can be modified for 40 ft (12.2 m) containers if necessary. The hold length is designed to suit the length and number of containers to be fitted into the hull, and to allow sufficient space for refrigeration coolers and coupling systems for those containers with perishable cargo. The accommodation and machinery on these vessels are usually located aft to leave a clear deck for cargo working and to allow the large crane an unrestricted region for operation. The shore container crane and its lifting spreader system will only lift standard containers and hatch lids with correctly designed corner fittings. All the holds have vertical guides to position the containers and to give support, especially to the lowest container which could distort under the load transmitted down from those above. The containers are placed in a fore and aft attitude as the cargo experiences less ship motion in this direction, and when lifted ashore they are more readily received by road and rail transport. One advantage of container vessels is that they can carry containers on deck, but the number of tiers depends on the strength of the hatch lids and the necessity of having a clear view from the wheelhouse. The stability of a vessel with a deck cargo must always be checked, as the centre of mass of the vessel will be raised and it may cause the ship to roll or capsize. All deck containers are lashed to the hatches with steel rods or wires having hooks and lashing screws to prevent them being lost at sea.

Above: a prefabricated section is lifted into position during construction of a merchant ship in Yokohama, Japan. Sections must be carefully positioned, and optical instruments are used to monitor alignment of the ship's hull and to check for possible distortion.
Opposite page, top: laying the wooden deck during construction of Cunard's liner Queen Elizabeth 2.
Opposite page, below: steel sections being prefabricated in a Finnish shipyard.

LASH vessels A lighter is a small barge which may be loaded with cargo. A LASH vessel is a mother ship which is capable of picking up loaded lighters at her stern and stowing them into large holds. The principle of the system is to collect together several loaded lighters at the same time into a rendezvous area with the LASH vessel ready for transportation over seas.

The LASH vessel has a single-strength deck, forward accommodation and a semi-aft engine room. The funnel uptakes are at the sides of the vessel to allow the massive gantry crane to pass down the deck on rails. Longitudinal bulkheads, steel divisions along the length of the vessel, and transverse bulkheads, steel divisions across the vessel, form holds within the ship to stow the lighters in cells. Vertical barge guides are provided in the holds and the double bottom is equipped with sockets to receive the barge corner posts.

Walkways are provided with interconnecting ladders in the holds for the inspection and maintenance of the lighters. The gantry crane is supported at the stern by two large cantilevers; its lifting capacity is in excess of 500 tons and it is capable of transporting the barge along the deck to the hold. Each lighter is handled in about 15 minutes, and at present the LASH vessel will carry about 80 lighters, each with a cargo capacity of approximately 400 tons, a length of 61 ft 6 in (18.8 m) and a width of 31 ft 2 in (9.5 m). As well as being stowed in the holds, lighters can also be stowed in tiers of two on top of the large single-piece pontoon-type hatch covers, which have metal fittings for keeping the lighters secure during heavy weather at sea. The crane is equipped with a hydraulically operated latching device to grip the lighters and hatch covers, and a swell compensator which holds the lighter steady at the stern irrespective of the relative movements of the ship and the lighter in the sea.

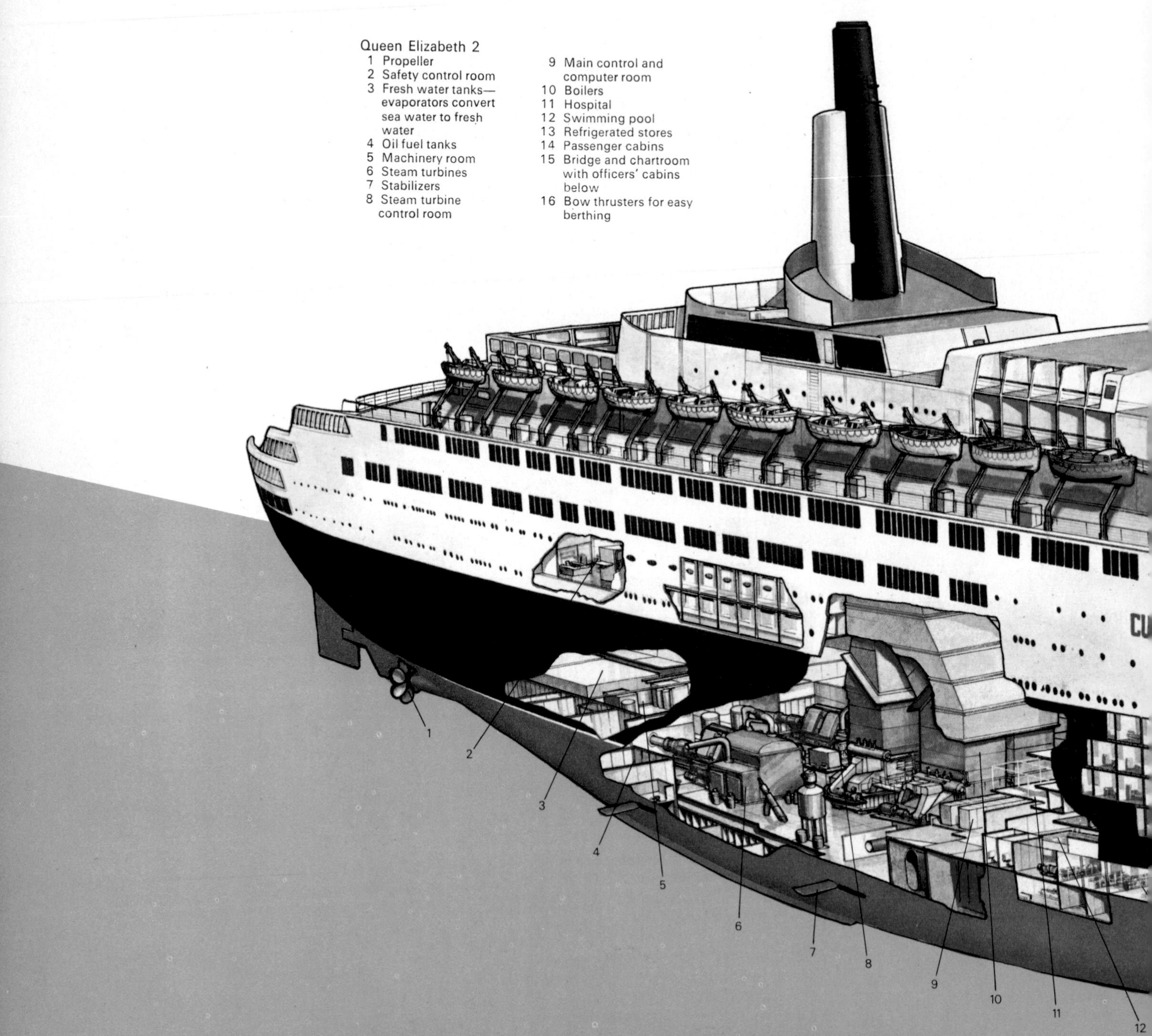

Queen Elizabeth 2
1 Propeller
2 Safety control room
3 Fresh water tanks—evaporators convert sea water to fresh water
4 Oil fuel tanks
5 Machinery room
6 Steam turbines
7 Stabilizers
8 Steam turbine control room
9 Main control and computer room
10 Boilers
11 Hospital
12 Swimming pool
13 Refrigerated stores
14 Passenger cabins
15 Bridge and chartroom with officers' cabins below
16 Bow thrusters for easy berthing

The advantages of the system are that cargo-handling operations can be carried out in parts of the world where large ships cannot be berthed as the depth of water is insufficient. Mixed cargoes can be handled simultaneously, and the lighters can be towed to various places up river after unloading, thus providing a virtual door-to-door service.

Passenger vessels In recent years the number of very large passenger liners has diminished in favour of the smaller vessel-capable of being converted for winter cruising.

Passenger vessels are more comprehensively subdivided than other merchant ships so that if several adjacent compartments are flooded, the ship will remain stable and stay afloat. If asymmetrical flooding occurs, the vessel has cross-flooding fittings to reduce the angle of heel.

Lifeboats are fitted port and starboard on the boat deck with sufficient capacity for the total number of passengers that the ship is certified to carry. Fire control is another important safety aspect, and the vessels are subdivided vertically into fire zones with steel bulkheads. In these zones the bulkheads must be capable of preventing the spread of a flame in a 30-minute standard fire test, and the accommodation must have an automatic fire alarm and detection system.

A gyroscopically controlled set of stabilizers or fins are a commor feature on most passenger vessels, to control the amount of roll and give a more comfortable crossing. For manoeuvring, these vessels are often fitted with bow thrusters, and they usually have twin-screw main propulsion.

The better cabins are located on the higher decks and the one, two or three-berth ordinary cabins on the lower decks. One of the most important areas in the accommodation is the foyer with reception desk, purser's office, main staircase and lifts. It should be centrally placed in order to receive the passengers so that their immediate needs can be dealt with

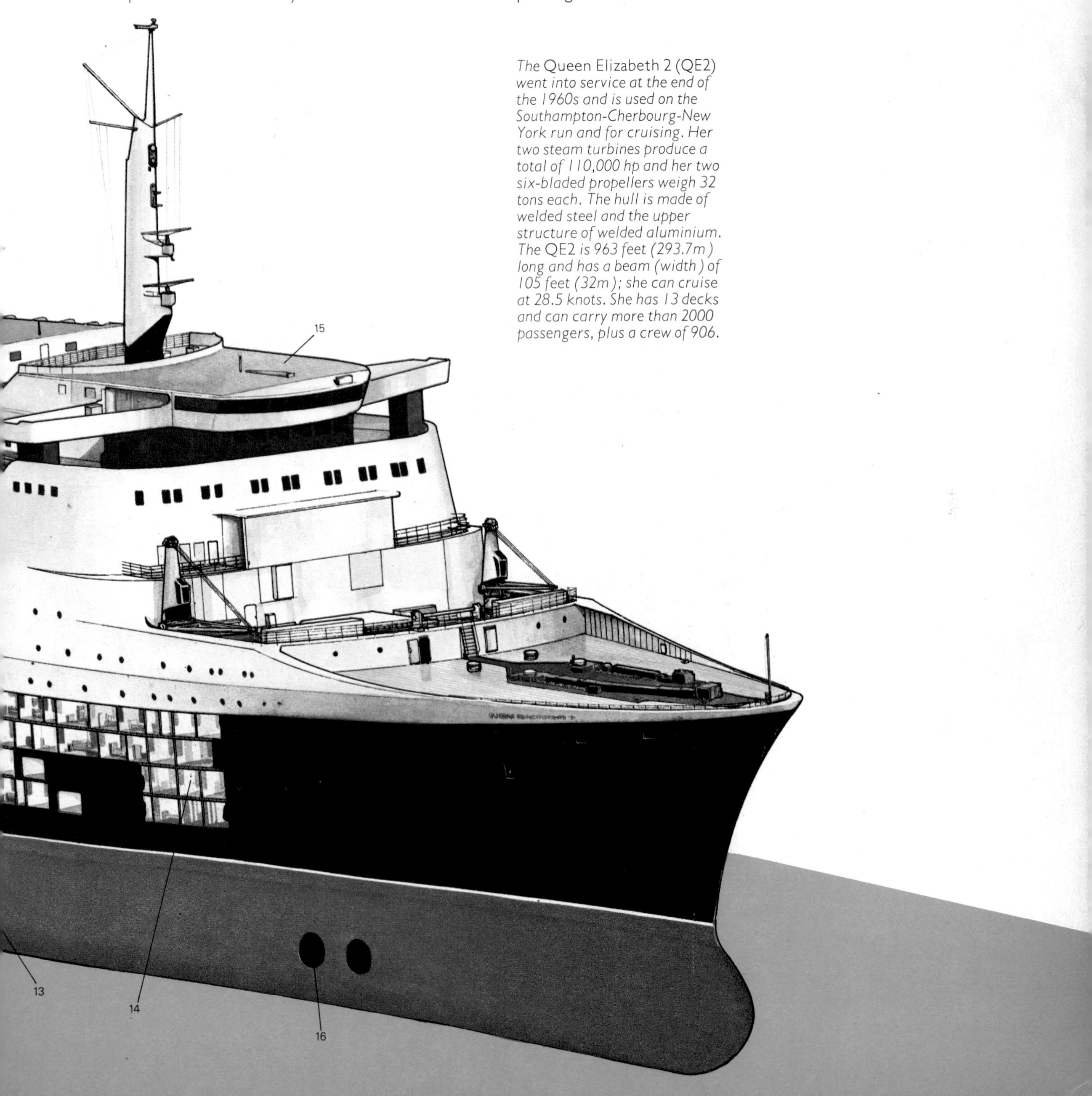

The Queen Elizabeth 2 (QE2) *went into service at the end of the 1960s and is used on the Southampton-Cherbourg-New York run and for cruising. Her two steam turbines produce a total of 110,000 hp and her two six-bladed propellers weigh 32 tons each. The hull is made of welded steel and the upper structure of welded aluminium. The QE2 is 963 feet (293.7m) long and has a beam (width) of 105 feet (32m); she can cruise at 28.5 knots. She has 13 decks and can carry more than 2000 passengers, plus a crew of 906.*

as soon as they embark. The following public rooms are quite common on most vessels: restaurants, ballroom, cinema, discotheque, shops, cocktail bars, clubrooms, banks and hairdressers. For recreation there will be a swimming pool and a deck area for games, the young children will have a nursery and there are playrooms for older children. The officers are berthed near to the bridge and the remaining crew and stewards have accommodation on a lower deck.

The vessel will normally comply with the regulations of all maritime countries, including those of the US coastguard. This will then allow the vessel to change to cruising at any time.

Oil tankers These vessels have a single main deck and a double bottom in the engine room only. Since tankers are divided into separate compartments, they are considered to be safe enough without having a double bottom along the full length of the ship. To reduce the risk of an explosion, the engines are fitted aft so that the shaft tunnel does not have to pass through any of the oil cargo tanks. At the extreme ends of the cargo tank range there is a cofferdam or bunker space to isolate the cargo from the other parts of the ship. Cofferdams are dry spaces across the vessel, preventing the possibility of any oil leaking directly into an adjacent compartment.

Pumps for discharging the cargo are fitted in a pump room in the bottom of the vessel. This pump room is often part of the cofferdam. The cargo pumps are usually driven by extended spindles from machinery in the main engine room. The oil is discharged from a tank by drawing it into the suctions at the end of a pipe leading from the main cargo pumps. It is then pumped vertically up from the pump room to the main deck where it passes along the deck pipelines until it reaches the deck crossover pipes. These crossovers are connected to the shore installations by hoses which are handled by the shore derrick. Oil tankers have small oiltight hatches with hinged lids giving access to the tanks by long steel ladders which reach to the bottom of the ship. The hatch coaming has a pipe leading vertically upwards to vent off vapour to the atmosphere if there should be a buildup of gas in the tank.

The oil tanks are subdivided by two longitudinal bulkheads into a centre tank and wing tanks, port and starboard. Sub-division of the oil tanks controls the movement of the cargo and prevents a large free surface across the ship which

The diagrams on the right are examples of structural features of common types of merchant ships.
Opposite page: a supertanker constructed in a Japanese shipyard was too large to be launched normally, so it was built in a drydock which was flooded when the ship was ready. The tanker has a capacity of 276,000 tons deadweight.

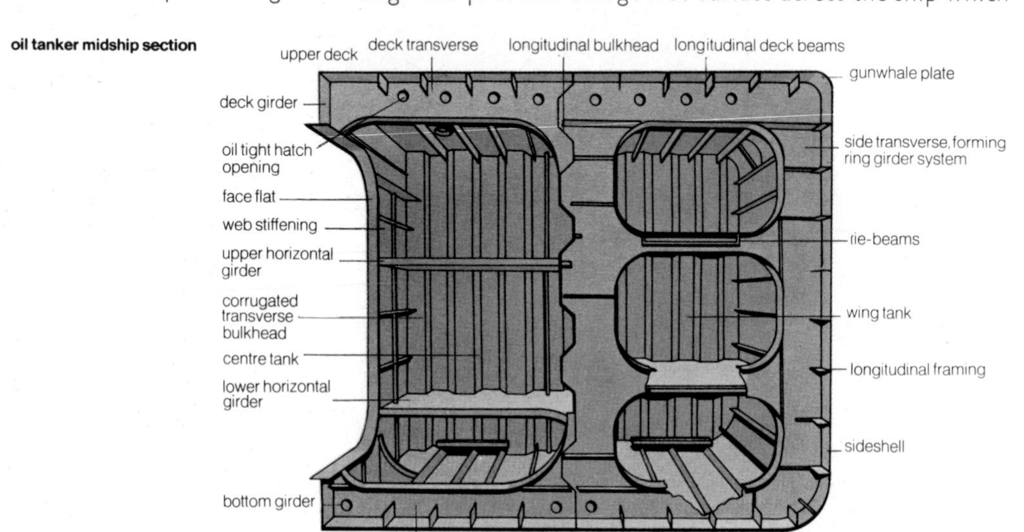

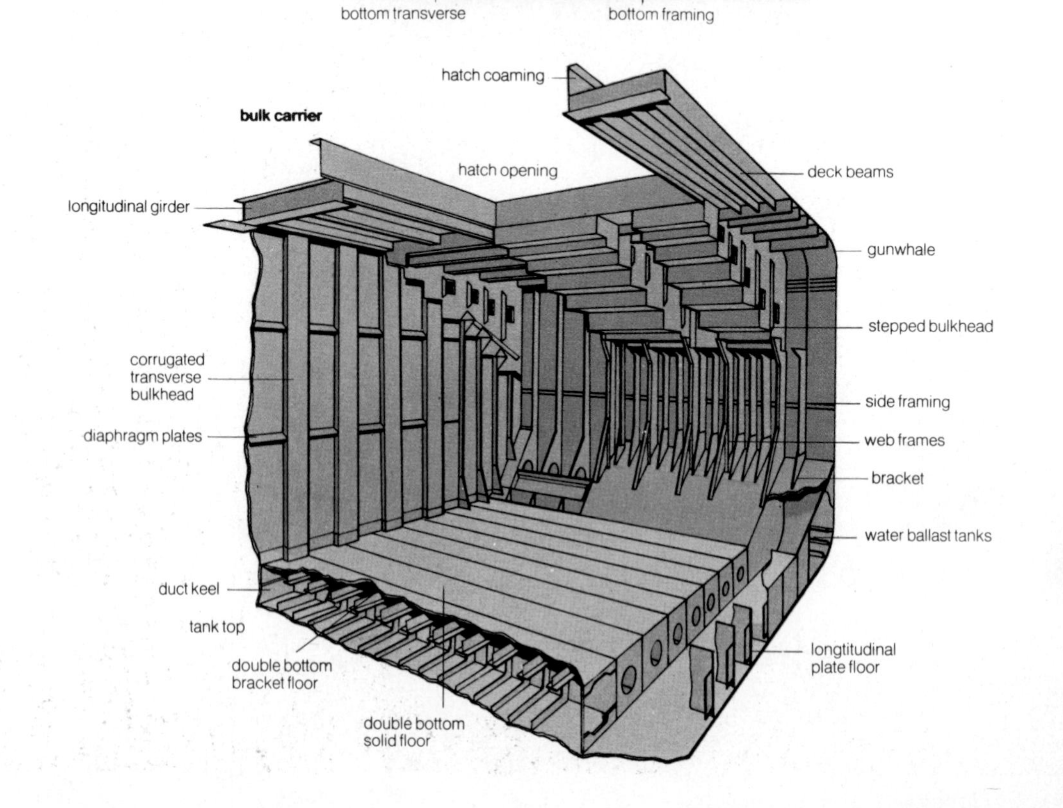

I H I 1 B D

would cause it to become unstable. Tank length is also important as oil in a partially filled tank will generate a wave caused by the movement of the ship. This wave will pass up and down the tank and could cause structural damage unless it is controlled by wash bulkheads or restricted by the length of the tank.

The engine power of a supertanker is very large, and a single propeller requires six blades so that the thrust is transmitted without overstressing the metal in any part of the blading. A bulbous bow is usually a standard part of the hull for a large tanker and it has the effect of modifying the flow of water at the bow, thereby reducing the power requirements from the ship's engines.

LPG vessels Vessels of this type are designed to carry propane, butane, anhydrous ammonia and other liquefied gases in specially designed tanks, which may be rectangular or hemispherical. A typical gas tanker has a design similar to a bulk carrier but it has gas tanks built into the hull which rest on chocks and are keyed to prevent movement when the vessel is rolling or pitching. The liquid gas temperature in the tanks may be well below zero; this will cause severe thermal stressing when the liquid moves and therefore the tanks will alter in shape owing to temperature changes. A void space between the gas tanks and the hull is filled with an inert gas to prevent the oxygen in the air and any leak of gas from the tank producing an explosive mixture. An inert gas unit in the engine-room is used to produce sufficient gas for the void space and to keep it topped up in case of leakage. The tanks are made from a low-temperature carbon steel which must withstand impact at low temperatures and thus not be susceptible to brittle fracture. In some vessels the gas tanks are not refrigerated but are insulated with four inches (10.1 cm) of polyurethane foam. When the liquid gas vaporizes it collects in domes at the top of the tank where it is drawn off and passed through a reliquefaction plant and then returned to the tank via condensers in liquid form. The domes at the top of each tank project three feet (0.91 m) through the main deck.

Alternatively there may be no tank refrigeration or reliquefaction plant on the ship and the gas is free to 'boil-off'. The vapour is then transferred to the engine room and used as bunkers for the main propulsion. This system makes

Wave conditions can create severe bending stress on the hull of a ship, in the form of alternating buoyancy (B) and weight (W) forces, depending on which part of the ship is in the wave and which in the trough. Other stresses on the steel plates of the hull also occur in rough water, according to the direction and depth of the waves. The longer the ship the more critical the stiffening and reinforcement of the hull.

the vessel less complex, cheaper on plant installation and bunkers but a percentage of the cargo is used over the voyage. Liquid-gas cargo is pumped from the bottom of the tanks using submerged pumps controlled from a room amidships. During pumping, a back pressure must always be maintained to prevent the gas boiling in the pump impeller and creating vapour in the riser when the tank is nearly emptied.

Air in the tanks is displaced before loading the liquid petroleum gas by using the inert gas system. Once the tanks are loaded they will always contain gas, so that the inerting procedure is not employed every time. Gas freeing of a cargo tank can be done by introducing inert-gas until the petroleum is diluted below the flammable limit and then blowing air into the tank and venting the gases at the top of the mast.

Prefabricated construction The way in which the vessel is constructed depends upon the type of ship and the technique adopted at the shipyard, and this will be influenced by the available yard machinery and cranes.

For example a bulk carrier will usually be constructed in the following way by most shipyards. Firstly the bottom shell and longitudinals will be laid on the building berth as a single unit after manufacture in the assembly shed, then the double-bottom unit will be lowered on to the bottom shell and welded into position. The wing tank unit is lifted into position, aligned and welded up, and a pair of bulkheads are erected the correct distance apart over the hold length, with an allowance made for their inclination to suit the declivity (downslope) of the building berth (necessary for launching). A side-shell panel can then be connected to the lower wing tank unit and bulkheads to form the sides of the hold. Then the upper wing tank is lowered into place and welded with the remaining deck panel finally completing the amidship structure. The ship is also built forward and aft of midships simultaneously. This technique, although not adopted by every shipyard, does allow an even spread of labour force. Working from midships gives a good reference structure for taking dimensions during the building of the vessel.

When each heavy unit is lifted on to the berth, the bottom of the vessel is checked for alignment by an optical system. Any distortion which may occur would affect the strength of the hull as well as hydrodynamic efficiency.

Oil Tanker
1 Mooring winches
2 Boiler
3 Crew mess and cabins
4 Radar
5 Engine rooms
6 Water inlet pipes
7 Wheel house, chartroom and radio room
8 Oil fuel bunker
9 Ballast tank
10 Pump room
11 Fire tower
12 Fire hydrant
13 Cargo tank
14 Mooring winches
15 Access for tank cleaning machinery
16 Tank hatches
17 Steam pipes for cleaning tanks
18 Derricks for lifting pipes on board
19 Discharge and loading points
20 Anchor winches
21 Mooring winches

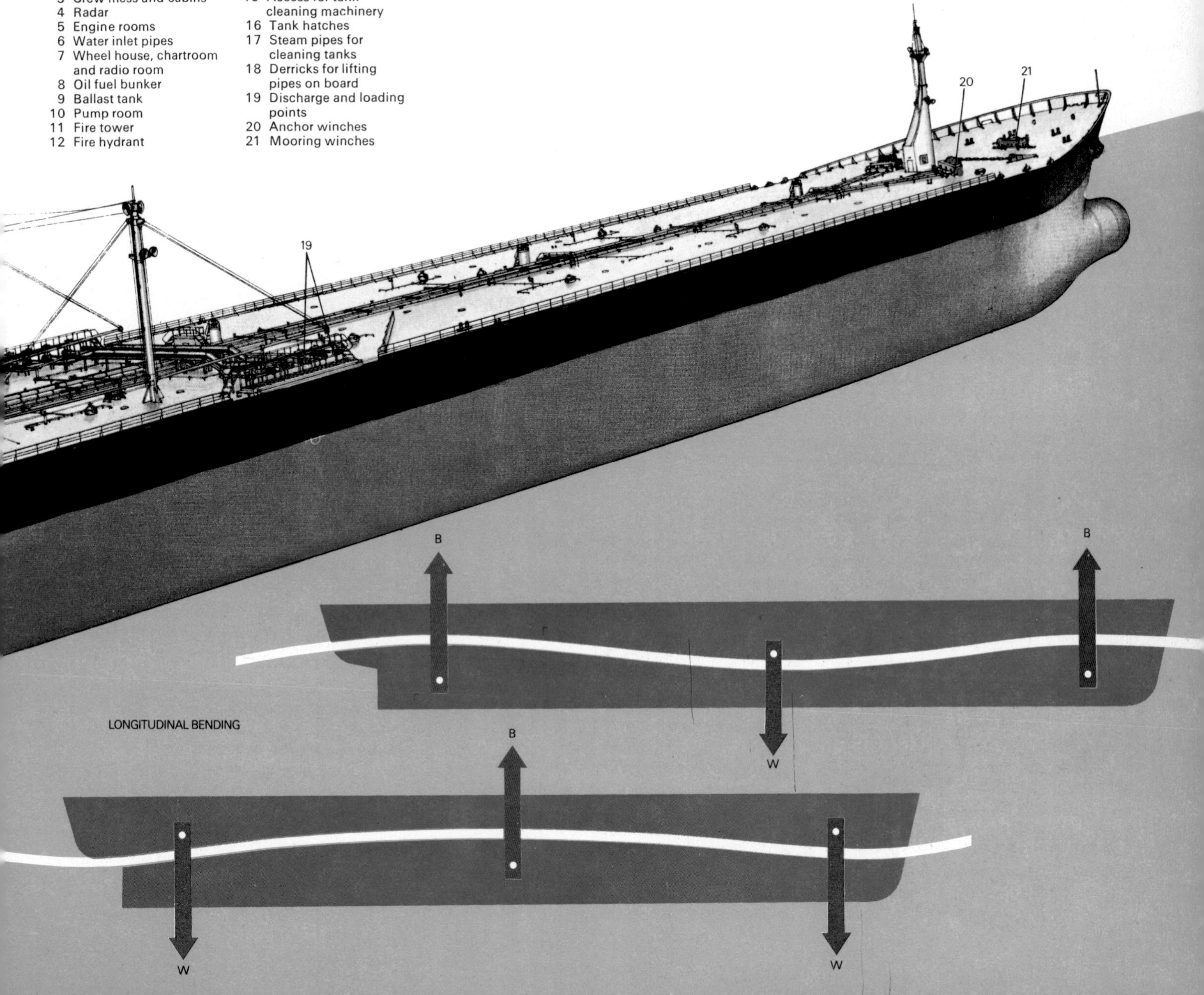

LONGITUDINAL BENDING

Warships

Up to World War II, the world's navies had fought in conventional naval battles, conducted at relatively close quarters with heavy guns. This had led to the favouring of large warships such as the battleship and the slightly smaller cruiser. These had a displacement of over 60,000 tons, carried vast 12 to 18 inch guns, and had an enormous thickness of armour plating—for example, the huge Japanese *Yamato* of World War II had 25 inch (63.5 cm) armour plating in parts of the turrets and superstructure. It became apparent during World War II that the nature of naval warfare had changed, with less emphasis given to sheer seaborne hitting power and much more to air power. A few torpedoes dropped from small, comparatively cheap aircraft could sink the largest and most expensive battleship. Aircraft range, performance and armament were increasing fast, and so was the technology of the aircraft carrier. Many of the 'naval' battles of the Pacific campaign were fought entirely by aircraft, without warships of either side coming anywhere near each other.

A modern navy consists of smaller ships, such as the frigate, which has anti-submarine and anti-aircraft duties as well as aircraft direction. The guided-missile destroyer (GMD) is somewhat larger, and approaches the size of the older conventional cruiser. The Royal Navy 'County' class of GMD has a displacement of 6309 tons; the 'Leander' class of frigate displaces 2200 tons. Compare these figures to the 45,000 tons of the USS New Jersey, the world's last operational battleship, which was launched in 1942 and was last used to shell North Vietnamese ports.

The construction features of a frigate can be seen as typical of warships in general. Heavy plating is not fitted to any modern warships. The side plating thickness of a frigate is less than that of many of the larger merchant vessels. Armour plating will not resist modern weapons, and it increases the displacement so that more power is required to propel the vessel at speed.

The hull is prefabricated and of all-welded construction with T-bars used for stiffening in the longitudinal direction. A grillage hull structure is the best design to resist underwater explosions and shock. It is formed by passing the T-bar longitudinally through the larger transverse framing, forming squares of stiffening to support the shell plates. The grillage structure gives the most efficient form of stiffening for minimum weight. Special quality steel is used in regions of the hull where there are high stresses, to reduce the possibility of cracks. To prevent corrosion, the steel is shot-blasted to remove mill scale and rust, and then it is painted before being used in the construction. Some parts of the vessel particularly susceptible to corrosion are shot blasted and zinc sprayed after construction. Weight can be saved by using aluminium for the superstructures, but it must be restricted to areas not likely to be subjected to blast, as it has a low melting point. Abrupt changes in the shape of superstructures, such as the forecastle and the long midship superstructure, cause a loss in strength, and for this reason gradual changes are made by sloping the ends of these structures.

Watertight subdivision is essential to keep the ship afloat in the event of damage; this is achieved by fitting a number of watertight transverse bulkheads, longitudinal bulkheads, decks and watertight flats. The transverse bulkheads are usually stiffened vertically and the plates comprising the bulkhead are welded together horizontally. The T-bar longitudinals are connected strongly to the bulkheads with stiffeners to help integrate the structure. Access through the bulkhead is often necessary and therefore watertight doors are fitted, but they must be above the waterline, and all compartments may be closed when the vessel is at sea.

A modern frigate is capable of running in a closed down condition in case of a nuclear attack; this means that special precautions are necessary to prevent contamination through the ventilation systems. In the upper structure, radar room, communications control room, computer rooms, electronic warfare office and an enclosed bridge are now common in the frigate. The No. 1 deck accommodation consists of the wardroom, officers' and petty officers' berths, galley, recreation rooms and sick bay. No. 2 deck is subdivided into gunbays, power rooms, gunners' stores, weapons spare gear store, the senior and junior ratings, dining hall, galley, scullery, 'Seacat' magazine, ship control centre, engineers' workshop and accommodation for junior ratings. The No. 3 deck contains the sonar instrument room, junior ratings' mess, refrigeration stores and fuel stores. Finally, in the bottom of the vessel, are the magazine and diesel oil tanks; there is also provision for the helicopter fuel, lubricating oil and the sonar space.

Aircraft carriers Aircraft carriers are complex ships with very special requirements. Their decks must be well above the waterline so that aircraft can land in bad weather, and must also be clear, with the bridge and funnel on the starboard side forming an 'island'. The hangar deck must be a clear space with a depth extending through two decks. To avoid passing the boiler uptakes through the hangar deck, they are led across to the starboard side and then up to the funnel.

The flight deck must be long enough to give sufficient landing area aft and catapult length forward to accelerate the heaviest aircraft. Angled flight decks are required to give the necessary deck length to land aircraft in rapid succession. Once the aircraft has been arrested, it is moved to the starboard forward corner of the flight deck to give a clear space to allow any aircraft that has missed the arresting wires to fly off the end of the angled deck and to return for another landing attempt.

Two aircraft lifts serving both ends of the hangar for transportation to and from the flight deck are operated electrically using a chain driven system. To maintain the aircraft, several workshops are arranged around the hangar on the port and starboard sides of the ship. Aircraft munitions and aircraft fuel are stored in lower regions of the vessel to give maximum protection. A long-range radar, radio communication and computers are necessary for tracking high-speed aircraft, and space must be provided for this equipment along with briefing, aircraft control, hangar control and flying control position rooms.

At the after end of the catapult, protection from the jet blast is necessary to prevent following aircraft from being damaged by hot exhaust gases. The deck panels must also be water cooled in this region to avoid overheating.

Firefighting arrangements on these vessels are extremely important and the hangars can be quickly drenched by a sprinkler system operated from remotely controlled pumps.

The submarine

Although the principle of the diving bell has been known for over 2000 years, and Leonardo da Vinci produced drawings for underwater craft, it was not until William Bourne, an English naval officer, produced a treatise on the principles of under-water ballasting in 1580 that the practicalities were examined. It was then realized that, by applying Archimedes' principles of a floating body displacing a quantity of water equal to its own weight, it was possible to build a craft into which water could be admitted, thereby increasing its weight sufficiently for it to submerge. Conversely, the expulsion of the added water would allow the craft once more to float upon the surface. It was on this basic plan that Cornelius van Drebbel, a Dutch mechanic, built what is believed to have been the first boat which not only went below the surface but also returned at will. His craft, which was demonstrated to King James I in the Thames off Westminster, used an expanding leather bulkhead which could be screwed back into position to expel the ballast water. Propulsion was by means of oars and a breathing tube led to the surface. Drebbel successfully navigated the first submersible from Westminster to Greenwich.

Early submarines During the 17th and 18th centuries a number of attempts were made to design more efficient underwater craft—in 1653 the Frenchman de Son produced a submersible propelled by a clockwork-driven paddle-wheel, which was not a success. Other schemes were tried for adding the necessary ballast to dive a boat and two Englishmen built practical craft, the first, Symons, using leather

Opposite page: the world's last operational battleship, the USS New Jersey *was launched in 1942. Of 45,000 tons displacement, her main armament is nine 16-inch guns capable of throwing a 2700 pound (1225kg) shell 30 miles (50km). She was last used for shelling ports in North Vietnam.*
Below: the large 'County' class guided missile destroyer, HMS Hampshire. *She is armed with two types of missile as well as four 4.5-inch guns.*

The USS Nautilus *was the first nuclear submarine. In a 1955 test she sailed 60,000 miles (96,560km) without refueling. In this photograph she sails into New York harbour not long after completing a trans-polar voyage under the arctic ice.*

bottles which filled with water. In 1773 Day sailed into Plymouth sound in a boat which relied on external ballast in the shape of rocks to destroy its surface buoyancy. These were to be released from within to regain the surface, but Day was the first submarine casualty. With no means of controlling his depth and possibly experiencing difficulty in jettisoning his ballast, it seems likely that his craft was crushed by the water pressure.

Only three years later David Bushnell, an American seeking methods of attacking British ships during the War of Independence, built the *Turtle*. She was a one-man, egg-shaped craft driven by a propeller operated by a crank, and she carried a charge of 150 lb (68 kg) of gun powder with a timing mechanism fitted. In 1776 the first submarine attack on a warship was carried out, being thwarted only by the fact that the screw securing the charge to the flagship's hull struck metal and would not penetrate. In 1800 a larger and more ambitious submersible, the *Nautilus*, was built by another American, Robert Fulton. This succeeded in sinking ships in two trial attacks but was never used in action. With only hand propulsion, no means of accurate depth control and with a weapon which required contact with the target, the submersible was not yet suitable for operational use.

It was not until 1850 that any progress was made, but from this point onwards the development of submersible craft was to advance rapidly. In that year Wilhelm Bauer, a Bavarian, produced a boat with a cast iron hull and the germ of a trimming system in which weights were moved fore and aft. She was equipped with a hand driven screw. In 1863, when the French produced the *Plongeur* with compressed air stored in bottles to operate the engine and blow the ballast tanks, the first major mechanical aids were incorporated in submarine design. In the same year, steam driven craft, the *Davids*, were used in the American Civil War.

From this time various methods of propulsion—heat storage, steam engines, electric motors and petrol [gasoline] engines—were employed but the greatest advances were made by John P. Holland, an Irishman who had emigrated to the USA, and Laubeuf of France. The series of craft designed and built by the former and the handsome *Narval* which the latter completed in 1899 were the true forerunners of today's submarine fleets.

By the early years of the 20th century all the major naval powers were forced into an interest in this new form of warfare, and submarine design followed the general principles which have continued to the present day.

Modern design The strength of the main pressure hull determines the depth to which the boat can dive. Of steel construction, its plates were originally riveted together but this has now been replaced by welding. The usual description of 'cigar-shaped' is approximately accurate—the cross section should be circular to obtain maximum strength and should be pierced by as few holes as possible. These are normally confined to the access hatches, the torpedo tube openings, the sleeves for periscopes and masts, escape hatches (normally secured shut in wartime), the engine-exhaust system and snort-mast (snorkel) leads, some of the trim tanks (see below) and the log (speed indicator). Each is equipped with a method of closure which is tested to full diving-depth pressure.

Outside the pressure hull are built the conning tower, the casing and the ballast tanks, which are either of the saddle tank type, great bulges hung from the main structure, or of the double hull type in which a whole skin is wrapped about the pressure hull. Some submarines, notably the British 'U' and 'V' class, were fitted with internal ballast tanks in which the tanks were fitted within the presure hull, but this arrangement, requiring additional penetration of the pressure hull, seriously reduced the diving-depth and is no longer employed.

Each ballast tank must have two openings: one at the bottom to let in the water ballast required on diving and through which that water is expelled by compressed air on surfacing, and one on the top through which air may escape to allow the water to enter. The lower opening can be merely a hole at the bottom of the tank, in which case the tank is known as free-flooding, or it may have a valve fitted, known as a Kingston valve. The valve on the top, which retains the air under slight pressure within the tank, is the main vent.

Diving The act of diving the submarine is achieved by having all Kingston valves open and then opening the main vents, thus releasing the air pressure which has previously kept the tanks dry. This reduces the submarine to the state known as neutral buoyancy, in other words the slightest downward force will cause the craft to sink.

This fact, by itself, will not take the submarine below the surface. Without other aids she would wallow in a dangerous state on the surface. She must be driven below the surface, and this is achieved by the thrust of the propellers forcing her forward and the hydroplanes, horizontal rudders placed two forward and two aft, directing her downwards.

Once below the surface it is necessary to 'catch a trim'. This means that, having achieved neutral buoyancy, she must have the weights within her so adjusted that, at the desired depth, she will remain static and horizontal. This state is achieved by transferring water into and out of trim tanks situated at either end and amidships. The deeper the submarine goes, the greater the compression of her hull and the less becomes her displacement. As a result, more water must be expelled from her trim tanks to ensure that her displacement always equals her weight. Only then will the submarine be in the state known as stopped trim, when she can hover at the desired depth without any mechanical assistance from her screws or hydroplanes.

Propulsion The means of propulsion for submarines was not satisfactorily solved until the diesel engine was incorporated in the design. Steam propulsion, though occasionally used in later years, presented the major problems of the necessity for a funnel, the delay in shutting-down the boilers before diving and the subsequent latent heat from those boilers. Petrol [gasoline] engines produced fumes which were occasionally explosive and presented the difficulty of storing a highly volatile fuel. The heavy oil engine, followed by the diesel, overcame most of these problems. It provided a means of propulsion using low flash-point fuel, an instant ability to dive even though the engines were running, and

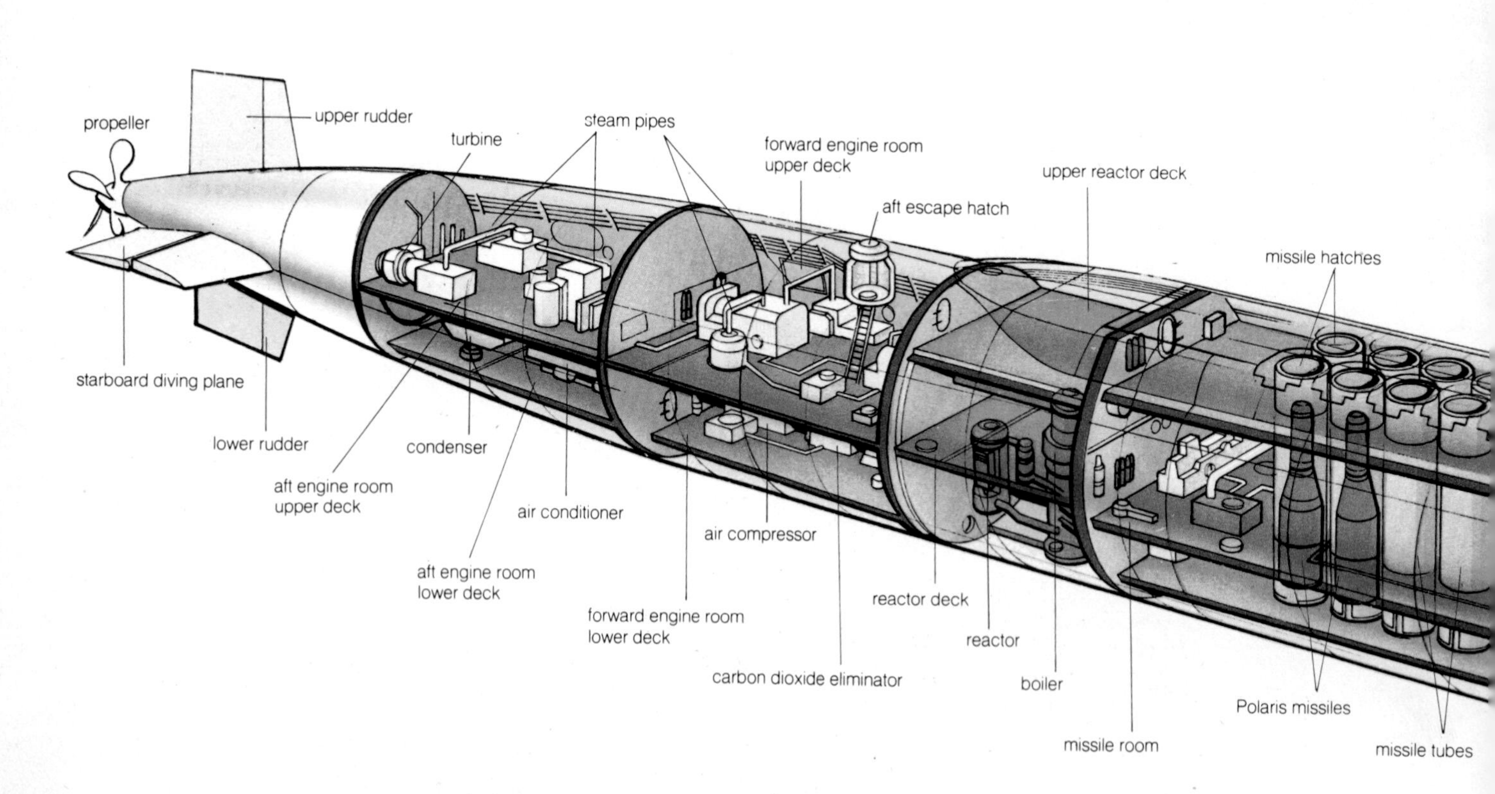

simpler and more dependable machinery. Today, in the normal patrol submarine, diesel engines provide the power required on the surface for propulsion and for charging the batteries. As the majority of such boats are now equipped with snort masts, which take air from the surface at periscope depth (up to 50 ft, 15.2 m), both these functions can now be carried out below the surface. Propulsion on the surface or at periscope depth can be either by direct drive, in which the diesel is coupled through a clutch to the main electric motors and through a tail clutch to the screw, or by diesel-electric drive in which the diesel drives a generator which provides power for the main electric motors to rotate the propeller.

When below periscope depth, propulsion is by means of the main motors, which draw their power from large storage batteries below the main deck. These are charged, when the diesels are running, either from the generators in a diesel-electric boat or from the main motors acting as generators in a direct-drive boat.

In nuclear submarines a coolant liquid is pumped in a closed circuit between a nuclear reactor, where it takes up heat from the radioactive core, and a boiler heat-exchanger, where it gives up the heat to a water feed, thereby generating steam. The steam is conducted through valves to the main propulsion turbines and auxiliary turbines which generate power for subsidiary systems. The main propulsion turbine can be connected to the submarine's screw either directly through a speed reduction gear or by means of a turboelectric drive (like diesel-electric drive, but with a steam turbine replacing the diesel engine).

Subsidiary systems are needed for many purposes. Hydraulic power operates the periscopes, the raising and lowering of the snort and other masts, the opening and shutting of the torpedo tube bow caps, the control of the main vents on the ballast tanks and many other minor yet

The missile compartment of a Polaris submarine. The vessel carries 16 nuclear missiles which can be launched while submerged and have a range of 2800 miles (4500km).

A Polaris submarine, because it is nuclear powered, can remain submerged for many months.

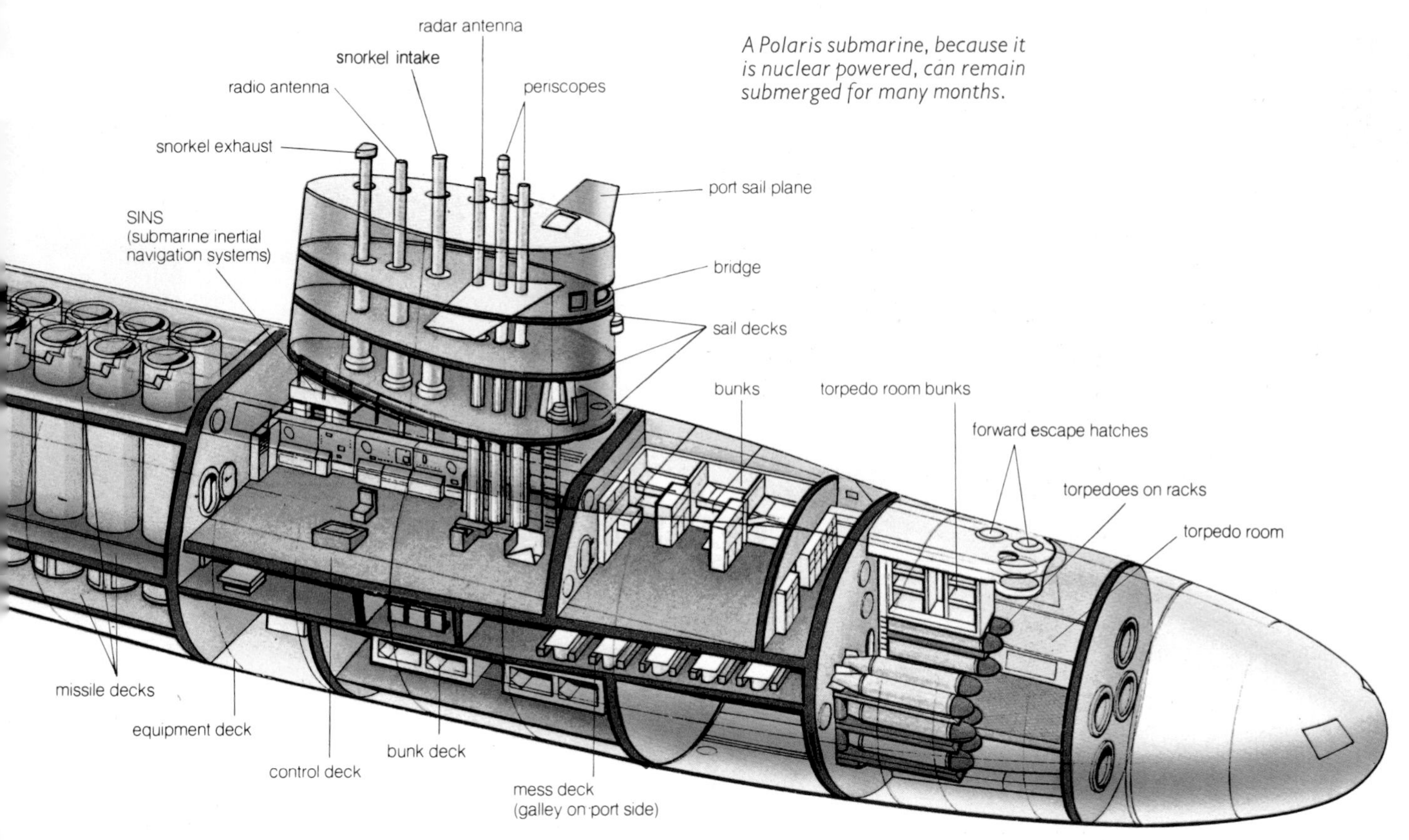

vital items. Few things these days are hand operated and most systems are electronically controlled.

Navigation In the early days, navigation of submarines was rightly described as 'by guess and by God'. The early submarines had no periscopes and, when these were introduced, some projected an inverted picture on a ground glass screen. Later improvements provided for increased ability to examine the sky as well as the sea; these included radar ranging and low light television scanning in addition to periscope sextants for taking Sun and star sights. Today, the need for a point of land for a periscope fix, or the tedious business of surfacing to allow the navigator to take a Sun, Moon or star sight with his sextant are gone. SINS (Submarine Inertial Navigation System), a complex of gyroscopes, now produces a remarkably accurate plot which requires 'up-dating' from external fixes only every few days. Assisted by the well-tried navigation aids of radar and echo sounding, the submarine navigator is today as well placed to know his exact position in in the ocean as anyone.

Armament The submarine's weapons have improved beyond recognition over the years. The original screwed charge was replaced by the spar torpedo (an explosive charge secured to the end of a long spar). Whitehead perfected his mobile torpedo in the 1870s, and this soon became the natural equipment for underwater craft. Originally dropped from collars on the upper deck, they were soon ejected from tubes projecting from the pressure hull which, provided the submarine was fitted with efficient bow caps, allowed reloading to take place. Methods of aiming these torpedoes progressed from 'eye shooting' to the modern computer controlled system. The gun was for many years a valuable weapon, but it was eventually defeated by the ability of radar to detect a submarine the moment it surfaced. Now the gun's function of striking from above the sea has been taken over by the cruise missile and the submarine has an overall armoury outranging and outperforming any surface ship except in the sphere of aircraft defence. With the possibility of fitting nuclear heads to any of their weapons, various classes of submarines in the major fleets now carry ballistic missiles such as *Polaris* (range about 2800 miles, 4500 km) and its successor *Poseidon* (range about 3000 miles, 4800 km) for long range strikes, cruise missiles covering from 250 miles (400 km) to short range, and torpedoes equipped with passive or active homing heads as well as wire guidance for close range. Information to programme and control this wide armoury of weapons comes from shore control, aircraft reconnaissance, the submarine's own radar and sonar, as well as the well-tried sighting through the periscope.

The whole aspect of submarine operations was changed when the *USS Nautilus* got underway on nuclear power in January 1955. From then on it was possible for a true submarine, as opposed to her submersible predecessors who were dependent on the atmosphere for their support, to operate for years without refuelling, to manufacture her own air and fresh water, to travel at sustained speeds hitherto unknown in the submarine world, and to remain totally submerged for periods impossible for diesel driven boats.

Modern submarines still carry torpedoes in case they encounter enemy surface ships or other submarines. This is the torpedo tube compartment of HMS Resolution.

The hydrofoil

A hydrofoil boat is analogous in principle to an aircraft. It comprises a boat-shaped hull to which are attached 'wings' or hydrofoils which generate lift as they travel through the

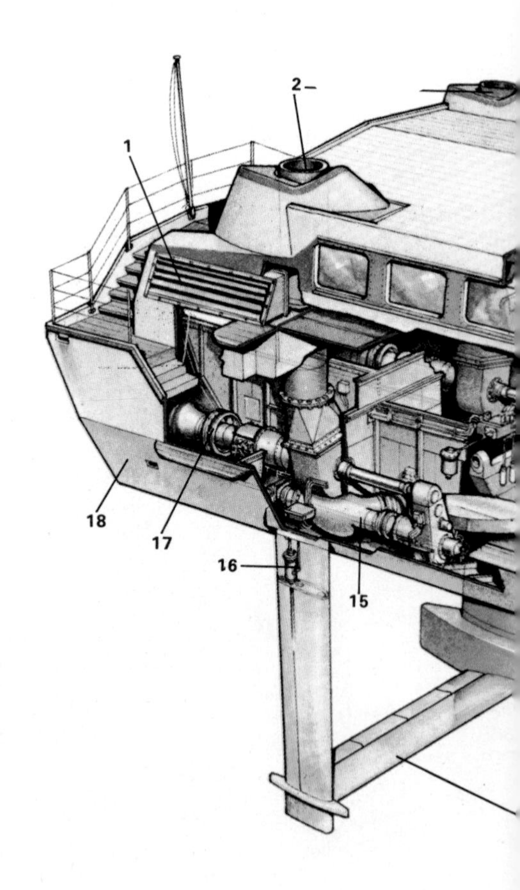

water, in the same way that the aerofoil design of the aircraft wing provides lift in the air. When the hydrofoil boat attains speed, the lift provided by the flow of water over the hydrofoil is sufficient to lift the hull entirely clear of the water. Once out of the water, the hull no longer suffers resistance from friction with the water, or from waves in rough water, so that higher speeds and a more stable ride can be attained.

The hydrofoil is not a new idea; a patent was issued to the Frenchman Farcot in 1869. Around 1900 the Italian Forlanini was building successful hydrofoil boats. The Wright brothers experimented with them before their successful aircraft was flown at Kitty Hawk in 1903; Alexander Graham Bell designed one; and the Russian designer Alexeyev and the German Hans von Schertel have been among the most important hydrofoil engineers since World War I.

Since water is several hundred times as dense as air, the hydrofoil can be much smaller than the wing of an aircraft. There are two principal types of hydrofoil boat design: the fully submerged hydrofoil and the surface-piercing hydrofoil.

Surface-piercing hydrofoil This design is by far the most popular for civilian use because of its simplicity. If lift is lost due to the hydrofoil breaking through waves, the craft sinks deeper into the water, thereby immersing more of the foil or foils which generates more lift. Large numbers of passenger ferries built under licence to designs of the Swiss Supramar company are now in service, and all except the largest of these are of this type.

A 200 ton naval hydrofoil, *Bras D'Or*, built in Canada, has demonstrated that the surface piercing design is suitable for operation on the open sea, where rough water can be expected. Foilborne speeds in excess of sixty knots have been achieved. The Russians, who are by far the biggest operators of hydrofoils, mostly on lakes and rivers, build several types of shallowly submerged hydrofoil craft, closely related to the surface-piercing type. Examples are the *Kometa*, *Meteor* and *Raketa* types.

Fully submerged hydrofoils In this type of design only vertical struts pierce the water surface, and changes in lift are effected by changing the angle of attack of the foil to the water. This is achieved by hydraulic or pneumatic cylinders activated by electrical signals from a sonar device pointing ahead of the craft, which reads the oncoming wave height and selects a suitable hydrofoil angle to give the necessary lift.

The principal user of this type of craft is the United States, which has built large naval craft, some having retractable hydrofoils which can swing up out of the water to allow the boat to operate in the conventional way in shallow channels. One of the most successful designs of this type is the *Tucumcari*, built by Boeing. Another system of stabilization, used on smaller craft, uses mechanical sensor-floats on arms ahead of the craft, mechanically linked to the fully submerged foils.

An early variation of the hydrofoil configuration was the 'ladder' type, the more 'rungs' being immersed the greater

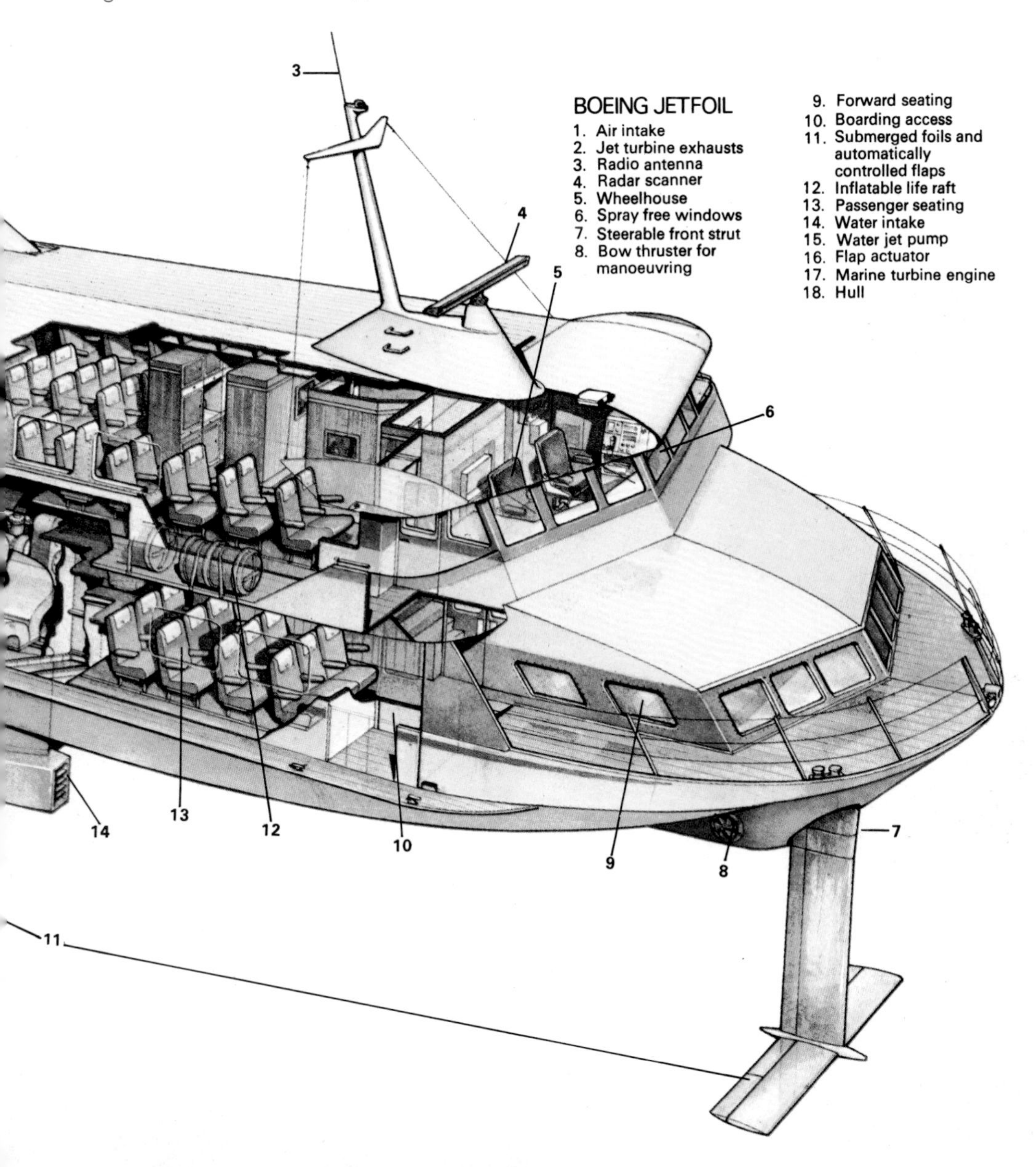

The Boeing Jetfoil has foil struts which swing up and out of the way for operation in shallow water. The waterjet propulsion system draws in water and discharges it at high pressure by means of pumps to push the craft forward. The Jetfoil is 90 feet (27m) long and has a beam (width) of 31 feet (9.5m); it cruises smoothly in waves up to 12 feet (3.7m) high, carrying 250 passengers or 25 tons of cargo.

the lift generated. An early hydrofoil boat of this type was the Bell-Baldwin HD4 which in 1919 attained a speed of sixty knots on the Bras D'Or lakes in Canada. (1 knot=1 nautical mile=6080ft=1852m per hour.)

In some designs the large part of the hull weight, usually between 80 and 90%, is supported by a large foil or foils located forward on the hull; in others the large foils are located aft. The remaining hull weight is supported by a smaller foil which can be turned to provide a steering function while the craft is foilborne.

Propulsion Power units for commercial craft usually consist of marine diesel engines which drive propellers at the end of long inclined shafts. In military craft, sometimes designed for anti-submarine duties towing sonar equipment, and where light weight and high speed are considerations but cost is not, gas turbines are used. Often an entirely separate propulsion system is also fitted for when the craft is operated at low speeds in the hullborne way. Water jets have also been used.

Limitations Above about fifty knots, hydrofoils have to be specially designed to cope with increased turbulence, in the same way that wings of supersonic aircraft have to be designed to cope with breaking the sound barrier. Hydrofoils will not replace the largest ocean-going vessels for the same reason that limits the practical size of aircraft: the cube-square law. As the size of a hydrofoil (or an aerofoil) increases, its weight increases in cubic increments, while the area available to support the weight only increases in square increments.

The air cushion vehicle

Air cushion vehicles—also known as hovercraft or, in the United States of America, as ground effect machines (GEM)—are vehicles which when in motion are supported by a layer of air, rather than by wheels or other direct means of contact with the ground over which they pass.

This absence of contact with the surface has brought the advantages of both adaptability and speed: the latter is particularly well demonstrated when the ACV is compared with the conventional ship. For example, the world's largest passenger ACV, the 190-ton SR-N4, is capable of speeds of nearly 80 knots (150 km/h) and cruises at around 50 (90 km/h). The top speed for a crack liner is around 35 knots (65 km/h).

There are a number of reasons why this is so. First, in a conventional ship that area of the hull which is normally submerged is subjected to drag as a result of the viscosity of the water through which it travels. Drag absorbs a good deal of engine power.

Second, wave formations are set up at bow and stern of a ship when it is under way. Again, this wave making process means a drain on the power supply. Although this is less important than drag at low speeds, as speed increases it takes over as the major power wastage problem.

Finally, there are the natural phenomena of currents and of windage on the exposed areas of hull and superstructure. These may sometimes work adversely.

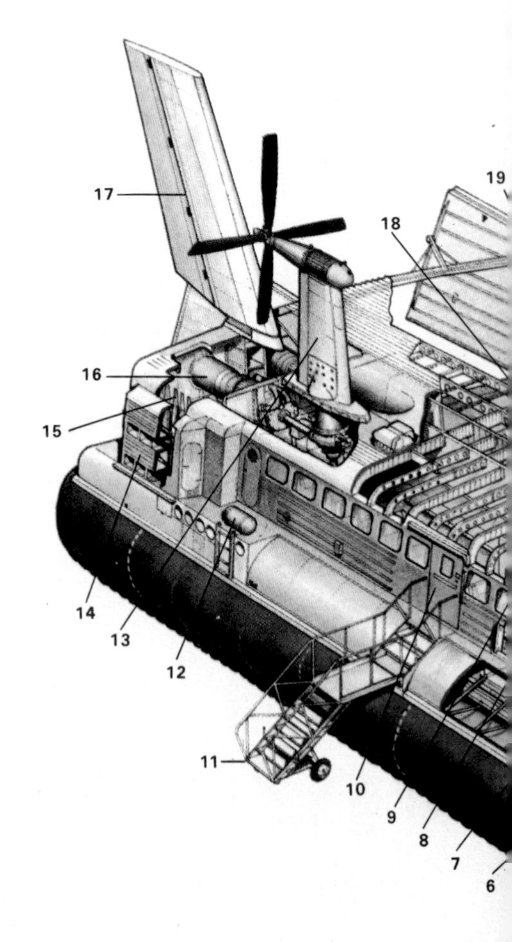

A US Navy patrol air cushion vehicle leaves on a mission in the Mekong Delta.

Considering the first two factors alone, it can be appreciated that the bigger and faster the ship, the larger the amount of energy wasted. There comes a point when the cost of deriving more speed from a ship outweighs the advantages —unless there are special factors like military or research requirements.

Since none of the ACV is immersed it has none of these problems. At low speeds a wave making process is set up, but at cruising speeds this disappears. So though the ACV is affected by adverse winds, it is generally faster not only than a conventional vessel of the same size but also larger ships.

In principle the ACV works as follows. The hull can be thought of as being something like an upturned tea tray with raised edges. If such a structure were placed carefully on the surface of water, a quantity of air would be trapped beneath it, retained by the edges which would now be jutting downwards. If, however, you attempted to propel the tray through the water, the air would escape and the tray would sink. Even if that did not happen, the submerged portions of the edges would be subjected to friction and would set up waves.

The pioneer designers were faced with two problems: how to raise the craft clear of the water, and how to keep the air cushion permanently in place.

They overcame the first by ducting air into the cushion compartment at pressure a little higher than atmospheric, and the second by arranging a system of air jets around the edge to provide a curtain of air which slowed down the rate of leakage from the cushion. This system was improved by the addition of a flexible skirt around the edge of the vessel.

It has been calculated that a pressure of only about 60 lb per square foot (300 kg/m²) is required to raise an ACV of 100 tons or more to a height of one foot (30 cm). The pressure required to inflate car tyres is a good deal greater.

Types of ACV Several variations of the basic ACV principle have been evolved. The simplest is called the air-bearing system. Air is blown through a central orifice in the undersurface and leaks away outwards under the flexible retaining skirt.

The plenum chamber vessel has a concave undersurface, and the cavity forms the upper section of a cushion chamber which is completed by the sea or ground surface. Again the air leaks away under the edges.

In the momentum-curtain system a ring of air jets is set around the circumference of the underside of the ACV. The air from these jets is directed downwards and inwards to retain the air cushion. This system has been further developed to include two rows of peripheral jets, one inside the other. The retaining air is blown out through one set, sucked up by the other after it has done its job, and then recirculated. This makes for greater efficiency, since it slows down the rate of air escape.

ACV propulsion systems have also been varied. The most popular for large vessels has been the airscrew or propeller. In the earliest machines, the fans that provided lift also drove air through a system of ducts to the stern where it was ejected for propulsion. In the SR-N4, the four engines that drive the lift fans also drive external airscrews for propul-

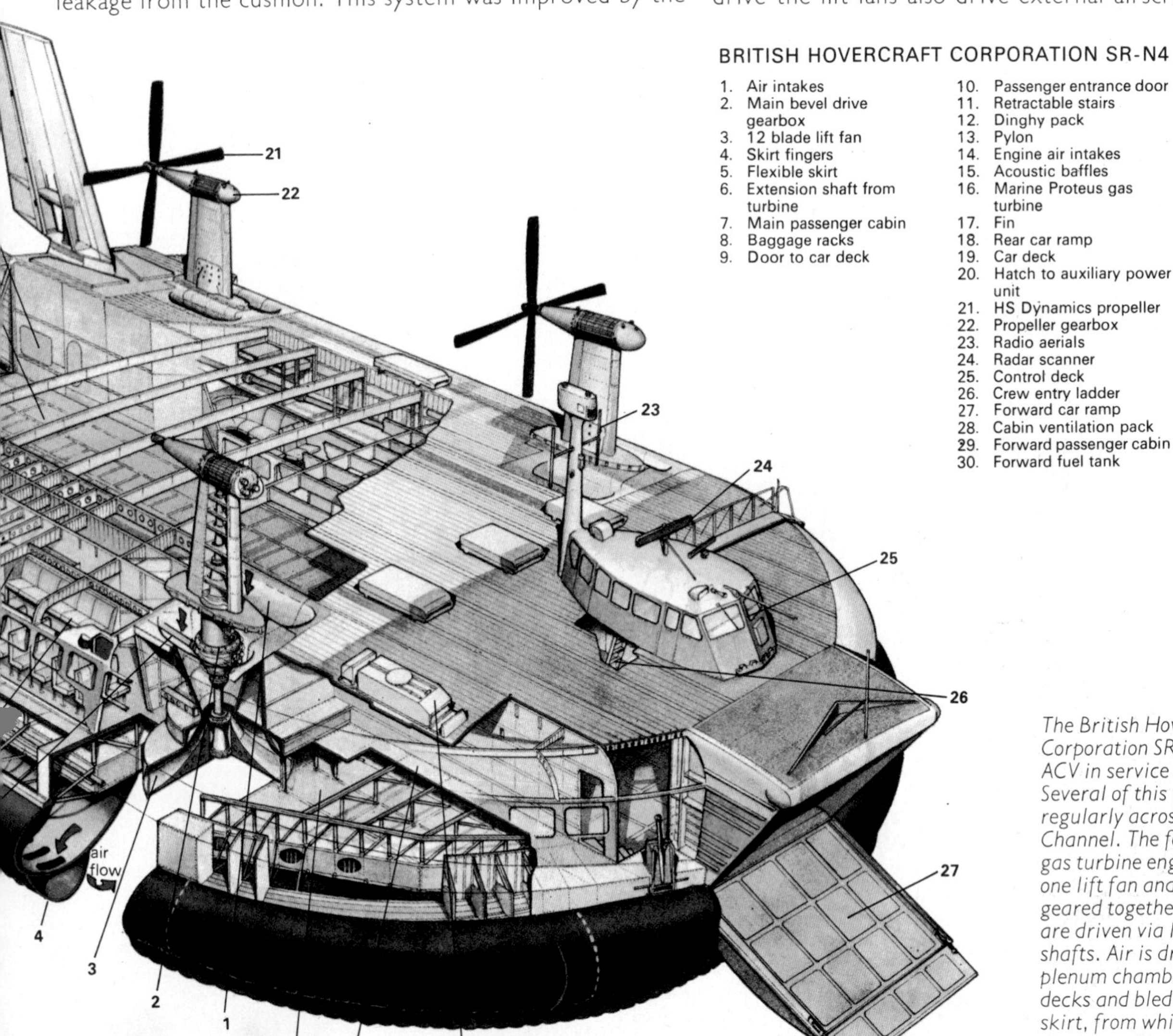

The British Hovercraft Corporation SR-N4 is the largest ACV in service at over 160 tons. Several of this type are operated regularly across the English Channel. The four rear-mounted gas turbine engines each drive one lift fan and one propeller, geared together. The front ones are driven via long extension shafts. Air is drawn into a plenum chamber under the decks and bled to the tubular skirt, from which it is discharged inward to form a stable cushion.

sion. In many other types, however, the lift and propulsion systems are separately powered. Some ACVs even have water propellers. These cannot go on land.

The problems of steering an ACV are very similar to those of steering an aircraft. As the ACV has no contact with sea or land, there is a danger of drift during turns. The helmsman overcomes this by banking, or tilting, his machine like an aircraft. He does so by reducing the pressure from the air jets on the side which he wants to dip. Directional control is exerted by varying the power of the airscrew, by using aero rudders, or with both systems at once.

Development The air cushion principle has fascinated designers for many years. Pioneering attempts at its use were made, for example, as far back as the 1930s, in both the United States and Finland. But it was not until after World War II that the real breakthrough came.

The inventor of the first successful ACV was Britain's C S Cockerell. Trained as an engineer and in electronics, he later turned his attention to the problems of boat design.

He tried at first to retain an air cushion under a boat by fitting hinged flaps at the bow and stern of his craft between side keels. Except for the hinging, it was almost exactly the same as the tea tray example.

Finding this technique to be ineffective, he replaced the flaps with sheets of water pumped vertically downwards. Air containment was still not very efficient, and finally he struck on the idea of using peripheral air jets for the purpose.

The world's first hovercraft, the SR-N1, was unveiled in 1959 when it travelled from the Isle of Wight to mainland England. Only a few weeks later it crossed the English Channel in two hours, and in 1965 the world's first regular passenger service was set up between the Isle of Wight and the mainland. Now a fleet of SR-N4s carries passengers and cars regularly to and from France, and in 1975 captured 29 per cent of the total traffic.

Uses of ACVs The ACV has truly arrived as a means of providing high-speed transport over a variety of terrains.

Because the air cushion acts not only as a form of support but also as an effective spring, the modern ACV can cope with waves of up to ten feet (3 m) and can operate over rough ground as well. It has been used for military purposes by the USA in Vietnam and elsewhere in the Far East. It was even employed on one occasion to carry a British expedition to the upper reaches of the Amazon.

The advantages of this type of craft in naval warfare are considerable. It is not only speedy, but the larger types can deliver torpedoes and other missiles with telling effect. At the same time, since they are not in contact with the sea themselves, they are immune to torpedo attack. The missile simply passes harmlessly beneath them.

But even that presupposes that an enemy submarine can find them. The insulation of the air cushion makes it impossible for submarine listening gear to pick up the sound of the ACV's propulsion system.

The United States Navy is planning a 2200 ton test craft for launching in 1976, and a 10,000 ton version for the late 1970s which will carry jet fighters of the vertical take-off type.

For both civil and military purposes, the ACV has the advantage that it can make the transition from sea to shore with relative ease. No expensive docking facilities are required—just an area of gently shelving beach.

There is also the whole area of ACV application on dry land. The concept has been used in the design of several devices including a type of lawnmower, a 'hoverpallet' for transporting heavy loads around the factory, and enormous craft like the American ACT 375, designed to carry a 375-ton payload across the Arctic wastes. Work on 'hovertrains' is going on in the United States, France, Japan and the Soviet Union. The air cushion principle has even been applied to aircraft—the US Bell LA4 employs an air cushion instead of wheels for take off and landing.

Amphibians

Amphibious vehicles, or amphibians, can move both on land and in water under their own power. Development of amphibians for military and pleasure use has produced many different technical approaches, from the military DUKW to the American home-made swamp buggies.

The first true amphibious vehicle was probably the French Fournier of 1906, which combined a boat-type hull with an automobile chassis. A shaft transmission drove both the rear axle and a single propeller. Modern pleasure amphibians still follow this basic pattern.

Most famous of all amphibians, and still one of the most versatile vehicles in use, is the American built General Motors Corporation DUKW ('Duplex Universal Karrier, Wheeled'), first built in 1942 and primarily used for ship-to-shore transport. Based on a truck chassis, it has a six cylinder 4.4 litre engine. All six wheels, which have rubber tyres, are driven.

The wheels all steer on land, and on water they assist a rudder. A single propeller is driven through a transfer case—a gearbox which enables it to be switched in and out. The six and a half ton amphibian can achieve 50 mile/h (80 km/h) on land and 6 mile/h (10 km/h) on water. Production ceased in 1945, but 'ducks' are still in use with armed forces all over the world.

Another major class of amphibian is known in the USA as the LVT, short for Landing Vehicle, Tracked. The LVT, nicknamed the 'Buffalo', started as a rescue vehicle. It was designed in 1932 by Donald Roebling for use in the Florida swamplands and later developed as a military vehicle for carrying men and materials over rivers or on sea-borne landings.

The LVT is driven on both land and water by tracks equipped with W shaped protruberances (grousers) to give greater thrust. Buffers set between the driving wheels support the tracks and prevent them from being forced inwards by water pressure. This gives a greater effective driving surface.

Later models were fitted with Cadillac engines and automatic gearboxes. Current LVTs have side screens along the top run of the tracks and a cowl over the front of them, so that water carried forward by the top of the tracks is directed towards the rear again and so contributes to the forward

Opposite page: the Voyager ACV was developed for use in Canada, where the variety of terrain includes lakes frozen in winter. Its capacity of 25 tons compares well with that of cargo aircraft. Below: the USS Nashville, *an assault ship which carries several large landing craft. They are released by flooding the stern until they float inside the mother ship.*
Inset: a tracked landing vehicle (LTV) coming ashore during Operation Deep Furrow in the Mediterranean, 1971. It carries 25 or 30 people and weighs about 38 tons.

thrust. Another gain in thrust comes from grilles at the back which channel the wash straight behind the vehicle. LVTs can cope with rocky beaches and heavy surf with equal ease.

During the Second World War, it was often necessary to convert a military vehicle into an amphibian, and Duplex Drive was devised by Nicholas Straussler to allow tanks to float into battle. The DD Sherman was so used on D Day in 1944. A platform of mild steel is welded round the water-proofed tank's hull and a raised canvas or plastic screen is erected round it to give buoyancy. This is generally supported by a series of rubber tubes inflated from cylinders of compressed air carried on the tank's superstructure.

Small propellers were originally driven by the tracks on early models, but the vulnerability of propellers on land led to the increasingly widespread use of water jet propulsion. A ducted propeller sucks in water from under the body and squirts it through steering valves at the back. Russia leads in this field, but the armies of many nations now have troop carriers and reconnaissance vehicles driven by water jets.

Flotation screens are still used on a few military vehicles like the Vickers 37 ton tank and the Ferret Mk V Scout car, but most modern designers prefer either to forgo amphibious qualities or to build true amphibians.

Developments in armour construction enable modern tanks to be lighter than their predecessors but just as well protected, so it is not difficult to produce light tanks that can float unassisted. A prime example is the Russian PT-76 amphibious tank, powered by a water jet.

The standard Russian technique allows tanks to submerge to cross rivers. This type of amphibian was pioneered during World War II by the Americans and Germans, and submersible tanks were to have been used in Hitler's abortive invasion of Britain.

They have snorkels (air tubes) extending to the surface, bringing air to both crew and engine. The French AMX30 tank has a 15 ft (4.6 m) long tube, wide enough to allow the commander to stand in the top and relay instructions to his submerged crew.

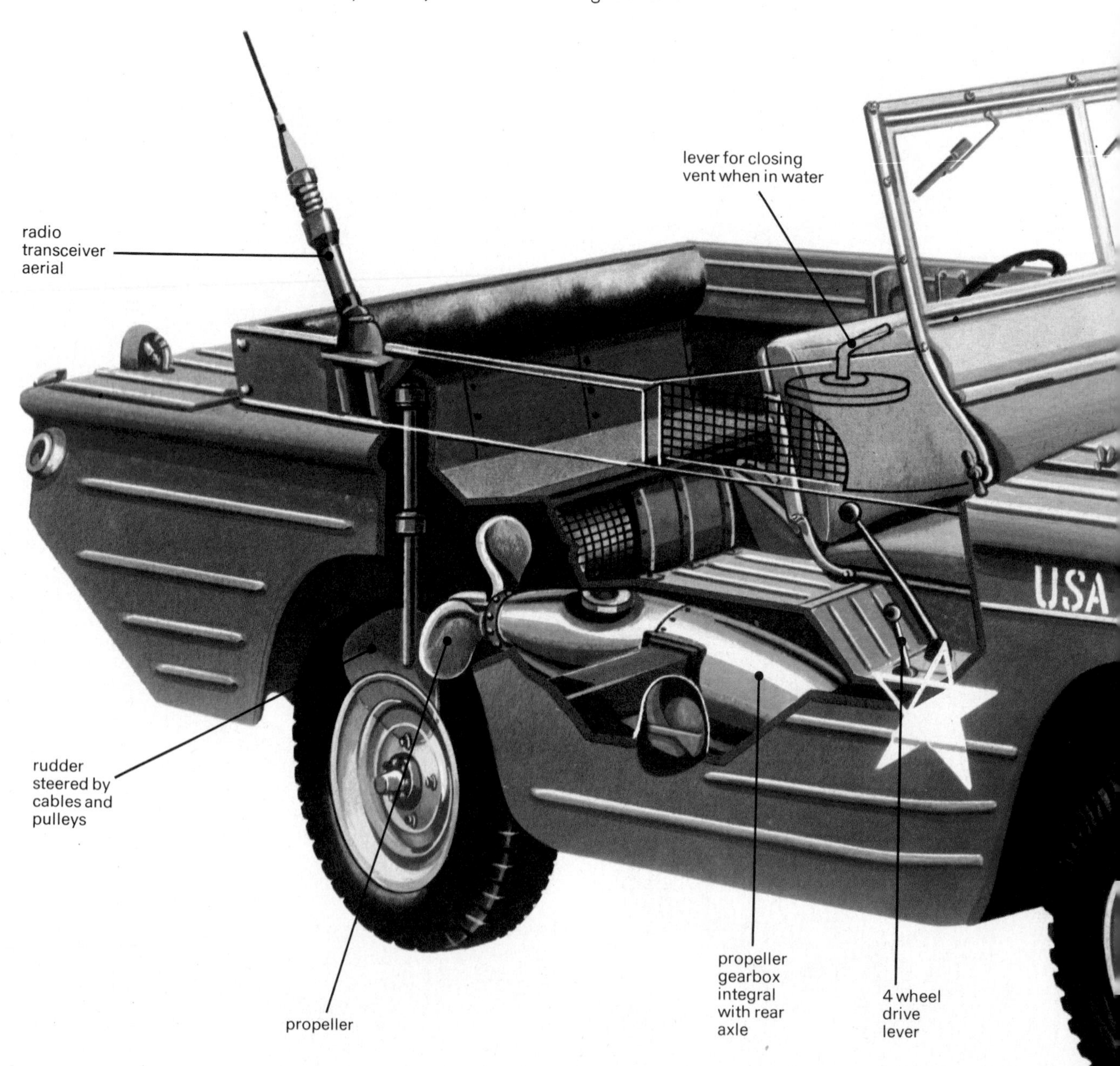

The air tube principle was also used on the British Austin Champ jeep. It had an extendable air pipe leading to the carburettor on the waterproofed engine. True amphibious jeeps were the World War II Volkswagen Schwimmwagen and the GPA ('General Purpose Amphibious') version of the American Jeep, which is still popular.

On the GPA Jeep, the watertight hull was constructed separately from the basic chassis to make replacement easy. The propeller was countersunk for protection, and was driven by a separate shaft mounted alongside the main drive to the rear wheels. It weighed more than a conventional Jeep, but could still reach 50 mile/h (80 km/h) on the road.

The bathyscaphe

For a long time underwater research into marine life and the nature of the ocean floor was limited by the equipment available. The development of the aqualung [scuba] has enabled a diver to dive to a maximum depth of about 165 ft (50 m). But Piccard's invention, in the late 1940s, of a diving vessel—the bathyscaphe—designed to withstand high pressure, enabled the scientist to descend several miles below the surface of the sea. The name comes from the Greek *bathys* meaning 'deep' and *scaphos* which means 'ship'.

The bathyscaphe is constructed of two main parts, a large, usually hull-shaped float which is a lighter-than-water container filled with fuel, and a spherical steel cabin suspended below it. In the float, the petrol is divided among several compartments, and the positive buoyancy that the craft has is balanced by quantities of iron shot. This ballast is situated in hoppers inside the float and secured by electro-magnets.

The sphere is the control centre of the craft. It houses the crew, the controls and any research equipment. As the sphere must withstand great pressures—at a depth of six miles (10,000 m) about eight tons per square inch (1240 bar) —the wall thickness varies from four to seven inches (10 cm

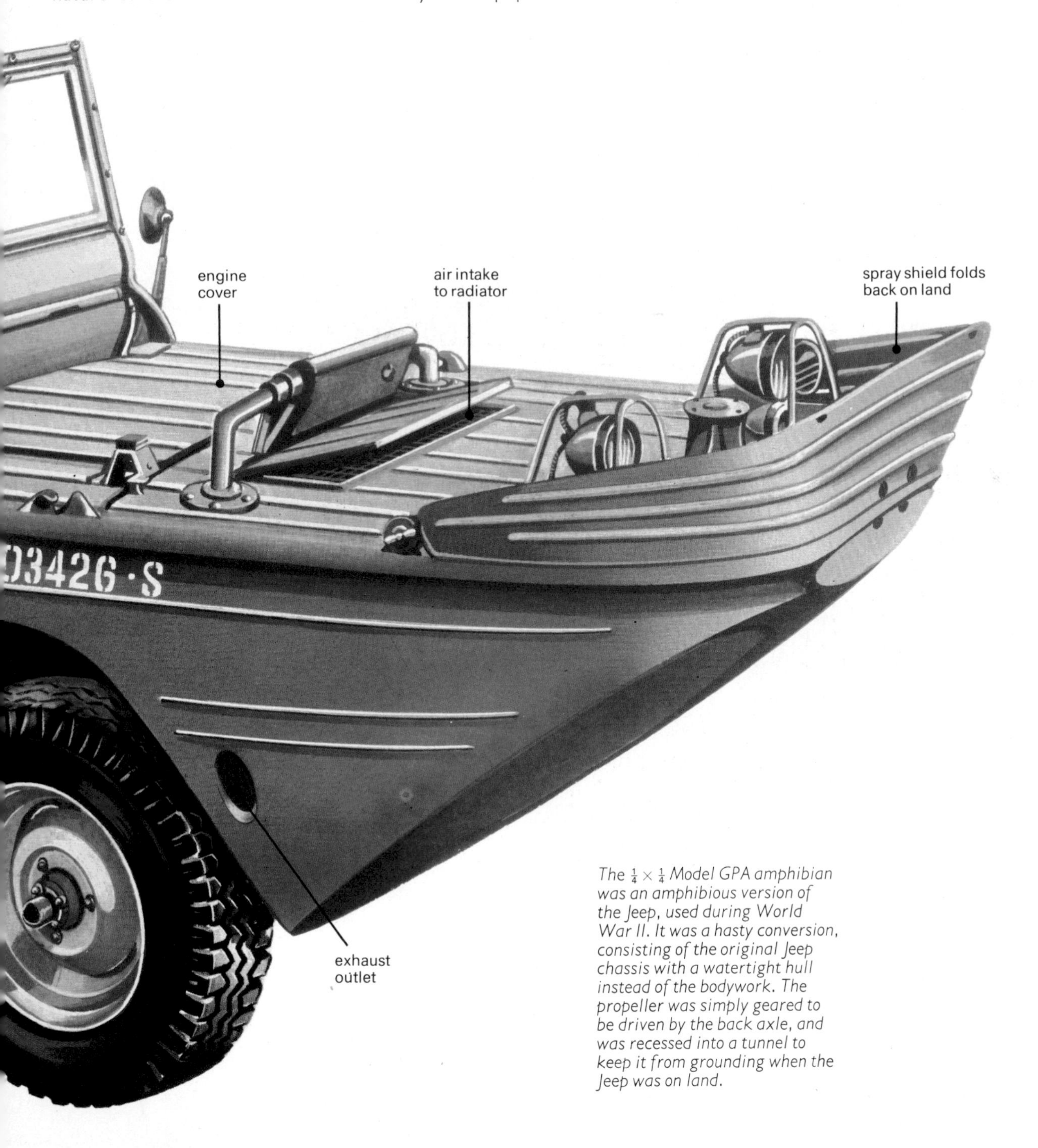

The $\frac{1}{4} \times \frac{1}{4}$ Model GPA amphibian was an amphibious version of the Jeep, used during World War II. It was a hasty conversion, consisting of the original Jeep chassis with a watertight hull instead of the bodywork. The propeller was simply geared to be driven by the back axle, and was recessed into a tunnel to keep it from grounding when the Jeep was on land.

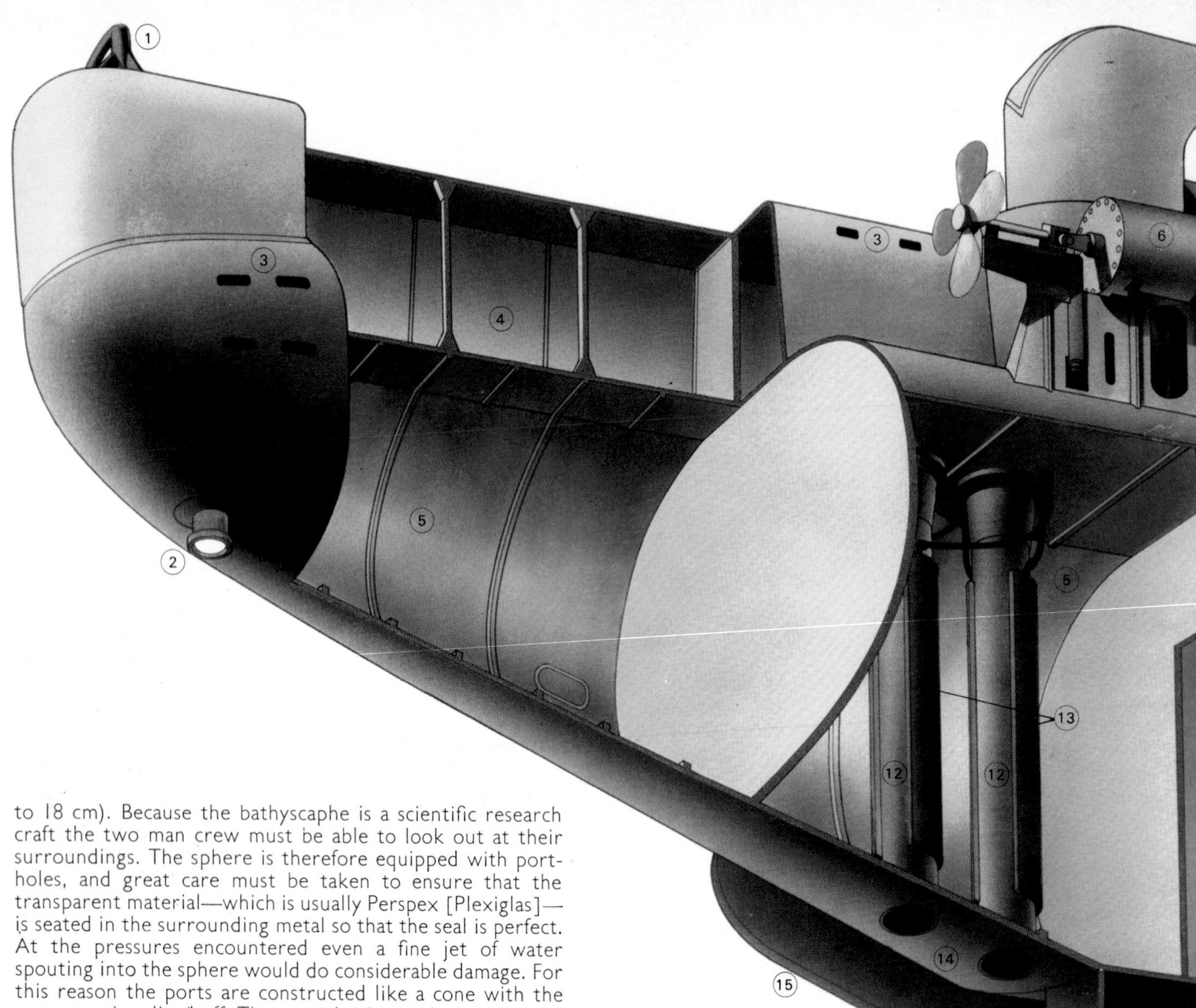

to 18 cm). Because the bathyscaphe is a scientific research craft the two man crew must be able to look out at their surroundings. The sphere is therefore equipped with portholes, and great care must be taken to ensure that the transparent material—which is usually Perspex [Plexiglas]—is seated in the surrounding metal so that the seal is perfect. At the pressures encountered even a fine jet of water spouting into the sphere would do considerable damage. For this reason the ports are constructed like a cone with the narrow point sliced off. They are glued into the sockets and inserted narrow end first into a matching hole, so mounting pressure merely pushes the port more securely into its bed. Similar ingenuity is required to stop leakages at the points where electrical conduits pass through the walls of the sphere, with the additional problem that the cables will flex when the craft is being used. The epoxy resin adhesive Araldite has been found to be an effective fixative and sealant, but experiments are going on with even tougher substances developed in the United States space programme.

The typical bathyscaphe sphere holds two men and provides air at normal atmospheric pressure, so that there is no need for the crew to decompress after a dive. Air cylinders are housed inside the sphere, and the supply usually lasts for about 24 hours. The sphere also contains the controls for the ballast and any external grabs, as well as sonar and television monitor.

Operating the bathyscaphe When the bathyscaphe is on the surface and preparing for a dive, some of the compartments in the float are filled with air. For diving, these compartments are flooded with sea water. As the vessel sinks, the outside of the float is subjected to increasing pressure. The bottoms of the compartments which contain petrol are open to the sea and the pressure of the water slightly compresses the lighter liquid, ensuring that at all times the pressure inside the float is the same as that outside. For this reason it is possible to construct the float of much lighter gauge metal than the sphere. Another effect of the compression of the petrol is that the deeper the bathyscaphe goes, the denser the petrol becomes until its density is even more than water. This makes the craft heavier and it sinks much faster. To control this, the crew can release small amounts of the ballast. The bathyscaphe is kept clear of the sea bed both by trimming the ballast so that the craft achieves neutral buoyancy at the desired level, and also by means of the guide chain.

This is a very simple device: simply a length of chain which hangs below the vessel. When the end touches the sea bed on the way down, the bathyscaphe is immediately lightened by the weight of the chain which has grounded. The bathyscaphe then slows up and when the weight of chain lying on the bottom reaches the right level the craft will have achieved neutral buoyancy. The chain also aids the stability of the vessel, acting as a drag line. Mobility on the bottom is limited: a massive structure is required to take two men to great depths and there is little room for anything more than small motors and electrical power packs. In practice, the crews of bathyscaphes frequently use undersea currents to help in the work of propulsion.

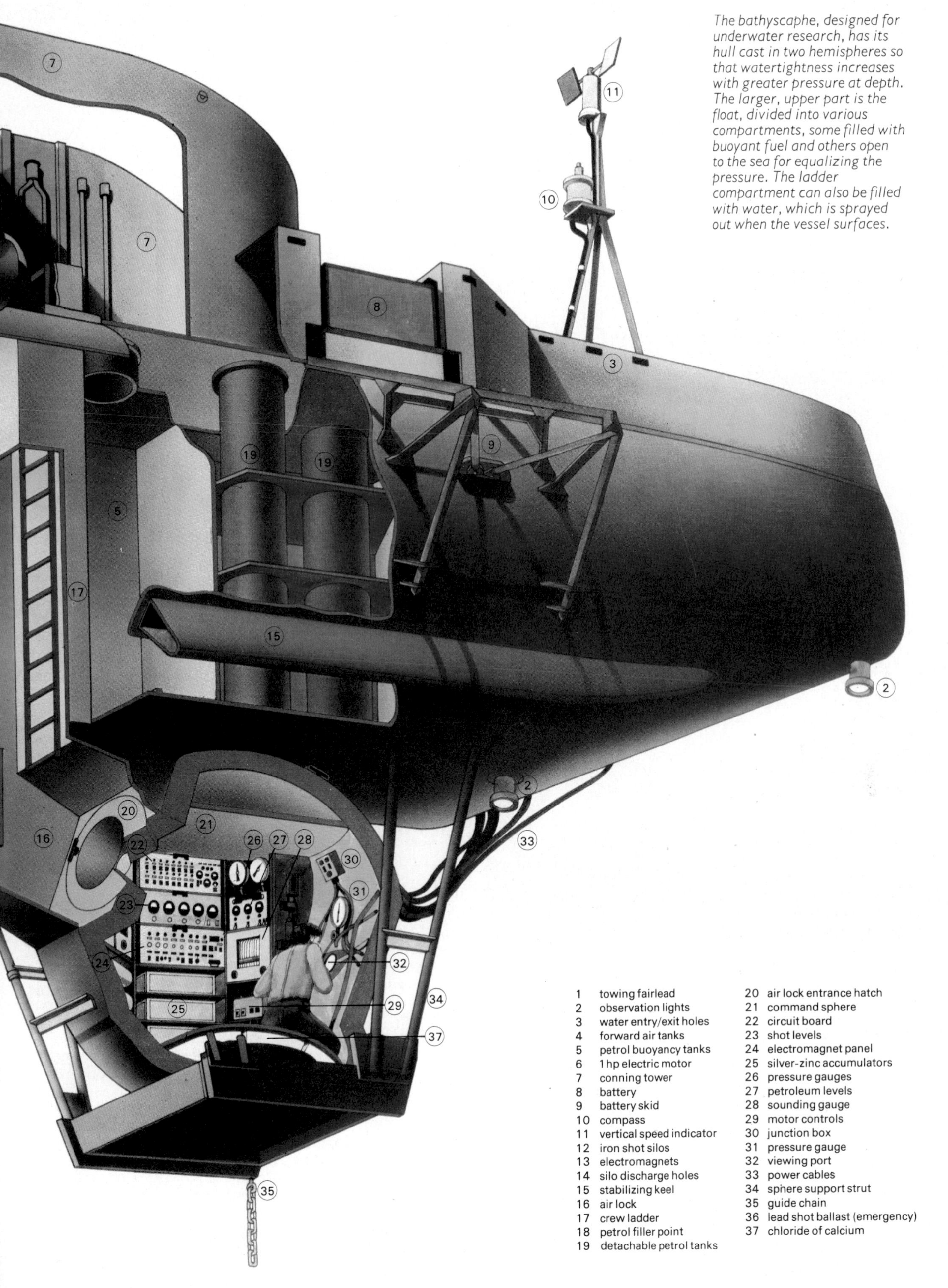

The bathyscaphe, designed for underwater research, has its hull cast in two hemispheres so that watertightness increases with greater pressure at depth. The larger, upper part is the float, divided into various compartments, some filled with buoyant fuel and others open to the sea for equalizing the pressure. The ladder compartment can also be filled with water, which is sprayed out when the vessel surfaces.

1 towing fairlead
2 observation lights
3 water entry/exit holes
4 forward air tanks
5 petrol buoyancy tanks
6 1 hp electric motor
7 conning tower
8 battery
9 battery skid
10 compass
11 vertical speed indicator
12 iron shot silos
13 electromagnets
14 silo discharge holes
15 stabilizing keel
16 air lock
17 crew ladder
18 petrol filler point
19 detachable petrol tanks
20 air lock entrance hatch
21 command sphere
22 circuit board
23 shot levels
24 electromagnet panel
25 silver-zinc accumulators
26 pressure gauges
27 petroleum levels
28 sounding gauge
29 motor controls
30 junction box
31 pressure gauge
32 viewing port
33 power cables
34 sphere support strut
35 guide chain
36 lead shot ballast (emergency)
37 chloride of calcium

To return to the surface the crew switches off the electromagnets, the rest of the shot empties away, and the light petrol 'balloons' the craft to the surface. In the event of a power failure exactly the same thing happens. This fail-safe system is augmented by other bunkers containing heavy scrap which can also be emptied and by the jettisoning of some pieces of heavy equipment secured by electromagnets.

Development Similar spheres called bathyspheres were used for very deep diving before the advent of the bathyscaphe after World War II. But these had the disadvantage of being suspended from a surface vessel by cables and limited in their usefulness because of this.

The inventor of the bathyscaphe was Professor Auguste Piccard. Famous in the 1930s for his ballooning exploits, this Swiss scientist decided to apply analagous ideas to deep sea exploration. The function of the float is the same as that of the balloon's gas bag. Ascent and descent in both are dependent on ballast trimming and, to some extent, on a drag cable. To enable the crew of either to breathe air at sea level pressure a self-contained air supply must be carried.

A large six-bladed fixed-pitch marine propeller is being cast. The largest solid propellers involve the melting of about 100 tons of metal. They are usually made from high-tensile non-ferrous alloys such as manganese bronze, and cast in moulds made of sand in a cement bond. The melting and pouring of the metal must be precisely controlled.

Piccard's work was supported initially by the Belgian National Foundation for Scientific Research (FRNS). Hence his famous balloon was called the FRNS 1. His first bathyscaphe was called the FRNS 2 and proved beyond doubt that his ideas worked. But it was the *Trieste*, named after the city where his backers came from, that provided his greatest triumph. Piloted by his son Jacques and a colleague, this vessel carried out a whole series of dives to ever greater depths, culminating in the present world record: a dive to 35,000 ft (11,000 m)—Mount Everest is only 29,000 ft (8800 m)—in 1960 in the Marianas Trench in the Pacific Ocean. Since then the *Trieste* has been successful in finding the remains of the sunken nuclear submarine *Thresher* and has managed to bring some of the wreckage to the surface.

Development work is now going on in France (FRNS III), the United States (the advanced *Trieste* II) and in the Soviet Union (the B-II, scheduled for launching in the late 1970s).

Propulsion and steering

The marine propeller The propeller, or screw, has been used in ships since about 1850. In its various forms the propeller remains the most efficient instrument for converting the power developed by a ship's machinery into the necessary thrust to push the ship through the water.

The propeller works by accelerating the water passing through it, exerting a forward thrust by the reaction from the increase in momentum of the accelerated flow. This momentum increase is achieved by giving either a low increase in speed to a large mass of water (large, slow-running propellers), or a high increase in speed to a small mass of water (small, fast-running propellers). The former represents the most efficient means of propulsion while the latter is less efficient, corresponding to what is commonly understood as jet propulsion.

The flow of water into the propeller is very significantly affected by the shape of the hull immediately ahead of it. In moving forward the ship drags along some of the surrounding water so that the relative speed of advance of the propeller through the water is actually less than the ship's speed. This field of flow affected by the ship forward of the propeller is called the wake field.

In addition, the axial speed of the water varies across the propeller 'disc' so that each blade in rotating will alternately pass through regions of high- and low-speed water. In general, when the blade is in the top of the aperture (at the 11 o'clock position) the relative water speed will be at its lowest and when the blade is in the bottom of the aperture the water speed will be at its highest. As a result the forces developed by the blades are cyclic in nature, fluctuating about the mean value and giving rise to problems with vibration, strength and cavitation.

Cavitation A propeller blade acts in a similar way to an aircraft wing in that a section through a blade has an aerofoil shape. The passage of the water sets up a pressure reduction on the forward side of the blades and a pressure increase on the aft side of the blades. The largest contribution to the propeller thrust comes from the pressure reduction, and if the pressure at any point falls to the pressure at which water vaporizes, then cavities of vapour are created in the water. This phenomenon is termed cavitation and can be harmful to the efficient operation of the propeller. The subsequent collapse of the cavities can also cause noise, erosion of the blade surfaces, and increased vibration.

The risk of cavitation is the main reason that a marine propeller differs from its airscrew equivalent in that it has much wider blades. A wide blade helps to restrict the level of the pressure reduction and so reduce the amount of cavitation. Nevertheless, because of the variations in water speed into the propeller it is often impossible to eliminate it completely.

Design The objective of the propeller designer is to produce a propeller that will convert the ship's power into thrust at the best possible efficiency, the propeller being strong enough to withstand all the forces involved without causing blade fracture and shaped in such a way that the harmful effects of cavitation are avoided. As such the propeller must be 'tailor-made' for the ship that it is to propel using a design procedure that is a blend of theory, model experiment and experience.

A very important requirement of any propeller design is to ensure that the ship's power is converted into thrust at the prescribed rate of revolutions, and this is dependent to a large extent on the propeller pitch. The pitch is the distance that the propeller would move forward in one complete revolution if the propeller was working in a solid substance. However, due to the fluid nature of water the propeller actually moves forward somewhat less than the pitch.

As the pitch of the propeller defines the angle of the blades to the flow, and thus the thrust and pressure forces that are developed by the action of the blades, it is also an important feature with regard to cavitation. Because of this the pitch is usually varied over the blade surface to suit the average water speed at each radius and so to minimize the risk of cavitation.

Types of propeller Although there are several types of propeller they all operate on the same basic principle of accelerating the water behind the ship.

The conventional propeller is a single-piece casting having two or more blades, although commonly four to six, designed specifically to operate at best efficiency at the ship's normal operating condition. Considerable effort is made to machine and finish these usually non-ferrous castings to a high degree of accuracy.

The controllable pitch propeller has its blades mounted separately on the hub so that, if required, they can be swivelled by a mechanism inside the hub which is operated hydraulically from inside the ship. By changing the pitch, or angle of the blades, this type of propeller can produce a variable amount of thrust to suit particular demands met in the ship's operation, such as when towing or manoeuvring. Complete control of the ship can thus be effected remotely from the bridge from full ahead to full astern without changing the direction of rotation of the engine, which would require more time.

The ducted propeller is either a conventional or controllable-pitch propeller fitted inside a 'nozzle' and because of its aerofoil shape the nozzle helps to accelerate the water flow into the propeller as well as producing a certain amount of thrust itself to propel the ship. This type of propeller has proved very efficient in special applications such as tugs and trawlers and is now being used on large bulk carriers.

Another type is the contra-rotating propeller system in which two conventional propellers are placed one in front of the other but rotating in opposite directions. The object of this is, in general, to reduce the loading per propeller, while recovering in the second propeller some of the rotational losses from the first.

Construction Propellers are made in a variety of materials but are normally manufactured in nickel-aluminium or manganese bronzes. For a conventional propeller, a mould is first prepared using a cement-bonded sand mixture; the metal is then cast, cooled under control and finally finished to the required complex shape.

The largest solid propeller so far produced has six blades, a diameter of 31 feet (9.4 m) and weighs 72 tons: it involved the melting of about 100 tons of metal. If the size of ships increases we may yet see propellers weighing 150 tons absorbing over 80,000 horsepower.

The world's largest propeller weighs 72 tons and is 31 feet (9.4m) in diameter. It is fitted to the 386,000 ton tanker Ioannis Coloctronis. *Extensive machining of the casting is necessary to obtain good balance and the precise shape of the propeller, which is designed for maximum efficiency at the ship's normal operating condition. This is why much more metal is melted to make the casting than is actually needed in the finished product. The larger ships of the future may have propellers weighing as much as 150 tons.*

The rudder Originally, steering of boats was achieved by suspending an oar over one or both sides of the craft at the stern. Around 200 AD, the Chinese realized that a better solution lay in setting the oar vertically through a shaft in the overhanging stern deck of their boats, and the rudder continued to be used in this form on Chinese junks. As a result of trade contact with the Chinese, Arab seamen began to hang the rudder like a hinged door on to the stern-post, and by the 12th century AD the rudder was in general use in the West.

Rudder action The passage of a ship through the water causes water to flow past the rudder, and the angle at which the rudder is inclined to the direction of flow is called the angle of attack. The steering action is dependent on the pressure distribution between the two hydrodynamic surfaces of the rudder. The pressure on the downstream side is less than the static pressure of the surrounding water, while the pressure on the upstream side is greater. The result of this is an outward force on the downstream side of the rudder, and this can be regarded as being made up of a lift force at right angles to the direction of flow and a drag force directly opposing the direction of flow. The variation of the lift and drag forces for different angles of attack is extremely important in rudder design as it is the lift force which creates the turning effect. At a certain angle of attack, called the critical angle, the rudder stalls: a phenomenon called burbling occurs and the rudder force is suddenly reduced. Burbling is caused by a breakdown in the streamlined flow on the downstream side of the rudder into a swirling irregular eddying flow. Rudders on merchant vessels are normally expected to operate up to an angle of 35° from the centreline to port or starboard, and so the critical angle is important as a reduction of rudder force would be undesirable within the working range.

Results from model tests with rudders in open water must be interpreted with care and cannot be directly applied to a ship as the rudder action is modified by the flow of water around the hull interacting with the propeller slip. Reliable results can only be obtained from full scale ship tests and then the model information corrected by a suitable factor for further rudder design.

When a ship's rudder is turned, the ship first moves a small distance sideways in the opposite direction to the intended turn and then moves around a circular path until it eventually faces the opposite direction. The distance moved forward from the point at which the rudder was turned to the point at which the ship is at right angles to its original direction is called the advance. Transfer is the sideways distance between these two points, and the diameter of the circular path followed by the ship is called the tactical diameter. During the turn the bow of the vessel lies always inside the turning curve, so that a drift angle is formed between the centreline of the vessel and the tangent to the turning curve. The tactical diameter is a measure of the ability of the rudder to turn the vessel and this is very important for a warship.

Other methods of steering A water jet unit may be used for propulsion and steering. The system is extremely useful for ferries and river craft where manoeuvring in confined waters is essential. Water is drawn into the unit and then discharged at high speed through a set of vanes which can be rotated to give thrust in any specified direction. Two units may be fitted to a vessel, one forward and one aft, to give maximum turning effect.

A bow thrust unit consists of a water tunnel at right angles to the centreline of a ship, fitted with a propeller whose blade angle or pitch can be varied by a hydraulic control. By altering the pitch, the amount of thrust can be adjusted to give the required lateral movement of the ship. The unit is controlled from the bridge of the vessel and the propeller is driven by a motor through a flexible connection and bevel

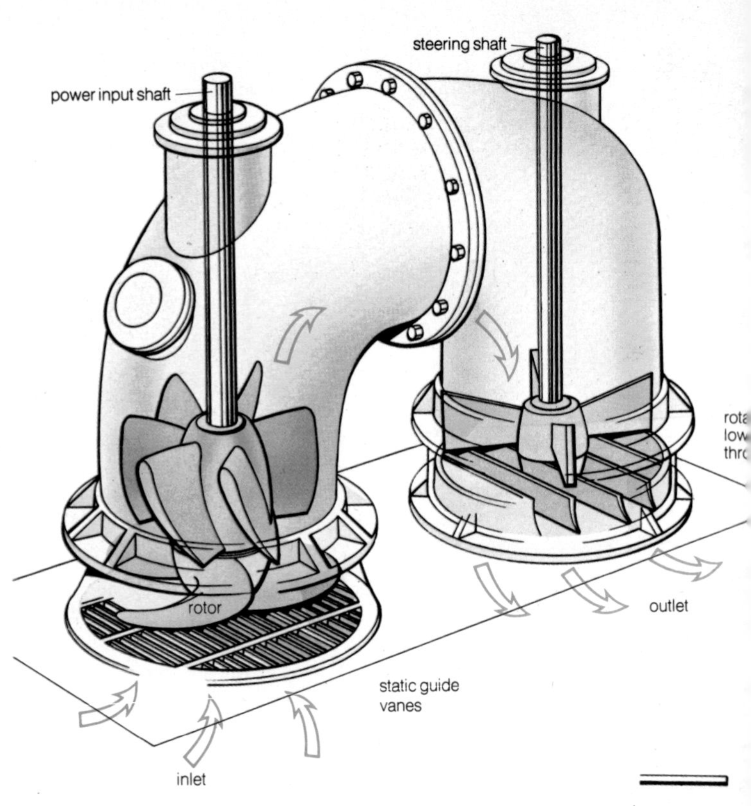

Various ship's steering arrangements are in common use today. In the jet propulsion unit (above) water is forced through the U-shaped tunnel and the outlet vanes deflect the water in the direction desired. The Kamewa bow thrust unit (centre) has a propeller in a tunnel at a right angle to the line of the ship; the propeller can drive water in either direction by altering the pitch of the blades. The Navyflux thruster (below) drives water out on either side of the ship by opening the appropriate shutters. The jet flap rudder drives water through a port or starboard aperture (lower right). At top right is a cutaway view of a conventional rudder for a modern ocean-going single screw vessel.

Navyflux Y-thruster in the bulbous bow of a vessel

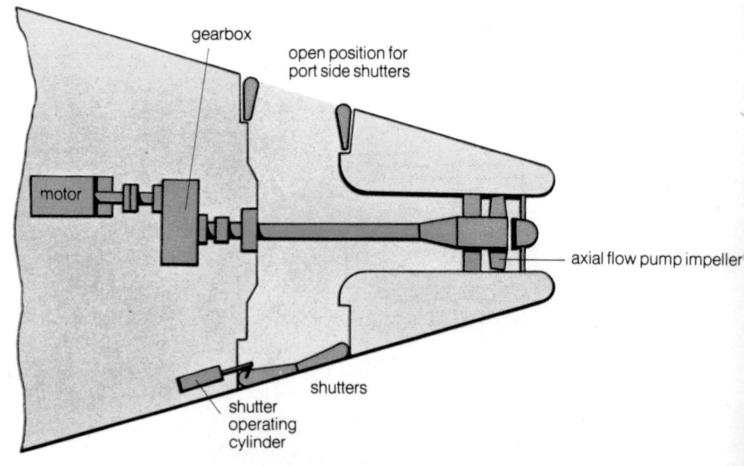

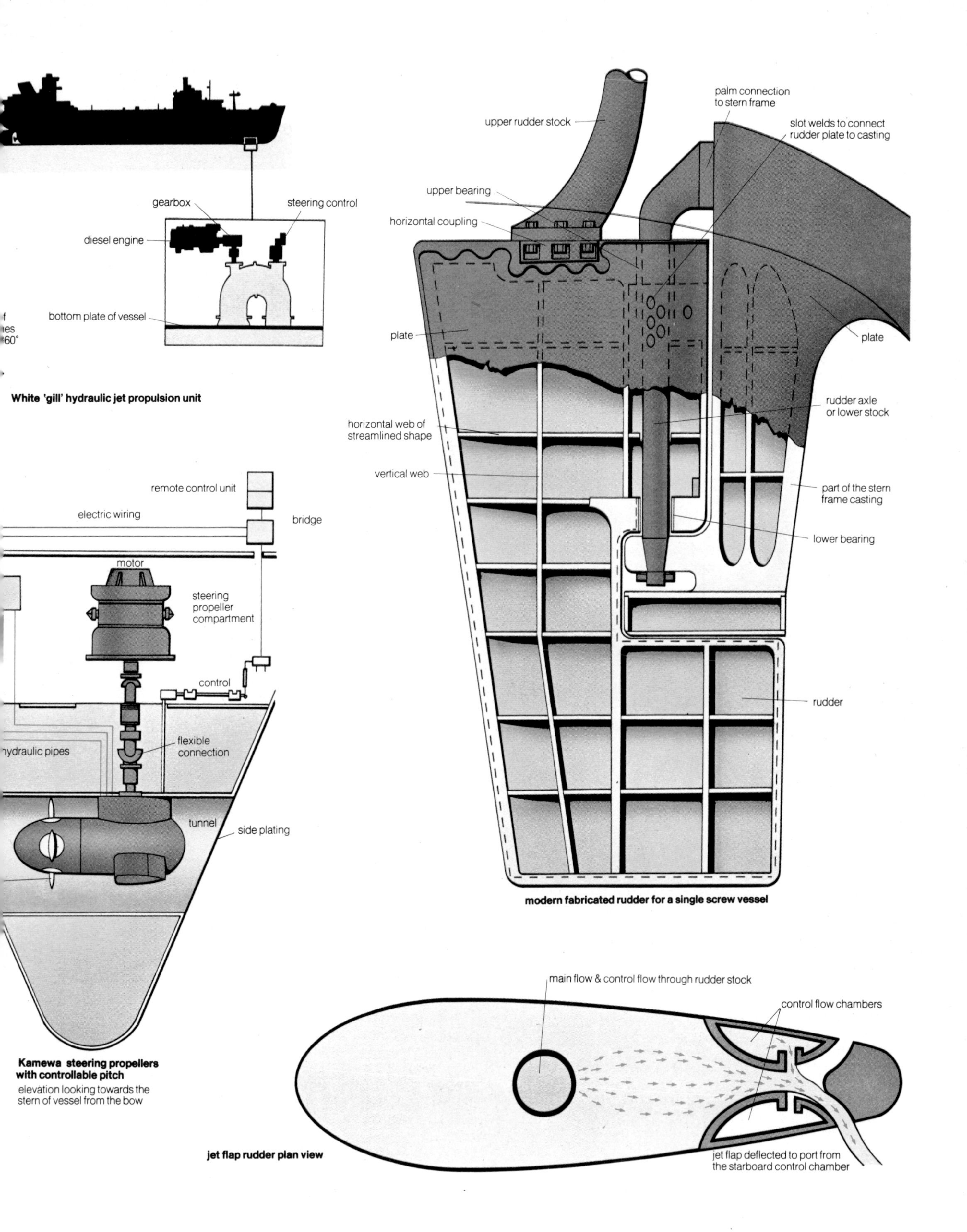

White 'gill' hydraulic jet propulsion unit

Kamewa steering propellers with controllable pitch
elevation looking towards the stern of vessel from the bow

modern fabricated rudder for a single screw vessel

jet flap rudder plan view

gearing. Bow thrusters are used on many ships such as oil tankers, bulk carriers and passenger vessels, and they are used for accurate course control for cable laying vessels.

A 'Navyflux' thruster, mounted in the bow of a ship, has a Y-shaped tunnel with openings at each side of the vessel and another directly forwards. An axial pump mounted in the front limb of the tunnel is used to discharge water out through the other two limbs. The openings of the two side limbs of the tunnel are controlled independently by shutters which are operated by a hydraulic mechanism. A lateral force is created by arranging the shutters so that different quantities of water flow out through the two side limbs of the tunnel; the maximum steering force is obtained when one of the shutters is entirely closed, the other is fully open and the pump is operated at maximum power. This type of thruster does not require forward motion of the ship for its operation, but at high forward speeds it will operate without the pump running.

An active rudder has a propulsion unit fitted into the rudder body and a fixed or variable pitch propeller at its trailing edge. When the rudder is turned, the propeller will produce thrust at an angle to the centreline of the vessel, causing a greater turning effect than the rudder alone. Here again, the rudder is not dependent on the forward motion of the ship for its operation.

An articulated rudder has a separate flap which can be turned through a greater angle than that of the rudder. This rudder flap alters the flow characteristics of the water over the rudder by increasing the camber of the surface. For low ship speeds the articulated rudder produces a greater rudder lift force and therefore a more positive turning effect. Articulated rudders are generally fitted on fishing vessels and tugs.

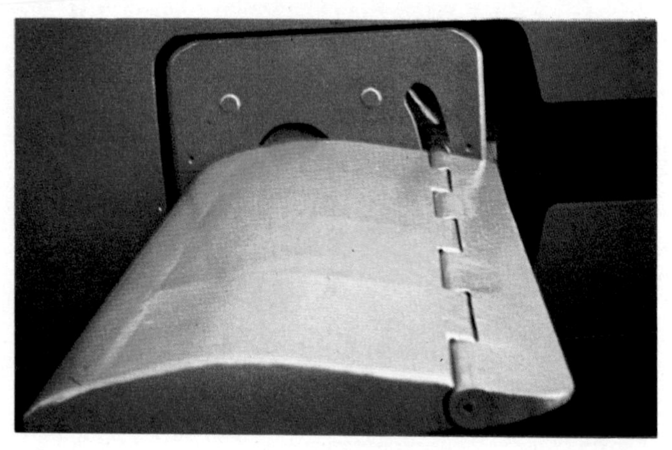

A stabilizer fin mounted on a ship's hull. The motion of the fin through the water generates upward or downward forces, according to the angle of attack, to counteract the rolling of the ship. The angle of the fin is altered automatically by a hydraulic motor controlled by a gyroscope which senses the ship's motion. When not in use, this type of fin is rotated forward into a watertight recess in the ship's hull.

A jet flap rudder has a similar action to the articulated rudder described above, but it is simpler mechanically. Water is pumped into the rudder, which is hollow, through the upper stock (the rudder turning linkage) and leaves through a vertical slot near to the trailing edge. The main jet of water emerging from this slot can be switched to the left or right by means of a control chamber which is also fed with water from the upper stock. The control chamber ejects a water jet into the main jet, causing the main flow to deflect through the desired slot according to which way the vessel is to be turned.

A cycloidal propeller combines the functions of a propeller and a rudder. It has vertical blades which can be turned through specific angles to produce an overall thrust in one direction. The correct angle of each blade is set by mechanical linkages and an eccentric which controls the direction of thrust for steering purposes. One of the advantages of this system is that it allows a ship to be turned in its own length.

A cylindrical rudder consists of a vertically mounted rotatable cylinder located behind the ship's propeller. The rudder operates by distorting the water flow around the cylinder, and this causes a lift force at right angles to the direction of fluid flow. To obtain a lift force on the other side of the cylinder, the direction of rotation is simply reversed. Conventional rudders can also be fitted with rotating cylinders to modify their flow characteristics and to improve the lift force. The cylinder may be fitted at the leading edge of the rudder or just in front of a rudder flap.

The stabilizer The ship's stabilizers are designed to reduce the rolling of the ship in rough water, in order to prevent cargo from shifting about and causing a list of the ship. Stabilizers also reduce problems for the catering services on ferries, ocean liners and cruise ships, as well as adding to passenger comfort. For accurate gunfire on warships, it is essential to keep motion as steady as possible, and stabilizers are fitted to many naval vessels to control the angle of roll. The stabilizer also reduces the stresses and strains on the ship's hull and internal framework caused by rolling. Various types of stabilizer have been tried with varying degrees of success; these include bilge keels, oscillating weights, anti-rolling tanks, gyroscopes and stabilizer fins.

Bilge keels These are normally fitted to the hull along both sides of the ship, and they extend for about one-third of the length of the vessel. They are riveted or welded to the shell where it curves to form the bilge at the bottom of the ship. Bilge keels are attached so that they offer minimum resistance to the forward motion, and they are not too strongly connected to the hull so that they will break off without damaging the shell if they strike an obstruction. They present resistance to motion in the rolling direction by impeding the fluid flow around the hull, and they are more effective when the ship is under way than when it is stationary. Bilge keels are fairly effective at damping out the angle of roll and they tend to increase the period of roll, that is the time taken to roll from one side to the other and back again.

Oscillating weights This system, now out of favour, involves moving weights from one side of a vessel to the other to counteract the motion created by the sea. The phase of the weight movement must lag 90 degrees behind the rolling motion of the vessel (the two movements must always be out of step) and thus the timing of the operation is critical. One method tried in an experimental installation was to move a truck on curved rails so that its weight produced a stabilizing force on the ship. Systems of this sort are no longer used partly because control in irregular waves is difficult and partly because they are noisy.

Anti-rolling tanks These were introduced very shortly after the first use of bilge keels. 'Frahm' anti-rolling tanks have been successfully used for a number of years on many ships. They are usually fitted near to amidships either in a tween

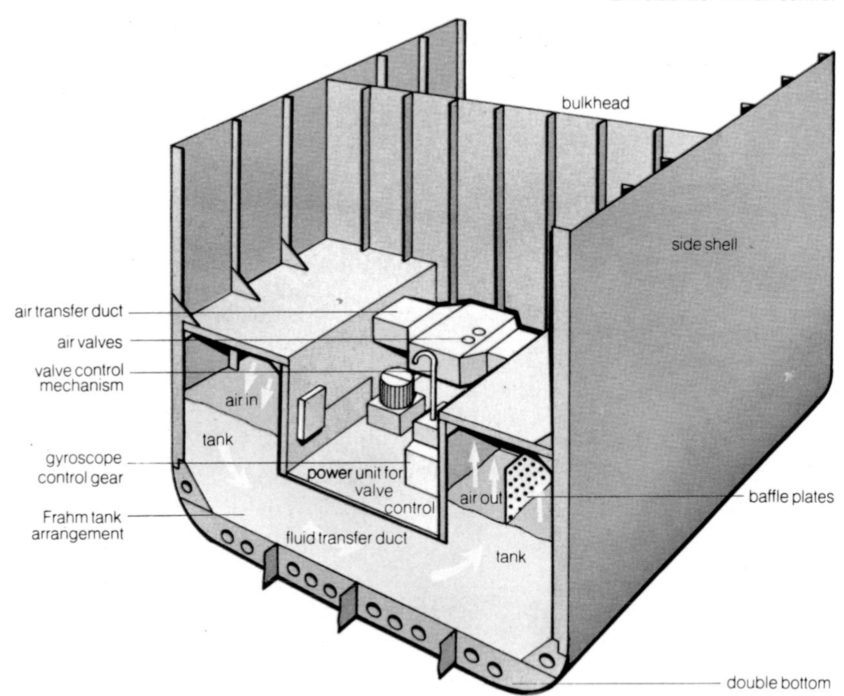

An activated tank stabilizer is a development of the Frahm stabilizer, in which water moves from one side of the tank to the other to compensate for the wave motion. In this more sophisticated design, the water distribution between the two sides of the tank is controlled by gyroscopically-controlled air valves, which regulate the volume of air above the water on each side.

deck space or lower down in the vessel above the double bottom. The arrangement is like a U-tube but with a larger cross-sectional area in the two vertical legs than in the horizontal leg. The relationship between these areas is important because the period of fluid oscillation when the ship rolls must be similar, but 90 degrees out of phase to that of the ship. There must be an air connection between the tops of the vertical tanks; otherwise the air in one tank will become pressurized and the other tank will have a partial vacuum when the liquid levels change. An air valve in the connection gives a means of fluid control if necessary.

In activated anti-rolling tanks a high capacity, low pressure air compressor supplies air to the upper part of the tanks, and by varying the pressure in each tank water can be moved from one side of the ship to the other to give a stabilizing effect. A gyroscope is used to stop and start the compressor and to operate the tank air valves as it senses the ship's motion.

Gyroscopic stabilizers Some vessels are fitted with large gyroscopes to control rolling. This technique reduces the average angle of roll by about 50%. Schlick in Germany was the first to use this system, and later Sperry stabilizers were introduced in the United States. The Schlick gyroscope was installed with the spin axis vertical, and the support frame axis horizontal. When the vessel rolls the gyroscope frame swings in its bearings in a fore and aft direction. This is called precession and is a function of the gyroscopic action. A roll to starboard would cause the top of the gyroscope frame to move aft if unresisted at the bearings. The opposite will occur for a roll to port. Because the rolling of the ship causes precession of the gyroscope, if precessional motion is resisted by applying brakes to the bearings of the support frame this will set up a stabilizing effect in opposition to the rolling of the ship. Brake control of the Schlick type of stabilizer is difficult to achieve, and the Sperry stabilizers were arranged with a precession motor meshed with a vertical ring gear to precess the gyroscope in a direction to oppose the rolling motion of the ship. Control of the precession motor is by a small pilot gyroscope sensitive to the transverse motion of the ship. The pilot gyroscope operates electrical contacts which power the precession motor in the required direction.

Flume stabilization A flume tank is placed transversely across the ship and comprises two side compartments and a centre compartment, which contain water. The motion of the fluid from one side of the vessel to the other is controlled by a restriction called a flume. Liquid depth is constant in the centre compartment during the transfer process. The tanks are carefully designed to tune the liquid frequency to the natural period of roll of the vessel and to maintain the 90 degree phase relationship necessary for stabilization. The flume prevents the liquid movement from coming into phase with the ship's movement and causing a disastrous increase in the rolling.

Fins Stabilizer fins project from the hull and produce a turning movement on the ship to oppose any rolling motion. As the ship moves through the water, the flow over the protruding fins, port and starboard, is deflected according to the angle of the fin, producing either an upward or a downward stabilizing force. As the ship rolls, the fin on the ascending side of the vessel will generate a downward force and the fin on the other side will produce an upward force. The magnitude of these forces depends on the angle through which the fins are rotated from the horizontal position, and the speed of the water over the fin surface. At low ship speeds the fins are not so effective as when the vessel is travelling at her designed cruising speed. When not in use, the fins may be retracted into a watertight box in the hull. An oil operated vane motor is used to turn the fins, the oil being delivered from a hydraulic pump controlled by a gyroscope which senses the ship's motion.

Navigation

Navigation is a business of motion. The Polynesians learnt to steer their catamarans by the stars and by the motion of the Pacific swells. The Mediterranean seamen tended to rely on prevailing winds until, around 1200, the magnetic compass appeared. It was soon found that the direction indicated was not true north but varied over the Earth, this variation now being printed on charts. When iron and steel ships were built, compensation for the consequent deviation became necessary, a problem solved in the last hundred years by the gyrocompass, which measures the direction of the Earth's axis without using magnetism.

In the Middle Ages, the navigator would cast over the stern a log tied to a knotted line which he would pay out, counting the knots while a half-minute sand-glass ran dry. Modern logs take the form of propellers or use water pressure, the latest types employing electronic principles. However, the word knot, meaning a nautical mile per hour, is also retained. The nautical mile is 6080 feet (1852 m), almost exactly a minute (one-sixtieth of a degree) of latitude.

Position finding from course, or direction of travel, and from speed, according to the time since leaving a starting point, is known as dead reckoning or DR. Were there neither tides nor winds the navigator at sea could rely on DR but, in practice, fixing or finding position by other means is necessary at intervals. On land, it is possible to pin-point a position on a map but at sea it is a matter of using distant objects, for example taking bearings on landmarks ashore. At night, these directions may be taken, with the aid of a compass, on a lighthouse.

In the wind oceans, dead-reckoning checks have traditionally depended on astro-navigation. Each point on the Earth's surface has a unique vertical and therefore a unique horizontal. The navigator measures the altitude or angle above the horizon of Sun, Moon or star by means of his sextant. According to where he thinks he is, he then calculates what the angle ought to be at the time. If the sextant attitude is greater than the calculated value the correction will be towards the star, the distance in nautical miles being the difference of altitude in minutes. A second star's altitude and a second correction will fix position.

Although astro-navigation is a worldwide aid, it is not 'all weather'. Cloud may cover the sky and fog will blot out the horizon which, in any event, may not be visible at all during the night so that star observations may have to be restricted to dusk and dawn.

The navigation desk of the frigate HMS Charybdis *is fitted with the Decca navigator above the table. This gives readings which can be plotted directly on an overprinted chart (opposite page). There are three dials; two readings are sufficient for an accurate fix. The wave motions from two transmitters intersect, like ripples on a pond. Decca operates in Europe and other major traffic areas;* Charybdis *and other ocean-going vessels also carry Loran for longer-range use.*

Electronics at sea In the early 1900s, Marconi successfully demonstrated radio communication which was used initially to broadcast time signals for checking the chronometers (accurate clocks) carried on ships for astro-navigation. It was soon found that an aerial wound in a loop could detect the direction from which a radio wave was coming. The navigator now had eyes which could 'see' a radio beacon at night or in fog.

By squeaking and timing the echoes, bats have learnt to fly inside pitch black caves. Radio waves travel nearly a million times as fast as sound waves, seven times round the Earth in a second. Nevertheless, shortly before World War II, echoes were being measured by radio. Using a directional aerial, generally in the form of a dish, distances and directions could be found and the results displayed on the face of a cathode ray tube. Thus radar enabled the mariner to see coastlines and other ships at night or in fog.

In recent decades, more money and effort has been expended on marine aids than in all the previous history of man. The lead line has been replaced by the echo sounder, which times the travel of sound waves from the hull of a ship to the seabed and back.

During World War II the Loran and Decca Navigator systems came into use. These set up patterns over the earth comparable to the intersecting ripples of water when stones are thrown simultaneously into a pond. More recently, Omega, an ultra-long-range, very-low-frequency system has been developed.

All these systems work on the same principle, though there are practical differences between them. If two radio transmitters send out the same signal, say a continuous tone, the two wave motions will coincide with each other—that is, they will be in phase—along certain lines. These lines form a hyperbolic pattern around the transmitters. A ship picking up the signals exactly in phase must therefore be on one of these lines, which are marked on charts. To give a precise fix, signals from another pair of stations must be compared, the hyperbolic lines from which are also marked. Combining the two gives a unique position, once the lines involved are known.

This is the principle of the Decca system, which uses additional lane-identification signals transmitted from the ground stations every minute to give a unique fix. These drive automatic counters on board the ship, so that a continuous readout of position is displayed.

The basic Loran system uses pulses instead of continuous wave transmissions, giving rather less information but a greater accuracy at long distances; the Loran C system combines both. These systems are also used by aircraft. A limited system called Consol, giving direction details only, is used by yachtsmen.

The most recent system involves the use of satellites which orbit the Earth, broadcasting details of their orbits by means of a signal code which is changed by remote control from the ground as the orbit alters. As the satellite moves across the sky, the rate of change of its Doppler shift varies—its distance from the ship varies most rapidly when it is close to the horizon, and least when it is overhead. The orbit and Doppler information are fed into a computer on board the ship, which calculates the ship's position to great accuracy.

Asdic (called after the Allied Submarine Detection Investigation Committee) is a device developed originally for detecting submarines by means of sound waves travelling through water. Its usefulness has now been extended to include detecting submerged wrecks and shoals of fish, as a navigational aid for ships, measuring the depth of water under a ship, and for research purposes. Sonar stands for sound navigation ranging, and this term has largely replaced the term Asdic. Another term, 'echo sounding', implies the principle of the system. Sound is normally thought of as being air vibrations, but water can also transmit vibrations. By listening for echoes, a picture of the area below the surface can be obtained.

In the basic technique, a short pulse of sound, which may be within the range audible to humans or may be on a much higher frequency, is transmitted from the bottom of the ship, usually by means of a transducer mounted in a hole cut in the ship's bottom. (A transducer is any device that converts electrical power into another form, such as sound, or vice versa.) The sound pulse travels downwards until it strikes the sea bottom or a submerged object, which reflects it back to the ship. There, it is picked up by another transducer, and the time taken for the pulse to travel down and back is measured. The speed at which sound travels in water depends on temperature, but it is roughly 4800 feet per second (1460 m/s), over four times its speed in air. The distance the pulse has travelled can be determined from the measurement of the time it took to travel that distance.

Each pulse lasts anything from a few thousandths of a second to a few seconds depending upon the range of the particular sonar (the longer the range, the longer the pulse). The pulses are emitted at intervals ranging from a fraction of a second to a few seconds or even minutes, again depending upon range. A typical shipboard echo sounder used for navigation has two range scales. On its shorter range it will measure depths up to 20 feet (6 m) in steps of a fraction of a foot. On its longer range it will measure depths up to 100 or more fathoms (1 fathom = 6 feet = 1.8 m).

The time interval, and therefore the range, can be measured by a rotating disc carrying a neon tube which flashes when the reflected sound pulse (the echo) is picked up. The disc rotates at a constant speed, carrying the neon tube past a fixed circular scale graduated in terms of distance.

Docks

For many centuries sea traders relied solely on the shelter afforded by natural harbours, inlets and river estuaries in order to load or discharge, victual or repair their ships. While lying at anchor, their vessels were at the mercy, not only of wind and tide, but of bands of marauders to whom they were easy prey. The need for protection from such threats led to the establishment of basins or wet docks where sailing ships could be fitted out in safety and where their cargoes could be dealt with in relative security.

The word 'dock', which to this day is used fairly loosely to describe a variety of places where ships are berthed, was first used to describe 'an artificial basin filled with water and enclosed by gates' during the second half of the sixteenth century—a period of considerable expansion in maritime trade. One of the first recorded enclosed dock basins was the Howland Great Wet Dock which was built on the south bank of the River Thames in the seventeenth century. Only in the

This hand-held radio locator is tuned to a coastal radio beacon whose bearing is then read off. Two such bearings, selected with the aid of an almanac, give a fix on a chart.

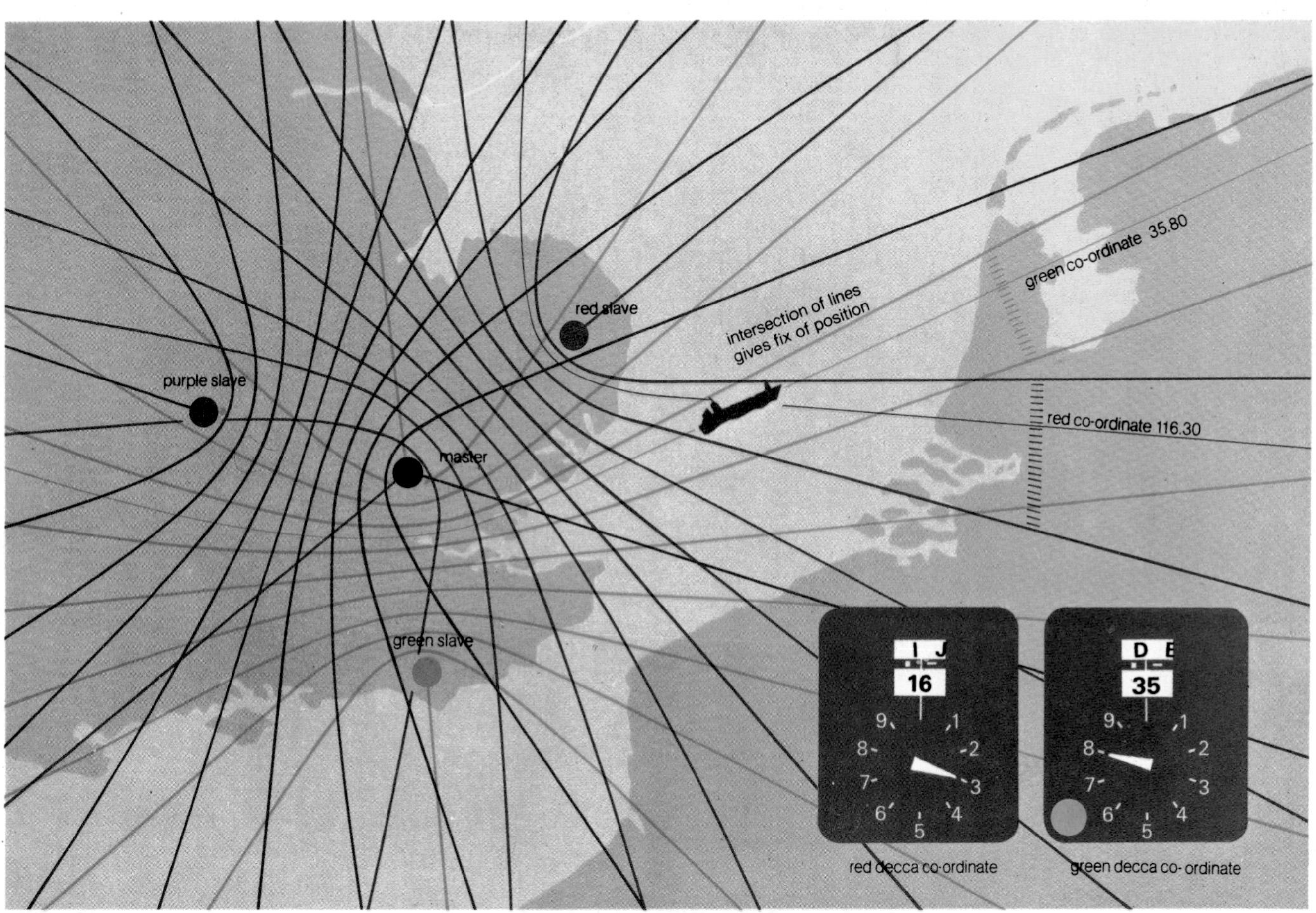

late eighteenth and nineteenth centuries did the great dock-building period begin in earnest, often closely associated with canal and railway-building ventures. This was also the period of the first iron steamships but the tremendous growth in ship sizes since those days has made many early docks obsolete.

The provision of gates at dock entrances is necessary because of the large tidal range which would otherwise cause the basins to have insufficient depth of water at low tide. In many countries the rise and fall of the tide is so insignificant that docks can be completely tidal. For example, in Melbourne, Australia, spring tides (those with the greatest range) rise less than 3 ft (1 m), in Rotterdam about 6.5 ft (2 m); and in Boston, USA, about 10 ft (3 m).

In Britain, however, with its large tidal ranges, more major dock systems are enclosed. The most notable exception to this is at Southampton where all dock berths are tidal, and where the effects of a 13 ft (4 m) tidal range are minimized by a phenomenon known as the 'double tide', which gives six hours of high water a day. An extreme example of a tidal range which makes enclosed docks imperative occurs in the Severn Estuary, where Bristol, for instance, experiences a maximum variation of almost 49 ft (15 m) between high and low water.

Lock entrances The dimensions of the lock-pit inevitably govern the maximum size of vessel which can enter an enclosed dock. With the trend towards larger ships in recent years, the constraints of existing entrance locks have become a problem. The largest lock in Britain, at Tilbury, is 1000 ft (305 m) long and 110 ft (33.5 m) wide, with a depth of 45½ ft (14 m), whereas the largest container ships, now operating between Europe and the Far East, are 950 ft (290 m) long overall, 106 ft (32 m) in beam, and have a maximum draught of 42½ ft (12 m). New entrance locks are being built to cater for vessels even larger than this: at the new West Dock at Bristol a lock measuring 1200 ft (366 m) long and 140 ft (43 m) wide is being constructed to take ships of 75,000 tons deadweight, and developments at Le Havre in France include a new lock 1312 ft (400 m) in length and 219 ft (67 m) wide, which is claimed to be the world's largest and is capable of accommodating a tanker of 500,000 tons deadweight.

The operation of an entrance lock is basically simple. By using a system of culverts and sluices, water is allowed to pass from the dock into the lock with both inner and outer gates shut. The water level, and with it any ship in the lock, rises until it reaches dock level, when the inner gates open and the ship moves into dock. A departing vessel can then be penned in the lock and lowered by allowing water in the lock to escape through the outer sluices.

Dock layouts Although certain cargoes such as coal or bulk grain require specialized handling facilities, dock berths have traditionally been multipurpose and vary little in design, layout and equipment. Usually the quay apron (the working area alongside ship) is equipped with rail tracks both for cranes and railway wagons, and is flush-surfaced to give access to road vehicles. Quay cranes of three to five tons capacity at 80 ft (24 m) radius are usually adequate for break-bulk general cargo operations (that is, where individual packages, drums, bales, and so on are handled piecemeal using cargo trays, nets, slings, or hooks), but cranes of greater capacity are installed where heavier cargo, for example steel traffic, is frequently dealt with. For even heavier items many ports are equipped with floating cranes, often with lifting capacities exceeding 100 tons.

Transit sheds adjacent to the quay apron give temporary covered accommodation to cargo prior to its loading aboard ship or collection by road or rail vehicles. Modern sheds have the maximum possible unobstructed floor area so that mobile equipment such as fork lift trucks, platform trucks and mobile cranes can be used to carry and stack cargo. To the rear of the sheds, loading bays with both road and rail access serve for the delivery of goods.

Container docks The dramatic changes that have occurred in cargo transportation over the past ten years have, however, completely transformed the layout of modern terminals. These new techniques include containerization—the carriage of general goods in large containers of internationally standardized dimensions—and 'roll-on roll-off' employing vessels with bow, stern, or side doors through which wheeled freight is loaded and discharged.

A typical container-handling dock has a large area of land serving each berth, ideally 20 to 25 acres (8 to 10 hectares), for container marshalling. It does not normally have covered

accommodation, except where container stuffing (packing) and unstuffing or Customs examination are carried out, although container warehouses have been constructed with their own internal gantry cranes for stacking.

Two or three giant gantry cranes, with lifting capacities of up to 40 tons, and capable of working a three-minute cycle (that is, loading and unloading 20 containers an hour) may be provided to a berth. For large ocean-going container ships at least 1000 ft (305 m) of quay is allocated for each berth. Mobile handling equipment may include van carriers, which straddle, lift, carry and stack containers three high, tractors and trailers, or side- or front-loaders, each with similar lifting capacities. Alternatively, the gantry cranes may span the entire stacking area, carrying out all movements between ship, container stack, and inland transport.

Ferry terminals Although many roll-on roll-off ferry terminals cater for passengers as well as freight and have passenger facilities of varying degrees of refinement, roll-on roll-off terminals consist mainly of a ramp, shore bridge, or link-span on to which the ferry can open its doors, and a large marshalling area for the vehicles it carries. In some cases a simple concrete ramp built out from the quay wall is all that is necessary, but most shore bridges are tailor-made for the individual vessel using them, with electrically operated machinery able to compensate for the ferry's changing draught during loading operations. Like all very successful ideas roll-on roll-off is a simple concept and it has revolutionized the carriage of cargo on short sea routes.

Bulk terminals The economics of transportation are resulting in the building of increasingly large vessels for bulk handling of raw materials but arrangements must be made to accommodate them. *Globtik Tokyo* and *Globtik London*, the largest tankets afloat, are 477,000 tons deadweight, 1243 ft (382 m) long, and have a draught of 92 ft 6 in (28 m). Oil tankers are usually brought to jetties sited in deep water but a relatively new system of loading and unloading uses what is known as a single-point mooring buoy or monobuoy mooring, linked by pipeline to the shore installations, and which is placed as far out to sea as is necessary.

Dry docks and floating docks At regular intervals all ships need to be inspected 'in the dry', and sometimes repaired. For this reason, most major ports are equipped with dry or graving docks, slipways being used for smaller ships.

Dry docks, which usually take one ship at a time, are simply basins which are capable of being pumped dry to leave a ship supported by an arrangement of 'keel blocks', so that work can be carried out on the hull, propellers, or rudder. The procedure for drydocking a ship is a precise affair and may take several hours; with the dock flooded the gate is opened and the ship enters, then the gate is closed and pumping begins. Accurate positioning is vital as the ship settles on to the blocks, prearranged to fit her hull, and to facilitate this, modern dry docks are usually fitted with guidance systems. In many international ports dry docks are being provided for tne largest tankers afloat or planned: the port of Rotterdam already has a dry dock 1350 ft (412 m) long, which can accommodate 500,000-ton tankers, and the construction of a super dry dock for 700,000-tonners will be carried out at Kiel over the next two years. The purpose of a floating dock is the same as that of a conventional dry dock, only the method of getting the ship out of the water differs. Ballast tanks are used to raise the submerged dock towards the surface and with it the ship to be repaired.

Far left: minesweepers to be repaired in Brest, France, are carefully positioned on keel blocks; then the water is pumped from the dry-dock.
Left: this floating dry-dock, in Hamburg, can lift ships of 70,000 tons deadweight. It uses ballast tanks to lift the dock as well as the ship.
Above left: this view of the container handling operation at Tilbury Docks, London, looks down the long boom of the crane.
Above: the traditional method of dockside freight handling uses slings.

Canals and locks

The term canal may be defined as any completely man-made open-water channel, whether it is used for irrigation, land drainage, water supply or navigation. When we use the word nowadays, however, it is usually in the context of navigation by boats, although many canals may serve a secondary function as a drainage or water supply channel.

Canals are not to be confused with river navigations. Such rivers may range from natural channels which, because of their size, are navigable far inland, to those which have been improved and made navigable by the construction of artificial 'cuts' to bypass acute bends, shallows and rapids, together with the construction of many locks.

A boat on a canal or river, or entering a dock, may have to be passed from one level to another. The simplest device which can do this is a pound or chamber lock. The general form of the modern lock is that of an open-topped chamber, with watertight gates at each end. Once the boat is inside, the gates are shut and the water level in the lock then rises or falls (depending on whether the boat is ascending or descending). When the water in the chamber has reached the right level, the appropriate gate is opened to allow the boat to leave the lock.

Locks have a history of at least two thousand years, although the earliest were not pound locks. One factor limiting the navigability of rivers is the depth of water and to increase this, dams were often built. Part of each dam was removable, permitting boats to be winched upstream through the gap, or swirled downstream by the current.

After the 1840s British canals were allowed to fall into disuse. Economic conditions forced families to live on their boats. This picture was taken in 1913.

The removable part is called a flash lock; the entire structure, a navigation weir. Its use was two-fold: the impounded water not only increased upstream depths, but could also be released in a surge to assist vessels in shallow water downstream. Locks of this type existed in China in 70 AD, and their use later spread as they were built into water-mill weirs as well. Although this reduced disputes between millers and boatmen, these could not be settled until the adoption of the pound lock, whose closed chamber minimizes the water loss from the higher level. In 984, the first known example of a pound lock was built in China, and was operated by raising or lowering gates at each end (now called guillotine gates). Europe's earliest pound lock which can be dated precisely was built in 1373 at Vreeswijk, Holland, and also had guillotine gates.

Most modern locks have an improved type of gate invented by Leonardo da Vinci in the fifteenth century for a canal in Milan, Italy. These gates turn on hinges, like doors, and each end of the lock has two such gates which meet to form a vee pointing upstream, giving them the name mitre gates. England's first pound locks with mitre gates were built in 1567 on the Exeter Canal.

One advantage of mitre gates is that since the vee shape points upstream, they are self sealing. When there is a difference in water level between one side and the other, the pressure holding the gates together is at its greatest.

Ancient locks were often built entirely of wood, but stone or brick chambers later became standard. Gates were usually of wood, lasting up to 50 years. To fill and empty the lock, hand-operated sluices were fitted to the gates, but it was later found that mounting these in conduits bypassing the gates gave a smoother water flow.

Where steep rises have to be negotiated, locks are built contiguously, with the upper gates of one acting as the lower

Below: the 'staircase' of locks built in 1825 on the Grand Union Canal at Foxton, Leicestershire, is still used.
Bottom of page: a cross section of a canal lock. To fill the lower chamber the lower gates are closed and the upper gates opened; the rising water level raises the boat so that it can move to the next chamber.

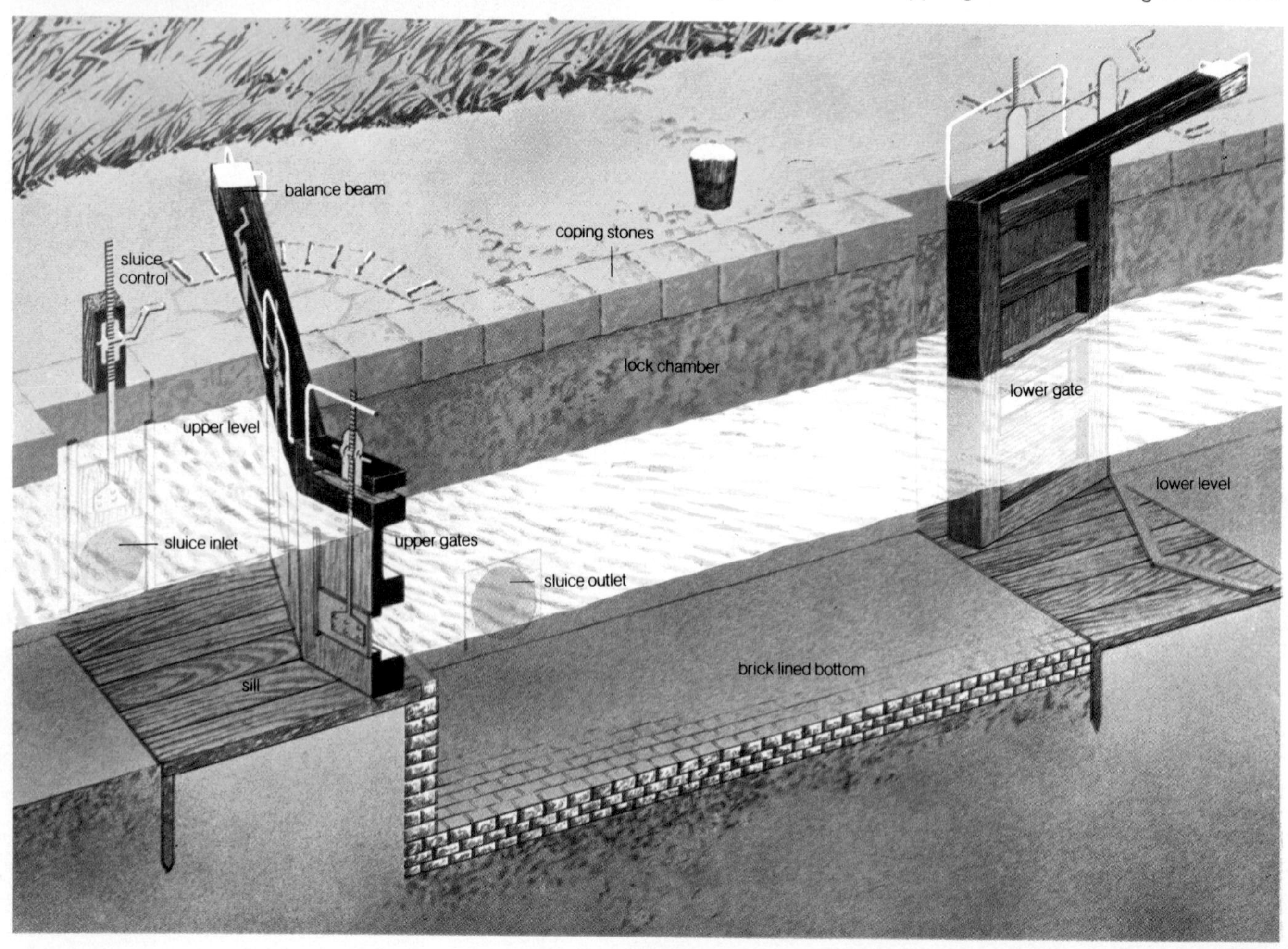

gates of the next. The resulting structure is called a staircase and many examples exist from 'two-steps' (three sets of gates) to the mighty 'Neptune's Staircase' on the Caledonian Canal, Scotland, which consists of eight locks condensed into one structure with nine sets of gates.

Construction Modern lock construction can readily be seen to represent a refinement of older types, with concrete usually used for the chamber and welded steel for the gates. Hydraulic power or electricity is used to operate the gates and sluices. As locks have grown in size to admit larger vessels, so, on canals, the problem of the water lost from higher to lower levels has increased. To overcome this, pumping may be employed, or economizer locks built. These have small reservoirs alongside the chamber to store some of the water emptied from it. This water is later used to refill the lock. The outstanding examples of this type are on the Rhine-Main-Danube Canal which will link Germany's waterways to those of south-eastern Europe. Here, multiple side-reservoirs allow savings of up to 60% of the lock water. The locks are 623 ft (190 m) long, 39 ft (12 m) wide and have rises up to 81 ft (25 m).

In the mid-nineteenth century, locks rarely had rises above 30 ft (9 m), but today's largest has a rise of 138 ft (42 m). This is at Ust-Kamenogorskiy, USSR, and lifts 1600 ton vessels past a hydroelectric dam on the River Irtish. The chamber, 328 ft (100 m) long and 56 ft (17 m) wide, takes about half an hour to fill or empty. The lower gate does not extend for the full height; it moves vertically to seal a 'tunnel' through which ascending vessels pass to enter the lock.

The Panama Canal also has large locks, especially remarkable as it was opened in 1914. These have mitre gates up to 82 ft (25 m) high and can accommodate vessels 1000 ft (305 m) long and 110 ft (34 m) wide, taking only eight minutes to fill or empty.

Gates Early dock entrance locks closely resemble eighteenth century canal locks, having brick or masonry chambers, timber mitre gates and manual operation. A dock entrance lock rarely needs a rise exceeding 20 ft (6 m), but the draught of ocean-going vessels requires that the gates be very deep. Modern gates are of several types: many are mitre gates but others are hinged along their lower edge (flap gates), moved sideways on rails (traversing caissons), or are floated into place (floating caissons). All are of hollow welded steel construction, permanently ballasted with concrete. Operating ballast may also be required: this is provided by pumping water in or out of the interior of the gate to sink or float it.

Although deep-water terminals are replacing enclosed docks for the largest ships, many massive dock entrances exist. Enclosed docks are unnecessary if the tidal range is less than about 12 ft (4 m), so there are many major ports without entrance locks. Typical ports where locks are necessary are Bremerhaven, Germany, which has a lock 1220 ft (372 m) long and 148 ft (45 m) wide, and Antwerp, Belgium, which has one 1180 ft (360 m) by 148 ft (45 m). The time taken to operate these is more dependent on manoeuvring the large ships than on actually filling the lock and can be up to 90 minutes. Unless a suitable natural water supply exists at a higher level, water must be pumped back up into the dock to compensate for that lost in locking, seepage and evaporation.

Canal routes The use of modern–type locks on river navigations was widespread long before canals became popular. The reason for this was that to build a canal over a high watershed involved major civil engineering works for which there was neither the expertise nor the capital available. Many early canals are called contour canals because they follow a very devious course along the natural contours in an effort to avoid the cost of major earthworks, aqueducts or tunnels. Even so, such major works were impossible to avoid; equally unavoidable was the problem of supplying the summit, or highest level, of the canal with sufficient water to compensate for that used in lockage. Each boat, as it passes through such a canal, draws two lock chambers full of water away from the topmost level, one as it ascends and another as it descends, and more water will be required for larger lock chambers. To make the problem more difficult for the engineer, this demand for water arises at the precise point where it is most difficult to supply, being situated on an upland plateau.

The first true summit-level canal was built in France to link the River Loire with the River Seine. This was the Canal de Briare, 21 miles (34 km) long with 40 locks to carry it over a 266 ft (81 m) summit level. Hugues Cosnier was the brilliant engineer responsible and from the day it was opened in 1642 his canal was a commercial success. As a prototype it paved the way for another far more ambitious work, that of the Canal du Midi which was built under the direction of Pierre Paul Riquet between 1671 and 1685 to link the Garonne at Toulouse with the Mediterranean. It is 149 miles (240 km) long and there are 99 locks which lift the canal over a 620 ft (189 m) summit level. Riquet went to amazing lengths to ensure that this high summit level had an adequate supply of water. To this end he drove nearly 40 miles (64 km) of feeder channels and built a huge dam at St Ferreol in the hills of the Massif Central.

History When it was built in the 17th century, the Canal du Midi was the greatest work of civil engineering then existing and it served as a prototype for the world to copy. For example, Francis Egerton, Duke of Bridgewater, visited the Midi Canal while he was on a grand tour of the Continent, and what he saw there emboldened him to embark on the construction of the Bridgewater Canal from his estate at

The recently completed Montech waterslope in France uses a pool of water which is sealed by a moveable dam. The dam is pushed up the slope of 42 feet (13m) by locomotives, pushing the water with it.

Worsley to Manchester, later extending it to the Mersey at Runcorn. This began the era of canal construction in England which lasted roughly from 1760 to 1830. First, the four great rivers of England, Mersey, Trent, Severn and Thames, were linked by a 'cross' of canals engineered by James Brindley and his assistants using the old contour canal principle. These canals were later supplemented by new and more direct routes of which the last was Thomas Telford's Birmingham and Liverpool Junction Canal which was completed in 1834. In contrast to Brindley's contour canals, the latter was surveyed and built on the 'cut and fill' technique adopted by the railway builders whereby the amount excavated in cuttings equals that required to raise the embankments.

This canal system of the English Midlands as devised by Brindley and his successors played a key role in Britain's industrial revolution by providing an economical method of transport for goods in bulk, especially coal. Unfortunately, however, Brindley adopted as standard on his canals a lock chamber 70 ft (21 m) long but only 7 ft (2.1 m) wide which restricted their use to special canal 'narrow boats' of similar dimensions carrying only 30 tons. Doubtless it was the water supply problem that determined this slim gauge, but although it proved adequate for 18th century needs, it became quite inadequate in the 20th century. France, for example, considers a 350 ton barge the minimum economic size and when it is realized that both the 30 tonner and the 350 tonner cost the same in man hours to work through a canal, it becomes easy to understand the reason why the greater part of England's canal system today is suitable only for pleasure craft.

Just as the Duke of Bridgewater was influenced by the example of the Canal du Midi, so, in turn, Americans and Canadians were influenced by the spate of canal construction in England. There was a certain time lag, for whereas in England canal construction reached its peak in 1792, in North America it did not begin seriously until 1816 when a number of waterways were constructed to provide better communication between the East Coast, the Great Lakes and the Middle West.

Canals today With significant exceptions, the canal heyday in England and North America was brief. Perhaps because railways were specifically an English invention, the English speaking world seems to have lost interest in its canals with their coming in the 1830s. By assuming too readily that railways were the only answer to every transport need, many canals were allowed to fall into neglect and decay. This was not the case on the continent of Europe where both transport systems continued to develop side by side. It was realized that each form of transport was suited to certain traffic and that the particular role of the canal was in carrying low-value bulk cargoes over long distances, provided it was capable of passing craft of sufficiently large tonnage. In Europe, therefore, certain trunk canal routes have been progressively enlarged over the years.

For example, the Brussels-Charleroi Canal, originally suitable for 400 ton craft when it was opened in 1895, has now been completely reconstructed to accept 1350 ton barges. Again, the Great North Holland Ship Canal, constructed between 1819 and 1825, enabled craft of 800 tons to reach Amsterdam, but thanks to recent reconstruction this figure has been raised to 2000 tons.

Another example is the Rhine-Danube Canal which was built in 1846 at the instigation of King Ludwig of Bavaria with 100 locks to carry it over the Rhine-Danube watershed. It could only pass 130 ton barges so that, although it survived into the 1940s, it needed too much labour to compete in the modern world and it is now being replaced by a greater new waterway capable of passing 1350 ton barges.

In England the only equivalent of these great waterways is the Manchester Ship Canal, which is 36 miles (58 km) long with five locks. It was opened in 1894 to enable sea-going ships of up to 4500 tons gross to reach the new port of Manchester.

In North America an outstandingly successful inland

The Rochdale Canal, in Lancashire, once played an important role in the town's industries, but Britain's canal system has not been maintained or improved over the years. In France, waterborne traffic increased from 76 million tons to 106 million tons between 1963 and 1971; in Britain in 1972 freight carried on all inland waterways came to only 30 million tons, and was declining. An American conference of civil engineers recently estimated that one dollar will move one ton of cargo five miles by air, 15 miles by truck, 67 miles by rail and 335 miles by water. Systems such as container handling and LASH (Lighter Aboard Ship; see page 62) would make canals compatible with other methods of shipping. To modernize the most important 100 miles or so of existing commercial waterways in Britain would cost perhaps £20 million; an official at the Department of the Environment declined to say how much motorways cost.

waterway was the famous Erie Canal built by the State of New York between 1817 and 1825 to link the Hudson River with Lake Erie. With 82 locks in a course of 363 miles (584 km) it was the longest canal in the world at the time of its opening. It was widened and deepened throughout as traffic grew until, in 1918, it was replaced by the New York State Barge Canal which can accommodate 2500 ton craft. North of this is the new (1959) St Lawrence Seaway, part canal, part river, which enables seagoing vessels to travel from Montreal to Lake Ontario and from there via the Welland Canal into Lake Erie.

All the other canals in eastern North America have fallen into disuse, but there is the great network of 6000 miles (10,000 km) of inland waterways formed by the Mississippi, its canalized tributaries and their associated canals. These have a common depth of 9 feet (3 m) and the smallest locks measure 600 by 100 ft (180 by 30 m). In the Southern States and connected with the Mississippi system are the Intracoastal waterways. Lock-free and a mixture of natural channels and artificial canals, they provide waterways for big barges sheltered from the sea. The Gulf Intracoastal runs for over 1000 miles (1600 km) from near the Mexican border to Florida where it is extended by the Atlantic Intracoastal as far as Norfolk, Virginia.

Compared with Brindley's little 70 feet (21 m) by 7 feet (2.1 m) locks with an average fall of 8 feet (2.4 m), some modern locks are of immense size. The largest in length and breadth is at Ijmuiden on the North Sea Ship Canal in Holland, which measures 1320 by 165 ft (402 by 50 m). The lock with the greatest fall is claimed by Russia with 138 ft (42 m) for a lock on the Irtish river navigation, though the 113 ft (34 m) deep John Day lock on the Columbia River section of the Mississippi system does not lag very far behind.

All these large locks are either on, or connected to, large river systems so that water supply problems do not occur. But even on summit-level canals the availability of water no longer governs the size of locks as was the case in Brindley's day. The modern canal engineer thinks in terms of pumping the lockage water back electrically rather than building costly dams to store more water at the summit level.

Boat lifts Though water can be pumped back, a long flight of locks uses a great deal of another precious commodity —time. Hence from the earliest days engineers have schemed to save both water and time by substituting various forms of boat lift for lock flights. Of these the only one remaining in England is the Anderton Vertical lift of 1875, which lowers boats from the Trent and Mersey Canal to the River Weaver. Ironically, although this is the sole survivor in England and is no longer in commercial use, its English engineers went on to build four much larger lifts of similar type on the Canal du Centre in Belgium: these are still in use by 400 ton barges. Also in Belgium is the huge Ronquières inclined plane lift on the Brussels-Charleroi Canal, which must be the largest canal structure in the world. It is a mile (1.6 km) long and lifts 1350 ton barges 220 ft (67 m). Another ingenious modern form of lift has just been brought into commission at Montech in southern France on the Garonne Lateral Canal. This consists of an inclined trough of concrete up which the barge, floating on a triangular wedge of water, is pushed by a moving dam propelled by locomotives.

Sea canals Another type of canal is the sea-to-sea ship canal designed to shorten a long sea passage. The trouble with this type of canal is that it is doomed to obsolescence because of the rapid increase in the size of shipping. Telford's Caledonian Canal was driven through the Great Glen of Scotland to enable sailing ships to avoid the long and perilous passage around Cape Wrath, but it was soon made obsolete by the coming of steam power and the increasing size of shipping. Telford was also responsible for the construction of the very similar Göta Canal in Sweden which opened an inland passage for Swedish shipping from the Baltic to the North Sea and so avoided the tolls levied by Denmark in the Kattegat. But here again, ships soon outgrew it.

The most famous 19th century canal builder was the Frenchman Ferdinand de Lesseps who was responsible for the Suez Canal. Because this 100 mile (160 km) long canal through the desert required no locks, its progressive enlargement has been relatively simple.

De Lesseps also planned to drive a 50 mile (80 km) long lockless sea level canal through the Isthmus of Panama to unite the Atlantic with the Pacific. But although his army of workmen toiled from 1881 to 1889, the climate and the terrain defeated them. Malaria and yellow fever took a terrible toll of the work force and treacherous rock strata caused incessant slips in the vast cutting he planned through the spine of Panama. The present, locked, Panama Canal was built and opened by the Americans in 1914, but it is already becoming too small and too slow, and there is talk of replacing it with a new lock-free canal such as de Lesseps dreamed of.

The Corinth Canal, four miles (6km) long, saves a considerable journey around the Peloponnesus in Greece. Cut through solid rock at the end of the nineteenth century, it can take vessels up to ten thousand tons.

Lighthouses

Until the introduction of lighthouses the usual way to mark the entrance to harbours or the presence of rocks or sandbanks was by means of a beacon, made up of a pile of stones or wooden spars. These simple warning devices are still used in small harbours, and more elaborate structures such as masts and pillars are widely used.

The first lighthouse of which there is any record was the Pharos of Alexandria (in Egypt), a huge structure built by Ptolemy in the third century BC. It has been estimated that the tower was built on a base 100 feet (30.5 m) square and was 450 feet (137 m) high. The Pharos survived until about AD 1200, when it was destroyed by an earthquake; it gave its name to pharology, the science of lighthouse building.

The Romans built many notable lighthouses, such as that at Ostia (the chief port for Rome) and others in Spain and France and at Dover. Following the fall of the Roman empire, navigational aids, along with many other aspects of that civilization, fell into disuse. Once again it was left to individual ports to set up and maintain their own lights.

From about the eleventh century onwards the increase in sea trading led to a revival of interest in lighthouse construction. Progress was slow, but in England and Europe from about 1600 onwards there was an increase in lighthouse building culminating in the great era of lighthouse construction in the eighteenth and nineteenth centuries.

Lighthouse illumination did not become really efficient until the early 1780s when the Swiss engineer Aimé Argand invented the type of oil burner which bears his name. This lamp used a circular wick, surrounded by a glass chimney which created a central, upward draught of air to assist the burning. This lamp produced a steady smokeless flame of high intensity, and it remained the principal source of light for over 100 years.

The Argand burner was adapted for use in domestic gas lighting, and it was gas-lighting technology which contributed to the next major advance in oil burners for lighthouses. The Argand lamp used a wick from which the oil vaporized for burning, but in 1901 Arthur Kitson produced a burner in which the oil was vaporized in a copper tube placed above the mantle (adapted from a gas mantle). The vapour then passed from the coiled tube to be burnt in the mantle, like gas. The oil was vaporized by the heat from the mantle, and a blowlamp was used to heat the coil before the lamp was lit.

The Kitson design was improved by David Hood in 1921, and this type of burner is still widely used today where electric lighting is not practical. Many unattended lights burn acetylene gas.

Electric lighting was first tried in the South Foreland light on the Kent coast in 1858, using a carbon arc lamp, and this experiment was followed in 1862 by the installation of arc lamps at the Dungeness light, also in Kent. Arc lamps did not, however, prove satisfactory for lighthouses and little use was made of them.

The first use of electric filament lamps was also at the South Foreland light, in 1922, and many lighthouses are now operating with electric lights, either filament lamps or high pressure xenon lamps. The power is supplied from the local mains where possible, or else by diesel powered generators.

Optical systems The development of efficient light sources led naturally to the development of reflector systems, because without some form of beam projection much of the intensity of the light is wasted. The three main groups of optical systems used in lighthouses are the catoptric (reflective), dioptric (refractive) and the catadioptric (reflective and refractive).

The first parabolic reflector was designed in 1752 by William Hutchinson, made up of small squares of mirrored glass set in plaster of Paris. The parabolic reflector is placed

behind the light source and the rays of light are reflected parallel to the axis of the reflector and emerge as a beam of light. Reflectors made from a hand-beaten composition of copper and silver soon replaced the heavy glass reflectors, and by 1800 they were standard equipment in lighthouses.

The use of a reflector increased the power of the light signal by about 350 times, and as the problem of beam projection had thus been overcome the question of individual light characteristics for each lighthouse could be dealt with. This question arose because ship owners often complained that although lighthouses were useful, it was difficult to tell one from another as they all emitted the same signal.

This problem was solved by arranging reflectors in different positions on a frame and revolving it, producing group flashings (two or three flashes in quick succession, followed by a period of darkness, then the flashes again).

The most important development in lighthouse engineering was the fresnel lens, invented by Augustin Fresnel in 1822. This is a dioptric lens, and has a central 'bullseye' lens surrounded by concentric rings of prismatic glass, each ring projecting a little way beyond the previous one. The overall effect of this arrangement is to refract (bend) into a horizontal beam most of the rays of light from a central lamp. Further reflecting elements may be placed above and below the refracting prisms to form a catadioptric arrangement.

Sometimes two lenses are placed one above the other, with a light at the centre of each, and this is called a bi-form optic. Many refinements have been made to Fresnel's original design, but the basic principle is essentially the same today.

Construction Most lighthouses are built of stone or pre-

Far left: a lamp and part of the lens system at the South Stack lighthouse. The circular parts are Fresnel lenses. The concentric rings of prismatic glass bend the light rays from the lamp into a horizontal beam. Without some such arrangement, most of the power of the lamp would be lost.
Left: a cross-section of an offshore lighthouse built of interlocking stone blocks. The detail of the service room and the lantern gallery shows the way the stones fit together. This type of construction gives a strong, stable structure which can withstand the action of strong winds and high seas.

cast concrete, and some are built many miles off-shore. The Eddystone light, for example, stands on a rock in the English Channel some 13 miles (21 km) from Plymouth. The present structure is the fourth to be built on the site, and was completed in 1881. The original was a wooden tower, built in 1698 by Henry Winstanley who was at the light in 1703 when it was washed away during a severe storm. The second one, also of wood, was built by John Rudyerd in 1708 and survived until 1755 when it burnt down.

The third and most famous Eddystone light was designed by John Smeaton and completed in 1759, using a hydraulic cement invented by Smeaton, and with the stone blocks dovetailed together for strength. Erosion of the rock on which it was built necessitated its replacement by the present light which was built nearby, using an even more complex dovetailing arrangement than Smeaton's, which was dismantled and re-erected on Plymouth Hoe as a memorial to him.

Where a suitable rock foundation is unavailable, such as where the hazard to be marked is a sandbank or coral reef, the light may be built on steel piles or concrete-filled caisson foundations. The light towers of these are often of open steel frame construction.

Light vessels The first light vessel (lightship) to go into service was moored near the Nore buoy in the Thames Estuary in 1731, and was found to be of great benefit by ship owners, who willingly subscribed to her upkeep. The lighting consisted of two ships' lanterns mounted 12 feet (3.7 m) apart on a cross beam on the single mast.

The light vessel was put on station by Robert Hamblin, who thought that navigation suffered from the difficulty of distinguishing one lighthouse from another, and considered that light vessels should be moored at dangerous points around the coast, using different arrangements of lanterns to enable each station to be easily identified.

The Nore vessel was placed on station against the wishes of Trinity House, the England lighthouse authority, who considered that lights which were only candles or oil lights would be ineffective as a guide to shipping. Despite these objections the King granted a patent for the light vessel, to run for 14 years from July 1730.

Following the immediate success of the Nore vessel, Trinity House became worried that light vessels would become so numerous as to upset the lighthouse system and eventually succeeded in persuading the King to revoke the patent in May 1732. The vessel had proved such a success, however, that it was impossible to remove it. Trinity House therefore obtained a patent in perpetuity and granted a lease for 61 years to Hamblin. After this initial breakthrough light vessels became accepted as valuable aids to navigation, but no other country tried them until 1800, and the first to be used in the USA did not go on station until 1820.

The lighting system now used in light vessels is multi-catoptric, consisting of eight parabolic reflectors mounted in pairs, one above the other, with an electric filament lamp in each. These reflectors are mounted on a frame which is rotated by a small electric motor. By varying the angular position of the reflectors and the speed of rotation of the frame it is possible to achieve single, double, triple and quadruple flashing lights.

THE ROMANCE OF THE RAILWAY

The story of the social and economic advancement of the last two centuries is inseparable from the saga of the railway. In England, the rise of the middle classes was assisted by the commuter railways; in America, the West was conquered by the noisy, smoking contraption which the Indians called the 'iron horse'. In Russia, a Czarist official complained that if railways were constructed, anybody would be able to go anywhere they wanted . . .

SANTA FE

The history of railways

The railway is a good example of a system evolved in various places to fulfil a need and then developed empirically. In essence it consists of parallel tracks or bars of metal or wood, supported transversely by other bars—stone, wood, steel and concrete have been used—so that the load of the vehicle is spread evenly through the substructure. Such tracks were used in the Middle Ages for mining tramways in Europe; railways came to England in the 16th century and went back to Europe in the 19th century as an English invention.

English railways The first Act of Parliament for a railway, giving right of way over other people's property, was passed in 1758, and the first for a public railway, to carry the traffic of all comers, dates from 1801. The Stockton and Darlington Railway, opened on 27 September 1825, was the first public steam railway in the world, although it had only one locomotive and relied on horse traction for the most part, with stationary steam engines for working inclined planes.

The obvious advantages of railways as a means of conveying heavy loads and passengers brought about a proliferation of projects. The Liverpool & Manchester, 30 miles (48 km) long and including formidable engineering problems, became the classic example of a steam railway for general carriage. It opened on 15 September 1830 in the presence of the Duke of Wellington, who had been Prime Minister until earlier in the year. On opening day, the train stopped for water and the passengers alighted on to the opposite track; another locomotive came along and William Huskisson, an MP and a great advocate of the railway, was killed. Despite this tragedy the railway was a great success; in its first year of operation, revenue from passenger service was more than ten times that anticipated. Over 2500 miles of railway had been authorized in Britain and nearly 1500 completed by 1840.

Britain presented the world with a complete system for the construction and operation of railways. Solutions were found to civil engineering problems, motive power designs and the details of rolling stock. The natural result of these achievements was the calling in of British engineers to provide railways in France, where as a consequence left-hand running is still in force over many lines

Track gauges While the majority of railways in Britain adopted the 4 ft 8½ inch (1.43 m) gauge of the Stockton & Darlington Railway, the Great Western, on the advice of its brilliant but eccentric engineer Isambard Kingdom Brunel, had been laid to a seven foot (2.13 m) gauge, as were many of its associates. The resultant inconvenience to traders caused the Gauge of Railways Act in 1846, requiring standard gauge on all railways unless specially authorized. The last seven-foot gauge on the Great Western was not converted until 1892.

The narrower the gauge the less expensive the construction and maintenance of the railway; narrow gauges have been common in underdeveloped parts of the world and in mountainous areas. In 1863 steam traction was applied to the 1 ft 11½ inch (0.85 m) Festiniog Railway in Wales, for which locomotives were built to the designs of Robert Fairlie. He then led a campaign for the construction of narrow gauges. As a result of the export of English engineering and rolling stock, however, most North American and European railways have been built to the standard gauge, except in Finland and Russia, where the gauge is five feet (1.5 m).

Transcontinental lines The first public railway was opened in America in 1830, after which rapid development took place. A famous 4–2–0 locomotive called the *Pioneer* first ran from Chicago in 1848, and that city became one of the largest rail centres in the world. The Atlantic and the Pacific oceans were first linked on 9 May 1869, in a famous ceremony at the meeting point of the Union Pacific and Central Pacific lines at Promontory Point in the state of Utah. Canada was crossed by the Canadian Pacific in 1885; completion of the

Opposite page: Stephenson's Rocket *won a competition in 1829, proving that steam haulage was practical. In the first year of operation of the Liverpool and Manchester line, revenue from passenger traffic was ten times that expected. Below: when Brunel's broad-gauge system gave way to the standard gauge in 1892, over 620 broad-gauge locomotives had to be scrapped.*

railway was a condition of British Columbia joining the Dominion of Canada, and considerable land concessions were granted in virtually uninhabited territory.

The crossing of Asia with the Trans-Siberian Railway was begun by the Russians in 1890 and completed in 1902, except for a ferry crossing Lake Baikal. The difficult passage round the south end of the lake, with many tunnels, was completed in 1905. Today more than half the route is electrified. In 1863 the Orient Express ran from Paris for the first time and eventually passengers were conveyed all the way to Istanbul (Constantinople).

Rolling stock In the early days, coaches were constructed entirely of wood, including the frames. By 1900, steel frames were commonplace; then coaches were constructed entirely of steel and became very heavy. One American 85-foot (26 m) coach with two six-wheel bogies weighed more than 80 tons. New lightweight steel alloys and aluminium began to be used; in the 1950s the Budd company in America was building an 85-foot coach which weighed only 27 tons. The savings began with the bogies, which were built without conventional springs, bolsters and so on; with only two air springs on each four-wheel bogie, the new design reduced the weight from 8 to 2½ tons without loss of strength or stability.

In the 1880s, 'skyscraper' cars were two-storey wooden vans with windows used as travelling dormitories for railway workers in the USA; they had to be sawn down when the railways began to build tunnels through the mountains. After World War II double-decker cars of a more compact design were built, this time with plastic domes, so that passengers could enjoy the spectacular scenery on the western lines, which pass through the Rocky Mountains.

The adoption of narrow-gauge railways in small, mountainous countries, such as this one in Abergynolwyn, Wales, made construction simpler and less costly. It did not matter that the narrow gauge restricted speeds because distances were so short.

Lighting on coaches was by means of oil lamps at first; then gas lights were used, and each coach carried a cylinder of gas, which was dangerous in the event of accident or derailment. Finally dynamos on each car, driven by the axle, provided electricity, storage batteries being used for when the car was standing. Heating on coaches was provided in the early days by metal containers filled with hot water; then steam was piped from the locomotive, an extra drain on the engine's power; nowadays heat as well as light is provided electrically.

Sleeping accommodations were first made on the Cumberland Valley Railroad in the United States in 1837. George Pullman's first cars ran on the Chicago & Alton Railroad in 1859 and the Pullman Palace Car Company was formed in 1867. The first Pullman cars operated in Britain in 1874, a year after the introduction of sleeping cars by two British railways. In Europe in 1876 the International Sleeping Car Company was formed, but in the meantime George Nagelmackers of Liege and an American, Col William D'Alton Mann, began operation between Paris and Vienna in 1873.

Goods vans [freight cars] have developed according to the needs of the various countries. On the North American continent, goods trains as long as 1¼ miles are run as far as 1000 miles unbroken, hauling bulk such as raw materials and foodstuffs. Freight cars weighing 70 to 80 tons have two four wheel bogies. In Britain, with a denser population and closely adjacent towns, a large percentage of hauling is of small consignments of manufactured goods, and the smallest goods vans of any country are used, having four wheels and up to 24.5 tons capacity. A number of bogie wagons are used for special purposes, such as carriages for steel rails, tank cars for chemicals and 50 ton brick wagons.

The earliest coupling system was links and buffers, which allowed jerky stopping and starting. Rounded buffers brought snugly together by adjustment of screw links with springs were an improvement. The buckeye automatic coupling, long standard in North America, is now used in Britain. The coupling resembles a knuckle made of steel and extending horizontally; joining automatically with the coupling of the next car when pushed together, it is released by pulling a pin.

The first shipment of refrigerated goods was in 1851 when butter was shipped from New York to Boston in a wooden van packed with ice and insulated with sawdust. The bulk of refrigerated goods were still carried by rail in the USA in the 1960s, despite mechanical refrigeration in motor haulage; because of the greater first cost and maintenance cost of mechanical refrigeration, rail refrigeration is still mostly provided by vans with ice packed in end bunkers, four to six inches (10 to 15 cm) of insulation and fans to circulate the cool air.

Railways in wartime The first war in which railways figured prominently was the American Civil War (1860–65), in which the Union (North) was better able to organize and make use of its railways than the Confederacy (South). The war was marked by a famous incident in which a 4–4–0 locomotive called the *General* was hi-jacked by Southern agents.

The outbreak of World War I was caused in part by the fact that the mobilization plans of the various countries, including the use of railways and rolling stock, was planned to the last detail, except that there were no provisions for stopping the plans once they had been put into action until the armies were facing each other. In 1917 in the United States, the lessons of the Civil War had been forgotten, and freight vans were sent to their destination with no facilities for unloading, with the result that the railways were briefly taken over by the government for the only time in that nation's history.

In World War II, by contrast, the American railways performed magnificently, moving 2½ times the level of freight in 1944 as in 1938, with minimal increase in equipment, and supplying more than 300,000 employees to the armed forces in various capacities. In combat areas, and in later conflicts such as the Korean war, it proved difficult to disrupt an enemy's rail system effectively; pinpoint bombing was difficult, saturation bombing was expensive and in any case railways were quickly and easily repaired.

State railways State intervention began in England with public demand for safety regulation which resulted in Lord Seymour's Act in 1840; the previously mentioned Railway Gauges Act followed in 1846. Ever since, the railways have been recognized as one of the most important of national resources in each country.

In France, from 1851 onwards concessions were granted for a planned regional system for which the Government provided ways and works and the companies provided track and rolling stock; there was provision for the gradual taking over of the lines by the State, and the Société Nationale des Chemins de Fer Français (SNCF) was formed in 1937 as a company in which the State owns 51% of the capital and the companies 49%.

The Belgian Railways were planned by the State from the outset in 1835. The Prussian State Railways began in 1850; by the end of the year 54 miles (87 km) were open. Italian and Netherlands railways began in 1839; Italy nationalized her railways in 1905–07 and the Netherlands in the period 1920–38. In Britain the main railways were nationalized from 1 January 1948; the usual European pattern is that the State owns the main lines and minor railways are privately owned or operated by local authorities.

In the United States, between the Civil War and World War I the railways, along with all the other important industries, experienced phenomenal growth as the country developed. There were rate wars and financial piracy during a period of growth when industrialists were more powerful than the national government, and finally the Interstate

Below: a central Pacific locomotive on a turntable during construction of the railway in 1864.
Bottom of page: the completion of the transcontinental railway in the USA at Promontory, Utah, on 10 May 1869. A golden spike was driven at the meeting point of the Union Pacific and Central Pacific lines.

Commerce Act was passed in 1887 in order to regulate the railways, which had a near monopoly of transport. After World War II the railways were allowed to deteriorate, as private car ownership became almost universal and public money was spent on an interstate highway system making motorway haulage profitable, despite the fact that railways are many times as efficient at moving freight and passengers. In the USA, nationalization of railways would probably require an amendment to the Constitution, but since 1971 a government effort has been made to save the nearly defunct passenger service. On 1 May of that year Amtrack was formed by the National Railroad Passenger Corporation to operate a skeleton service of 180 passenger trains nationwide, serving 29 cities designated by the government as those requiring train service. The Amtrack service has been heavily used, but not adequately funded by Congress, so that bookings, especially for sleeper-car service, must be made far in advance.

The locomotive

Few machines in the machine age have inspired so much affection as railway locomotives in their 170 years of operation. Railways were constructed in the sixteenth century, but the wagons were drawn by muscle power until 1804. In that year an engine built by Richard Trevithick worked on the Penydarren Tramroad in South Wales. It broke some cast iron tramplates, but it demonstrated that steam could be used for haulage, that steam generation could be stimulated by turning the exhaust steam up the chimney to draw up the fire, and that smooth wheels on smooth rails could transmit motive power.

Steam locomotives The steam locomotive is a robust and simple machine. Steam is admitted to a cylinder and by expanding pushes the piston to the other end; on the return stroke a port opens to clear the cylinder of the now ex-

panded steam. By means of mechanical coupling, the travel of the piston turns the drive wheels of the locomotive.

Trevithick's engine was put to work as a stationary engine at Penydarren. During the following twenty-five years, a limited number of steam locomotives enjoyed success on colliery railways, fostered by the soaring cost of horse fodder towards the end of the Napoleonic wars. The cast iron plateways, which were L-shaped to guide the wagon wheels, were not strong enough to withstand the weight of steam locomotives, and were soon replaced by smooth rails and flanged wheels on the rolling stock.

John Blenkinsop built several locomotives for collieries, which ran on smooth rails but transmitted power from a toothed wheel to a rack which ran alongside the running rails. William Hedley was building smooth-wheeled locomotives which ran on plateways, including the first to have the popular nickname *Puffing Billy*.

In 1814 George Stephenson began building for smooth rails

This converted Royal Scot *4-6-0 passenger locomotive has a total boiler heating surface of 1862 square feet, including tubes, firebox and superheater. Its drive wheels are 6 feet 9 inches in diameter, and the total wheelbase of engine and tender is 54 feet $5\frac{1}{2}$ inches.*

STEAM LOCOMOTIVE

1. Cylinder
2. Smoke deflectors
3. Steam chest
4. Piston
5. Chimney
6. Valve gear
7. Connecting rod
8. Coupling rods
9. Sanding pipes
10. Water injector
11. Safety valve
12. Boiler tubes
13. Fire box
14. Regulator
15. Steam brake lever
16. Whistle
17. Vacuum brake lever
18. Reversing gear
19. Sand lever

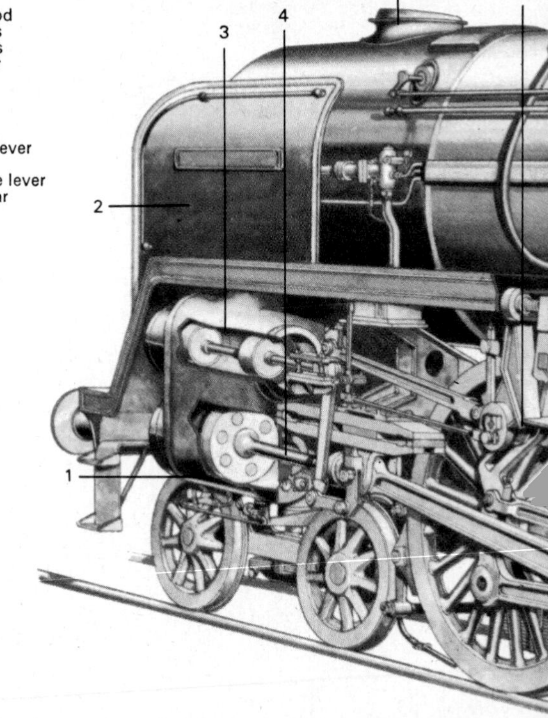

at Killingworth, synthesizing the experience of the earlier designers. Until this time nearly all machines had the cylinders partly immersed in the boiler and usually vertical. In 1815 Stephenson and Losh patented the idea of direct drive from the cylinders by means of cranks on the drive wheels instead of through gear wheels, which imparted a jerky motion, especially when wear occurred on the coarse gears. Direct drive allowed a simplified layout and gave greater freedom to designers.

In 1825 only 18 steam locomotives were doing useful work. One of the first commercial railways, the Liverpool & Manchester, was being built, and the directors had still not decided between locomotives and cable haulage, with railside steam engines pulling the cables. They organized a competition which was won by Stephenson in 1829, with his famous engine, the *Rocket*, now in London's Science Museum.

Locomotive boilers had already evolved from a simple flue to a return-flue type, and then to a tubular design, in which a nest of fire tubes, giving more heating surface, ran from the firebox tube-plate to a similar tube-plate at the smokebox end. In the smokebox the exhaust steam from the cylinders created a blast on its way to the chimney which kept the fire up when the engine was moving. When the locomotive was stationary a blower was used, creating a blast from a ring of perforated pipe into which steam was directed. A further development, the multitubular boiler, was patented by Henry Booth, treasurer of the Liverpool & Manchester, in 1827. It was incorporated by Stephenson in the *Rocket*, after much trial and error in making the ferrules of the copper tubes to give water-tight joints in the tube plates.

After 1830 the steam locomotive assumed its familiar form, with the cylinders level or slightly inclined at the smokebox end and the fireman's stand at the firebox end.

As soon as the cylinders and axles were no longer fixed in or under the boiler itself, it became necessary to provide a frame to hold the various components together. The bar frame was used on the early British locomotives and exported to America; the Americans kept to the bar-frame design, which evolved from wrought iron to cast steel construction, with the cylinders mounted outside the frame. The bar frame was superseded in Britain by the plate frame, with cylinders inside the frame, spring suspension (coil or laminated) for the frames and axleboxes (lubricated bearings) to hold the axles.

As British railways nearly all produced their own designs, a great many characteristic types developed. Some designs with cylinders inside the frame transmitted the motion to crank-shaped axles rather than to eccentric pivots on the outside of the drive wheels; there were also compound locomotives, with the steam passing from a first cylinder or cylinders to another set of larger ones.

When steel came into use for building boilers after 1860, higher operating pressures became possible. By the end of the nineteenth century 175 psi (12 bar) was common, with 200 psi (13.8 bar) for compound locomotives. This rose to 250 psi (17.2 bar) later in the steam era. (By contrast, Stephenson's *Rocket* only developed 50 psi, 3.4 bar.) In the 1890s express engines had cylinders up to 20 inches (51 cm) in diameter with a 26 inch (66 cm) stroke. Later diameters increased to 32 inches (81 cm) in places like the USA, where there was more room, and locomotives and rolling stock in general were built larger.

Supplies of fuel and water were carried on a separate tender, pulled behind the locomotive. The first tank engine, carrying its own supplies, appeared in the 1830s; on the continent of Europe they were confusingly called tender engines. Separate tenders continued to be common because they made possible much longer runs. While the fireman stoked the firebox, the boiler had to be replenished with water by some means under his control; early engines had pumps running off the axle, but there was always the difficulty that the engine had to be running. The injector was invented in 1859. Steam from the boiler (or latterly, exhaust steam) went through a cone-shaped jet and lifted the water into the boiler against the greater pressure there through energy imparted in condensation. A clack (non-return valve) retained the steam in the boiler.

Early locomotives burned wood in America, but coal in Britain. As British railway Acts began to include penalties for emission of dirty black smoke, many engines were built after 1829 to burn coke. Under Matthew Kirtley on the Midland Railway the brick arch in the firebox and deflector plates were developed to direct the hot gases from the coal to pass over the flames, so that a relatively clean blast came out of the chimney and the cheaper fuel could be burnt. After 1860 this simple expedient was universally adopted. Fireboxes were protected by being surrounded with a water jacket; stays about four inches (10 cm) apart supported the inner firebox from the outer.

Steam was distributed to the pistons by means of valves. The valve gear provided for the valves to uncover the ports at different parts of the stroke, so varying the cut-off to provide for expansion of steam already admitted to the cylinders and to give lead or cushioning by letting the steam in about $\frac{1}{8}$ inch (3 mm) from the end of the stroke to begin the reciprocating motion again. The valve gear also provided for reversing by admitting steam to the opposite side of the piston.

Long-lap or long-travel valves gave wide-open ports for the exhaust even when early cut-off was used, whereas with short travel at early cut-off, exhaust and emission openings became

Far left: the A-3 class Flying Scotsman, *built in the 1920s and now restored. The extra tender was added to carry water, because watering facilities no longer exist on the British Rail system.*
Left: Venus, *a British Rail 'Britannia' class locomotive. This 4-6-2 class was introduced in 1951; they were the first express passenger steam engines to be built by British Rail after nationalization.*

smaller so that at speeds of over 60 mph (96 kph) one-third of the energy of the steam was expanded just getting in and out of the cylinder. This elementary fact was not universally accepted until about 1925 because it was felt that too much extra wear would occur with long-travel valve layouts.

Valve operation on most early British locomotives was by Stephenson link motion, dependent on two eccentrics on the driving axle connected by rods to the top and bottom of an expansion link. A block in the link, connected to the reversing lever under the control of the driver, imparted the reciprocating motion to the valve spindle. With the block at the top of the link, the engine would be in full forward gear and steam would be admitted to the cylinder for perhaps 75% of the stoke. As the engine was notched up by moving the lever back over its serrations (like the handbrake lever of a car), the cut-off was shortened; in mid-gear there was no steam admission to the cylinder and with the block at the bottom of the link the engine was in full reverse.

Walschaert's valve gear, invented in 1844 and in general use after 1890, allowed more precise adjustment and easier operation for the driver. An eccentric rod worked from a return crank by the driving axle operated the expansion link; the block imparted the movement to the valve spindle, but the movement was modified by a combination lever from a crosshead on the piston rod.

Steam was collected as dry as possible along the top of the boiler in a perforated pipe, or from a point above the boiler in a dome, and passed to a regulator which controlled its distribution. The most spectacular development of steam locomotives for heavy haulage and high speed runs was the introduction of superheating. A return tube, taking the steam back towards the firebox and forward again to a header

Some Japanese trains travel at 150 miles an hour. This 'bullet' train was photographed at a platform in Tokyo.

at the front end of the boiler through an enlarged flue-tube, was invented by Wilhelm Schmidt of Cassel, and modified by other designers. The first use of such equipment in Britain was in 1906 and immediately the savings in fuel and especially water were remarkable. Steam at 175 psi, for example, was generated 'saturated' at 371°F (188°C); by adding 200°F (93°C) of superheat, the steam expanded much more readily in the cylinders, so that twentieth-century locomotives were able to work at high speeds at cut-offs as short as 15%. Steel tyres, glass fibre boiler lagging, long-lap piston valves, direct steam passage and superheating all contributed to the last phase of steam locomotive performance.

Steam from the boiler was also for other purposes. Steam sanding was introduced for traction in 1887 on the Midland Railway, to improve adhesion better than gravity sanding, which often blew away. Continuous brakes were operated by a vacuum created on the engine or by compressed air supplied by a steam pump. Steam heat was piped to the carriages, and steam dynamos [generators] provided electric light.

Steam locomotives are classified according to the number of wheels. Except for small engines used in marshalling yards, all modern steam locomotives had leading wheels on a pivoted bogie or truck to help guide them around curves. The trailing wheels helped carry the weight of the firebox. For many years the 'American standard' locomotive was a 4–4–0, having four leading wheels, four driving wheels and no trailing wheels. The famous Civil War locomotive, the *General*, was a 4–4–0, as was the New York Central *Engine No* 999, which set a speed record of 112.5 mph (181 kph) in 1893. Later, a common freight locomotive configuration was the *Mikado* type, a 2–8–2.

A Continental classification counts axles instead of wheels, and another modification gives drive wheels a letter of the alphabet, so the 2–8–2 would be 1–4–1 in France and 1D1 in Germany.

The largest steam locomotives were articulated, with two sets of drive wheels and cylinders using a common boiler. The sets of drive wheels were separated by a pivot; otherwise such a large engine could not have negotiated curves. The largest ever built was the Union Pacific *Big Boy*, a 4–8–8–4, used to haul freight in the mountains of the western United States. Even though it was articulated it could not run on sharp curves. It weighed nearly 600 tons, compared to less than five tons for Stephenson's *Rocket*.

Steam engines could take a lot of hard use, but they are now obsolete, replaced by electric and especially diesel-electric locomotives. Because of heat losses and incomplete combustion of fuel, their thermal efficiency was rarely more than 6%.

Diesel locomotives Diesel locomotives are most commonly diesel-electric. A diesel engine drives a dynamo [generator] which provides power for electric motors which turn the drive wheels, usually through a pinion gear driving a ring gear on the axle. The first diesel-electric propelled rail car was built in 1913, and after World War II they replaced steam engines completely, except where electrification of railways is economical.

Diesel locomotives have several advantages over steam engines. They are instantly ready for service, and can be shut down completely for short periods, whereas it takes some time to heat the water in the steam engine, especially in cold weather, and the fire must be kept up while the steam engine is on standby. The diesel can go further without servicing, as it consumes no water; its thermal efficiency is four times as high, which means further savings of fuel. Acceleration and high-speed running are smoother with a diesel, which means less wear on rails and roadbed. The economic reasons for turning to diesels were overwhelming after the war, especially in North America, where the railways were in direct competition with road haulage over very long distances.

The British Rail High Speed Train is the prototype for the next generation of inter-city passenger trains. In regular service its operating speed is 125mph (201kph); it holds the world speed record for diesel-electric propulsion of 143mph (231kph).

Electric traction The first electric-powered rail car was built in 1834, but early electric cars were battery powered, and the batteries were heavy and required frequent re-charging. Today electric trains are not self-contained, which means that they get their power from overhead wires or from a third rail. The power for the traction motors is collected from the third rail by means of a shoe or from the overhead wires by a pantograph.

Electric trains are the most economical to operate, provided that traffic is heavy enough to repay electrification of the railway. Where trains run less frequently over long distances the cost of electrification is prohibitive. DC systems have been used as opposed to AC because lighter traction motors can be used, but this requires power substations with rectifiers to convert the power to DC from the AC of the commercial mains. (High voltage DC power is difficult to transmit over long distances.) The latest development of electric trains has been the installation of rectifiers in the cars themselves and the use of the same AC frequency as the commercial mains (50 Hz in Europe, 60 Hz in North America), which means that fewer substations are necessary.

Railway systems

The foundation of a modern railway system is track which does not deteriorate under stress of traffic. Standard track in Britain comprises a flat-bottom section of rail weighing 110 lb per yard (54 kg per metre) carried on 2112 cross-sleepers per mile (1312 per km). Originally creosote-impregnated wood sleepers [cross-ties] were used, but they are now made of post-stressed concrete. This enables the rail to transmit the pressure, perhaps as much as 20 tons/in² (3150 kg/cm²) from the small area of contact with the wheel, to the ground below the track formation where it is reduced through the sole plate and the sleeper to about 400 psi (28 kg/cm²). In soft ground, thick polyethylene sheets are generally placed under the ballast to prevent pumping of slurry under the weight of trains.

The rails are tilted towards one another on a 1 in 20 slope. Steel rails may last 15 or 20 years in traffic, but to prolong the undisturbed life of track still longer, experiments have been carried out with paved concrete track (PACT) laid by a slip paver similar to concrete highway construction in reinforced concrete. The foundations, if new, are similar to those for a motorway. If, on the other hand, existing railway formation is to be used, the old ballast is sealed with a bitumen emulsion before applying the concrete which carries the track fastenings glued in with cement grout or epoxy resin. The track is made resilient by use of rubber-bonded cork packings 0.4 inch (10 mm) thick. British Railways purchases rails in 60 ft (18.3 m) lengths which are shop-welded into 600 ft (183 m) lengths and then welded on site into continuous welded track with pressure-relief points at intervals of several miles. The continuous welded rails make for a steadier and less noisy ride for the passenger and reduce the tractive effort.

Signalling The second important factor contributing to safe rail travel is the system of signalling. Originally railways relied on the time interval to ensure the safety of a succession of trains, but the defects rapidly manifested themselves, and a space interval, or the block system, was adopted, although it was not enforced legally on British passenger lines until the Regulation of Railways Act of 1889. Semaphore signals became universally adopted on running lines and the interlocking of points [switches] and signals (usually accomplished mechanically by tappets) to prevent conflicting movements being signalled was also a requirement of the 1889 Act. Lock-and-block signalling, which ensured a safe sequence of movements by electric checks, was introduced on the London, Chatham and Dover Railway in 1875.

Track circuiting, by which the presence of a train is detected by an electric current passing from one rail to another through the wheels and axles, dates from 1870 when William Robinson applied it in the United States. In England the Great Eastern Railway introduced power operation of points and signals at Spitalfields goods yard in 1899, and three years later track-circuit operation of powered signals was in operation on 30 miles (48 km) of the London and South Western Railway main line.

Day colour light signals, controlled automatically by the trains through track circuits, were installed on the Liverpool Overhead Railway in 1920 and four-aspect day colour lights (red, yellow, double yellow and green) were provided on Southern Railway routes from 1926 onwards. These enable drivers of high-speed trains to have a warning two block sections ahead of a possible need to stop. With track circuiting it became usual to show the presence of vehicles on a track diagram in the signal cabin which allowed routes to be controlled remotely by means of electric relays. Today, panel operation of considerable stretches of railway is commonplace; at Rugby, for instance, a signalman can control the points at a station 44 miles (71 km) away, and the signalbox at London Bridge controls movements on the busiest 150

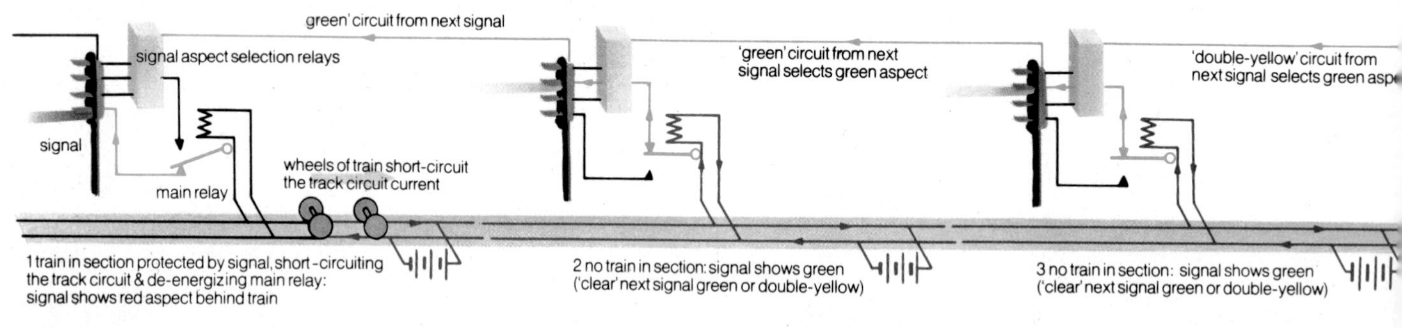

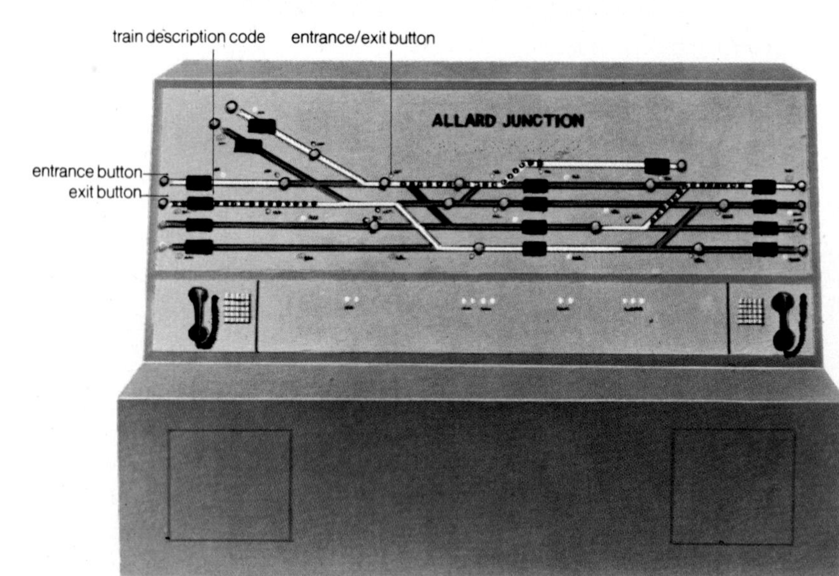

A modern 'entrance-exit' type of signal-box control console can control many miles of line and handle more than thirty trains at a time. The signalman 'sets up' a route for a train by pressing the appropriate entrance and exit buttons; this action sets the points and clears the signals along the track, and the route is automatically illuminated on the track plan in white when it has been cleared.

track-miles of British Rail. By the end of the 1980s, the 1500 miles (2410 km) of the Southern Region of British Rail are to be controlled from 13 signalboxes. In modern panel installations the trains are not only shown on the track diagram as they move from one section to another, but the train identification number appears electronically in each section. Computer-assisted train description, automatic train reporting and, at stations such as London Bridge, operation of platform indicators, is now usual.

Whether points are operated manually or by an electric point motor, they have to be prevented from moving while a train is passing over them and facing points have to be locked, and proved to be locked (or 'detected') before the relevant signal can permit a train movement. The blades of the points have to be closed accurately (0.16 inch or 0.4 cm is the maximum tolerance) so as to avert any possibility of a wheel flange splitting the point and leading to a derailment.

Other signalling developments of recent years include completely automatic operation of simple point layouts, such as the double crossover at the Bank terminus of the British Rail's Waterloo and City underground railway. On London Transport's underground system a plastic roll operates junctions according to the timetable by means of coded punched holes, and on the Victoria Line trains are operated automatically once the driver has pressed two buttons to indicate his readiness to start. He also acts as the guard, controlling the opening of the doors, closed circuit television giving him a view along the train. The trains are controlled (for acceleration and braking) by coded impulses transmitted through the running rails to induction coils mounted on the front of the train. The absence of code impulses cuts off the current and applies the brakes; driving and speed control is covered by command spots in which a frequency of 100 Hz corresponds to one mile per hour (1.6 km/h), and 15 kHz shuts off the current. Brake applications are so controlled that trains stop smoothly and with great accuracy at the

This manual switchbox at Sheffield Park, Sussex, is still in use, but most such equipment has been replaced by electronic centralized equipment, sometimes operated by computers.

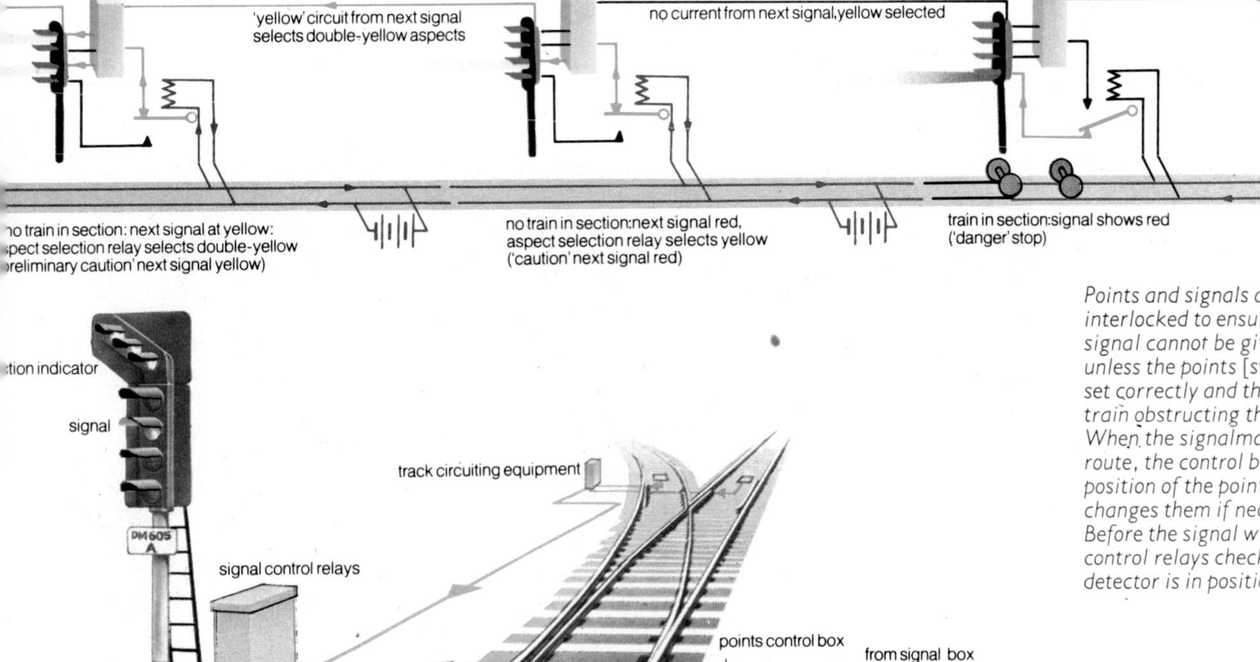

Points and signals are interlocked to ensure that a clear signal cannot be given to a train unless the points [switches] are set correctly and there is no train obstructing the route set. When the signalman sets up a route, the control box checks the position of the points and changes them if necessary. Before the signal will clear, the control relays check that the detector is in position and lock it.

desired place on platforms. Occupation of the track circuit ahead by a train automatically stops the following train, which cannot receive a code.

On British main lines an automatic warning system is being installed by which the driver receives in his cab a visual and audible warning of passing a distant signal at caution; if he does not acknowledge the warning the brakes are applied automatically. This is accomplished by magnetic induction between a magnetic unit placed in the track and actuated according to the signal aspect, and a unit on the train.

Train control In England train control began in 1909 on the Midland Railway, particularly to expedite the movement of coal trains and to see that guards and enginemen were relieved at the end of their shift and were not called upon to work excessive overtime. Comprehensive train control systems, depending on complete diagrams of the track layout and records of the position of engines, crews and rolling stock, were developed for the whole of Britain, the Southern Railway being the last to adopt it during World War II, having hitherto given a great deal of responsibility to signalmen for the regulation of trains. Refinements of control include advance traffic information (ATI) in which information is passed from yard to yard by telex giving types of wagon, wagon number, route code, particulars of the load, destination station and consignee. In 1972 British Rail decided to adopt a computerized freight information and traffic control system known as TOPS (total operations processing system) which was developed over eight years by the Southern Pacific company in the USA.

Although a great deal of rail traffic in Britain is handled by block trains from point of origin to destination, about one-fifth of the originating tonnage is less than a train-load. This means that wagons must be sorted on their journey. In Britain there are about 600 terminal points on a 12,000-mile network which is served by over 2500 freight trains made up of varying assortments of 249,000 wagons and 3972 locomotives, of which 333 are electric. This requires the speed of calculation and the information storage and classification capacity of the modern computer, which has to be linked to points dealing with or generating traffic throughout the system. The computer input, which is by punched cards, covers details of loading or unloading of wagons and their movements in trains, the composition of trains and their departures from and arrivals at yards, and the whereabouts of locomotives. The computer output includes information on the balance of locomotives at depots and yards, with particulars of when maintenance examinations are due, the numbers of empty and loaded wagons, with aggregate weight and brake force, and whether their movement is on time, the location of empty wagons and a forecast of those that will become available, and the numbers of trains at any location, with collective train weights and individual details of the component wagons.

A closer check on what is happening throughout the system is thus provided, with the position of consignments in transit, delays in movement, delays in unloading wagons by customers, and the capacity of the system to handle future traffic among the information readily available. The computer has a built-in self-check on wrong input information.

Freight handling The merry-go-round system enables coal for power stations to be loaded into hopper wagons at a colliery without the train being stopped, and at the power station the train is hauled round a loop at less than 2 mph (3.2 km/h), a trigger device automatically unloading the

The primary retarder at Temple Mills has clamps inside the rails to slow the wagons as necessary. This view looks toward the 'hump'; the artificial elevation can be seen in the background.

The signal boards in many modern marshalling yards are located high above the yard in a tower. This control tower is at British Rail's Tees yard.

wagons without the train being stopped. The arrangements also provide for automatic weighing of the loads. Other bulk loads can be dealt with in the same way.

Bulk powders, including cement, can be loaded and discharged pneumatically, using either rail wagons or containers. Iron ore is carried in 100 ton gross wagons (72 tons of payload) whose coupling gear is designed to swivel, so that wagons can be turned upside down for discharge without uncoupling from their train. Special vans take palletized loads of miscellaneous merchandise or such products as fertilizer, the van doors being designed so that all parts of the interior can be reached by a fork-lift truck.

British railway companies began building their stocks of containers in 1927, and by 1950 they had the largest stock of large containers in Western Europe. In 1962 British Rail decided to use International Standards Organisation sizes, 8 ft (2.4 m) wide by 8 ft high and 10, 20, 30 and 40 ft (3.1, 6.1, 9.2 and 12.2 m) long. The 'Freightliner' service of container trains uses 62½ ft (19.1 m) flat wagons with air-operated disc brakes in sets of five and was inaugurated in 1965. At depots 'Drott' pneumatic-tyred cranes were at first provided but rail-mounted Goliath cranes are now provided.

Cars are handled by double-tier wagons. The British car industry is a big user of 'company' trains, which are operated for a single customer. Both Ford and Chrysler use them to exchange parts between specialist factories and the railway thus becomes an extension of factory transport. Company trains frequently consist of wagons owned by the trader; there are about 20,000 on British railways, the oil industry, for example, providing most of the tanks it needs to carry 21 million tons of petroleum products by rail each year despite competition from pipelines.

Gravel dredged from the shallow seas is another developing source of rail traffic. It is moved in 76 ton lots by 100 ton gross hopper wagons and is either discharged on to belt conveyers to go into the storage bins at the destination or, in another system, it is unloaded by truck-mounted discharging machines.

Cryogenic (very low temperature) products are also transported by rail in high capacity insulated wagons. Such products include liquid oxygen and liquid nitrogen which are taken from a central plant to strategically-placed railheads where the liquefied gas is transferred to road tankers for the journey to its ultimate destination.

Switchyards Groups of sorting sidings, in which wagons [freight cars] can be arranged in order so that they can be detached from the train at their destination with the least possible delay, are called marshalling yards in Britain and classification yards or switchyards in North America. The work is done by small locomotives called switchers or shunters, which move 'cuts' of trains from one siding to another until the desired order is achieved.

As railways became more complicated in their system layouts in the nineteenth century, the scope and volume of necessary sorting became greater, and means of reducing the time and labour involved were sought. (By 1930, for every 100 miles that freight trains were run in Britain there were 75 miles of shunting.) The sorting of coal wagons for return to the collieries had been assisted by gravity as early as 1859, in the sidings at Tyne dock on the North Eastern Railway; in 1873 the London & North Western Railway sorted traffic to and from Liverpool on the Edge Hill 'grid irons': groups of sidings laid out on the slope of a hill where gravity provided the motive power, the steepest gradient being 1 in 60 (one foot of elevation in sixty feet of siding). Chain drags were used for braking the wagons. A shunter uncoupled the wagons in 'cuts' for the various destinations and each cut was turned into the appropriate siding. Some gravity yards relied on a code of whistles to advise the signalman what 'road' (siding) was required.

In the late nineteenth century the hump yard was introduced to provide gravity where there was no natural slope of

the land. In this the trains were pushed up an artificial mound with a gradient of perhaps 1 in 80 and the cuts were 'humped' down a somewhat steeper gradient on the other side. The separate cuts would roll down the selected siding in the fan or 'balloon' of sidings, which would end in a slight upward slope to assist in the stopping of the wagons. The main means of stopping the wagons, however, were railwaymen called shunters who had to run alongside the wagons and apply the brakes at the right time. This was dangerous and required excessive manpower.

Such yards appeared all over North America and north-east England and began to be adopted elsewhere in England. Much ingenuity was devoted to means of stopping the wagons; a German firm, Fröhlich, came up with a hydraulically operated retarder which clasped the wheel of the wagon as it went past, to slow it down to the amount the operator thought necessary.

An entirely new concept came with Whitemoor yard at March, near Cambridge, opened by the London & North Eastern Railway in 1929 to concentrate traffic to and from East Anglian destinations. When trains arrived in one of ten reception sidings a shunter examined the wagon labels and prepared a 'cut card' showing how the train should be sorted into sidings. This was sent to the control tower by pneumatic tube; there the points [switches] for the forty sorted sidings were preset in accordance with the cut card; information for several trains could be stored in a simple pin and drum device.

The hump was approached by a grade of 1 in 80. On the far side was a short stretch of 1 in 18 to accelerate the wagons, followed by 70 yards (64 m) at 1 in 60 where the tracks divided into four, each equipped with a Frohlich retarder. Then the four tracks spread out to four balloons of ten tracks each, comprising 95 yards (87 m) of level track followed by 233 yards (213 m) falling at 1 in 200, with the remaining 380 yards (348 m) level. The points were moved in the predetermined sequence by track circuits actuated by the wagons, but the operators had to estimate the effects on wagon speed of the retarders, depending to a degree on whether the retarders were grease or oil lubricated.

Pushed by an 0–8–0 small-wheeled shunting engine at $1\frac{1}{2}$ to 2 mph ($2\frac{1}{2}$ to 3 km/h), a train of 70 wagons could be sorted in seven minutes. The yard had a throughput of about 4000 wagons a day. The sorting sidings were allocated: number one for Bury St Edmunds, two for Ipswich, and so forth. Number 31 was for wagons with tyre fastenings which might be ripped off by retarders, which were not used on that siding. Sidings 32 to 40 were for traffic to be dropped at wayside stations; for these sidings there was an additional hump for sorting these wagons in station order. Apart from the sorting sidings, there were an engine road, a brake van road, a 'cripple' road for wagons needing repair, and transfer road to three sidings serving a tranship shed, where small shipments not filling entire wagons could be sorted.

British Rail built a series of yards at strategic points; the yards usually had two stages of retarders, latterly electro-pneumatically operated, to control wagon speed. In later yards electronic equipment was used to measure the weight of each wagon and estimate its rolling resistance. By feeding this information into a computer, a suitable speed for the wagon could be determined and the retarder operated automatically to give the desired amount of braking. These predictions did not always prove reliable.

At Tinsley, opened in 1965, with eleven reception roads and 53 sorting sidings in eight balloons, the Dowty wagon speed control system was installed. The Dowty system uses many small units (20,000 at Tinsley) comprising hydraulic rams on the inside of the rail, less than a wagon length apart.

Left: British Rail's experimental Advanced Passenger Train (APT-E) has been operated at 125mph over track which would limit a conventional train to 80mph. It has body-tilting and self-levelling soft air suspension. Such advanced technology will not, by itself, solve British Rail's problems. The nation's railways were nationalized in 1948, and many thousands of people commute to work by rail each day in London alone, yet the subsidies necessary to operate the system are resented. Opposite page: an aerial view of the Tees marshalling yard, opened 21 May 1963. The photo shows only part of the yard, which can sort thousands of wagons a day. The use of marshalling yards is falling off because of competition from motor haulage. Railway mileage is actually falling in many countries, despite the fact that the true cost of railway haulage is lower.

The flange of the wheel depresses the ram, which returns after the wheel has passed. A speed-sensing device determines whether the wagon is moving too fast from the hump; if the speed is too fast the ram automatically has a retarding action. Certain of the units are booster-retarders; if the wagon is moving too slowly, a hydraulic supply enables the ram to accelerate the wagon. There are 25 secondary sorting sidings at Tinsley to which wagons are sent over a secondary hump by the booster-retarders. If individual units fail the rams can be replaced.

An automatic telephone exchange links all the traffic and administrative offices in the yard with the railway control office, Sheffield Midland Station and the local steelworks (principal source of traffic). Two-way loudspeaker systems are available through all the principal points in the yard, and radio telephone equipment is used to speak to enginemen. Fitters maintaining the retarders have walkie-talkie equipment. The information from shunters about the cuts and how many wagons in each, together with destination, is conveyed by special data transmission equipment, a punched tape being produced to feed into the point control system for each train over the hump.

As British Railways have departed from the wagon-load system there is less employment for marshalling yards. Freightliner services, block coal trains from colliery direct to power stations or to coal concentration depots, 'company' trains and other specialized freight traffic developments obviate the need for visiting marshalling yards. Other factors are competition from motor transport, closing of wayside freight depots and of many small coal yards.

Modern passenger service In Britain a network of city to city services operates at speeds of up to 100 mph (161 km/h) and at regular hourly intervals, or 30 minute intervals on such routes as London to Birmingham. On some lines the speed is soon to be raised to 125 mph (201 km/h) with high speed diesel trains whose prototype has been shown to be capable of 143 mph (230 km/h). With the advanced passenger train (APT) now under development, speeds of 150 mph (241 km/h) are envisaged. The Italians are developing a system capable of speeds approaching 200 mph (320 km/h) while the Japanese and the French already operate passenger trains at speeds of about 150 mph (241 km/h).

The APT will be powered either by electric motors or by gas turbines, and it can use existing track because of its pendulum suspension which enables it to heel over when travelling round curves. With stock hauled by a conventional locomotive, the London to Glasgow electric service holds the European record for frequency and speed over a long distance. When the APT is in service, it is expected that the London to Glasgow journey time of five hours will be reduced to $2\frac{1}{2}$ hours.

In Europe a number of combined activities organized through the International Union of Railways included the Trans-Europe-Express (TEE) network of high-speed passenger trains, a similar freight service, and a network of railway-associated road services marketed as Europabus.

Mountain railways

Cable transport has always been associated with hills and mountains. In the late 1700s and early 1800s the wagonways used for moving coal from mines to river or sea ports were hauled by cable up and down inclined tracks. Stationary steam engines built near the top of the incline drove the cables, which were passed around a drum connected to the steam engine and were carried on rollers along the track. Sometimes cable-worked wagonways were self-acting if loaded wagons worked downhill, for they could pull up the lighter empty wagons. Even after George Stephenson perfected the travelling steam locomotive to work the early passenger railways of the 1820s and 1830s cable haulage was sometimes used to help trains climb the steeper gradients, and cable working continued to be used for many steeply-graded industrial wagonways throughout the 1800s. Today a few cable-worked inclines survive at industrial sites and for such unique forms of transport as the San Francisco tramway [streetcar] system.

Funiculars The first true mountain railways using steam locomotives running on a railway track equipped for rack and pinion (cogwheel) propulsion were built up Mount Washington, USA, in 1869 and Mount Rigi, Switzerland, in 1871. The latter was the pioneer of what today has become the most extensive mountain transport system in the world. Much of Switzerland consists of high mountains, some exceeding 14,000 ft (4250 m). From this development in mountain transport other methods were developed and in the following 20 years until the turn of the century funicular railways were built up a number of mountain slopes. Most worked on a similar principle to the cliff lift, with two cars connected by cable balancing each other. Because of the length of some lines, one mile (1.6 km) or more in a few cases, usually only a single track is provided over most of the route, but a short length of double track is laid down at the halfway point where the cars cross each other. The switching of cars through the double-track section is achieved automatically by using double-flanged wheels on one side of each car and flangeless wheels on the other so that one car is always guided through the righthand track and the other through the left-hand track. Small gaps are left in the switch rails to allow the cable to pass through without impeding the wheels.

Funiculars vary in steepness according to location and may have gentle curves; some are not steeper than 1 in 10 (10 per cent), others reach a maximum steepness of 88 per cent. On the less steep lines the cars are little different from, but smaller than, ordinary railway carriages. On the steeper lines the cars have a number of separate compartments, stepped up one from another so that while floors and seats are level a compartment at the higher end may be 10 or even 15 ft (3 or 4 m) higher than the lowest compartment at the other end. Some of the bigger cars seat 100 passengers, but most carry fewer than this.

Braking and safety are of vital importance on steep mountain lines to prevent breakaways. Cables are regularly inspected and renewed as necessary but just in case the cable breaks a number of braking systems are provided to stop the car quickly. On the steepest lines ordinary wheel brakes would not have any effect and powerful spring-loaded grippers on the car underframe act on the rails as soon as the cable becomes slack. When a cable is due for renewal the opportunity is taken to test the braking system by cutting the cable and checking whether the cars stop within the prescribed distance. This operation is done without passengers.

The capacity of funicular railways is limited to the two cars, which normally do not travel at more than about 5 to 10 mph (8 to 16 km/h). Some lines are divided into sections with pairs of cars covering shorter lengths.

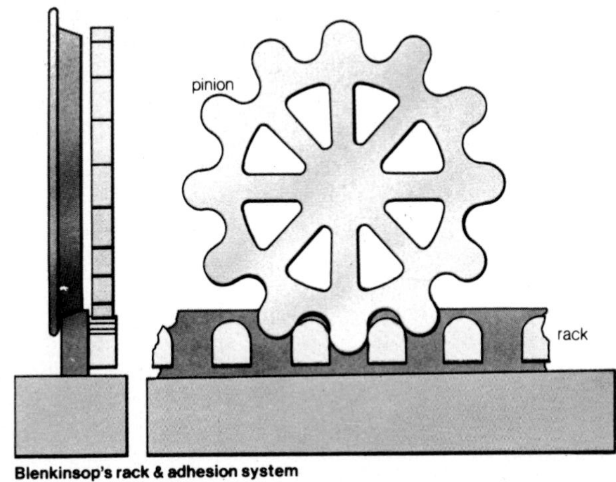

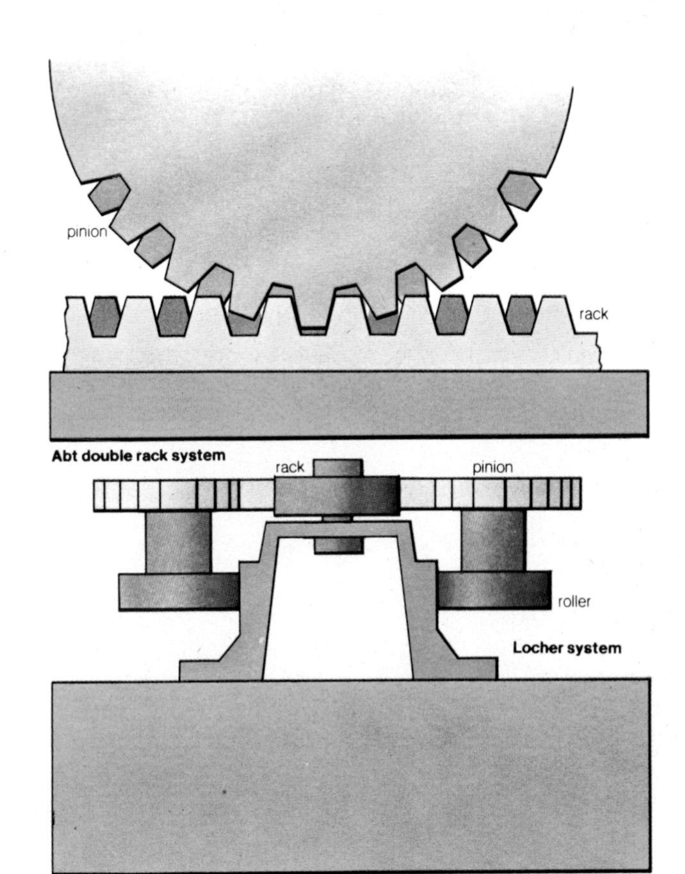

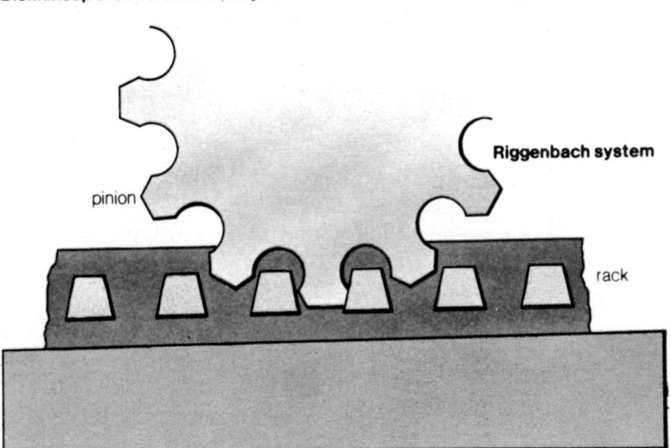

Rack railways The rack and pinion system principle dates from the pioneering days of the steam locomotive between 1812 and 1820 which coincided with the introduction of iron rails. One engineer, Blenkinsop, did not think that iron wheels on locomotives would have sufficient grip on iron rails, and on the wagonway serving Middleton colliery near Leeds he laid an extra toothed rail alongside one of the ordinary rails, which engaged with a cogwheel on the locomotive. The Middleton line was relatively level and it was soon found that on railways with only gentle climbs the rack system was not needed. If there was enough weight on the locomotive driving wheels they would grip the rails by friction. Little more was heard of rack railways until the 1860s, when they began to be developed for mountain railways in the USA and Switzerland.

The rack system for the last 100 years has used an additional centre toothed rail which meshes with cogwheels under locomotives and coaches. There are four basic types of rack varying in details: the Riggenbach type looks like a steel ladder, and the Abt and Strub types use a vertical rail with teeth machined out of the top. One or other of these systems is used on most rack lines but they are safe only on gradients no steeper than 1 in 4 (25 per cent). One line in Switzerland up Mount Pilatus has a gradient of 1 in 2 (48 per cent) and uses the Locher rack with teeth cut on both sides of the rack rail instead of on top, engaging with pairs of horizontally-mounted cogwheels on each side, driving and braking the railcars.

The first steam locomotives for steep mountain lines had vertical boilers but later locomotives had boilers mounted at an angle to the main frame so that they were virtually horizontal when on the climb. Today steam locomotives have all but disappeared from most mountain lines and survive in regular service on only one line in Switzerland, on Britain's only rack line up Snowdon in North Wales, and a handful of others. Most of the remainder have been electrified or a few converted to diesel.

Top of page: diagrams showing four rack-and-pinion systems.
Above: the Brienzer-Rothorn railway in Switzerland. The boiler is level on the steeper parts of the slope.
Opposite page: an old view of the Swiss Brünigbahn, with Riggenbach rack on the steeper slopes.

07
POWELL AND MARKET
AQUATIC PARK
MARITIME MUSEUM
HYDE AND BEACH
FISHERMAN'S WHARF
4 BLOCKS FROM TERMINAL
527
"I stop at the St. Francis"

Trams and trolleybuses

The early railways used in mines with four-wheel trucks and wooden beams for rails were known as tramways. From this came the word tram for a four-wheel rail vehicle. The world's first street railway, or tramway, was built in New York in 1832; it was a mile (1.6 km) long and known as the New York & Harlem Railroad. There were two horse-drawn cars, each holding 30 people. The one mile route had grown to four miles (6.4 km) by 1834, and cars were running every 15 minutes; the tramway idea spread quickly and in the 1880s there were more than 18,000 horse trams in the USA and over 3000 miles (4830 km) of track. The building of tramways, or streetcar systems, required the letting of construction contracts and the acquisition of right-of-way easements, and was an area of political patronage and corruption in many city governments.

The advantage of the horse tram over the horse bus was that steel wheels on steel rails gave a smoother ride and less friction. A horse could haul on rails twice as much weight as on a roadway. Furthermore, the trams had brakes, but buses still relied on the weight of the horses to stop the vehicle. The American example was followed in Europe and the first tramway in Paris was opened in 1853 appropriately styled 'the American Railway'. The first line in Britain was opened in Birkenhead in 1860. It was built by George Francis Train, an American, who also built three short tramways in London in 1861: the first of these ran from Marble Arch for a short distance along the Bayswater Road. The lines used a type of step rail which stood up from the road surface and interfered with other traffic, so they were taken up within a year. London's more permanent tramways began running in 1870, but Liverpool had a line working in November 1869. Rails which could be laid flush with the road surface were used for these lines.

A steam tram was tried out in Cincinatti, Ohio in 1859 and in London in 1873; the steam tram was not widely successful because tracks built for horse trams could not stand up to the weight of a locomotive.

The solution to this problem was found in the cable car. Cables, driven by powerful stationary steam engines at the end of the route, were run in conduits below the roadway, with an attachment passing down from the tram through a slot in the roadway to grip the cable, and the car itself weighed no more than a horse car. The most famous application of cables to tramcar haulage was Andrew S Hallidie's 1873 system on the hills of San Francisco—still in use and a great tourist attraction today. This was followed by others in United States cities, and by 1890 there were some 500 miles (805 km) of cable tramway in the USA. In London there were only two cable-operated lines—up Highgate Hill from 1884 (the first in Europe) and up the hill between Streatham and Kennington. In Edinburgh, however, there was an extensive cable system, as there was in Melbourne.

The ideal source of power for tramways was electricity, clean and flexible but difficult at first to apply. Batteries were far too heavy; a converted horse car with batteries under the seats and a single electric motor was tried in London in 1883, but the experiment lasted only one day. Compressed air driven trams, the invention of Major Beaumont, had been tried out between Stratford and Leytonstone in 1881; between 1883 and 1888 tramcars hauled by battery locomotives ran on the same route. There was even a coal-gas driven tram with an Otto-type gas engine tried in Croydon in 1894.

There were early experiments, especially in the USA and Germany, to enable electricity from a power station to be fed to a tramcar in motion. The first useful system employed a small two-wheel carriage running on top of an overhead wire and connected to the tramcar by a cable. The circuit was completed via wheels and the running rails. A tram route on this system was working in Montgomery, Alabama, as early as 1886. The converted horse cars had a motor mounted on one of the end platforms with chain drive to one axle. Shortly afterwards, in the USA and Germany there were trials on a similar principle but using a four-wheel overhead carriage known as a troller, from which the modern word trolley is derived.

Real success came when Frank J Sprague left the US Navy in 1883 to devote more time to problems of using electricity for power. His first important task was to equip the Union Passenger Railway at Richmond, Virginia, for electrical working. There he perfected the swivel trolley pole which could run under the overhead wire instead of above it. From this success in 1888 sprang all the subsequent tramways of the world; by 1902 there were nearly 22,000 miles (35,000 km) of electrified tramways in the USA alone. In Great Britain there were electric trams in Manchester from 1890 and London's first electric line was opened in 1901.

Above: a British horse-drawn tramcar, with a turntable in the foreground, photographed in the 1870s. The steel wheels on steel rails generated little friction, and the horse could pull twice the weight than it could with a conventional vehicle on the road. Opposite page: San Francisco's famous cable cars are clamped to a cable, which runs in a slot in the street and is pulled by a stationary engine.

Except in Great Britain and countries under British influence, tramcars were normally single-decked. Early electric trams had four wheels and the two axles were quite close together so that the car could take sharp bends. Eventually, as the need grew for larger cars, two bogies, or trucks, were used, one under each end of the car. Single-deck cars of this type were often coupled together with a single driver and one or two conductors. Double-deck cars could haul trailers in peak hours and for a time such trailers were a common sight in London.

The two main power collection systems were from overhead wires, as already described—though modern tramways often use a pantograph collecting device held by springs against the underside of the wire instead of the traditional trolley—and the conduit system. This system is derived from the slot in the street used for the early cable-cars, but instead of a moving cable there are current supply rails in the conduit. The tram is fitted with a device called a plough which passes down into the conduit. On each side of the plough is a contact shoe, one of which presses against each of the rails. Such a system was used in inner London, in New York and Washington DC, and in European cities.

Trams were driven through a controller on each platform. In a single-motor car, this allowed power to pass through a resistance as well as the motor, the amount of resistance being reduced in steps by moving a handle as desired, to feed more power to the motor. In two-motor cars a much more economical control was used. When starting, the two motors were connected in series, so that each motor received power in turn—in effect, each got half the power available, the amount of power again being regulated by resistances. As speed rose the controller was 'notched up' to a further set of steps in which the motors were connected in parallel so that each received current direct from the power source instead of sharing it. The controller could also be moved to a further set of notches which gave degrees of electrical braking, achieved by connecting the motors so that they acted as generators, the power generated being absorbed by the resistances. An American tramcar revival in the 1930s resulted in the design of a new tramcar known as the PCC type after the Electric Railway Presidents' Conference Committee which commissioned it. These cars, of which many hundreds were built, had more refined controllers with more steps, giving smoother acceleration.

The decline of the tram springs from the fact that while a tram route is fixed, a bus route can be changed as the need for it changes. The inability of a tram to draw in to the kerb to discharge and take on passengers was a handicap when road traffic increased. The tram has continued to hold its own in some cities, especially in Europe; its character, however, is changing and tramways are becoming light rapid transit railways, often diving underground in the centres of cities. New tramcars being built for San Francisco are almost indistinguishable from light railway vehicles.

The lack of flexibility of the tram led to experiments to dispense with rails altogether and to the trolleybus, or trackless tram. The first crude versions were tried out in Germany and the USA in the early 1880s. The current collection system needed two cables and collector arms, since there were no rails. A short line was tried just outside Paris in 1900 and an even shorter one—800 feet (240 m)—opened in Scranton, Pennsylvania, in 1903. In England, trolleybuses were operating in Bradford and Leeds in 1911 and other cities soon followed their example. America and Canada widely changed to trolleybuses in the early 1920s and many cities had them. The trolleybuses tended to look, except for their collector arms, like contemporary motor buses. London's first trolleybus, introduced in 1931, was based on a six-wheel bus chassis with an electric motor substituted for the engine. The London trolleybus fleet, which in 1952 numbered over 1800, was for some years the largest in the world, and was composed almost entirely of six-wheel double-deck vehicles.

The typical trolleybus was operated by means of a pedal-operated master control, spring-loaded to the 'off' position, and a reversing lever. Some braking was provided by the electric motor controls, but mechanical brakes were relied upon for safety. The same lack of flexibility which had condemned trams in most parts of the world also condemned the trolleybus. They were tied as firmly to the overhead wires as were the trams to the rails.

Right: electric tramcars coupled together in Zurich.
Opposite page: a rubber-tyred trolley bus, which takes its power from overhead lines. The word trolley comes from an early current-collecting device called a 'troller'. Trolleys are more manoeuvrable than trams because they do not run on rails, but routes still cannot readily be changed because of the expense of moving the overhead lines. Three types of current-collecting devices are also shown: two trollers and a type of plough, which ran in a conduit, or slot in the street.

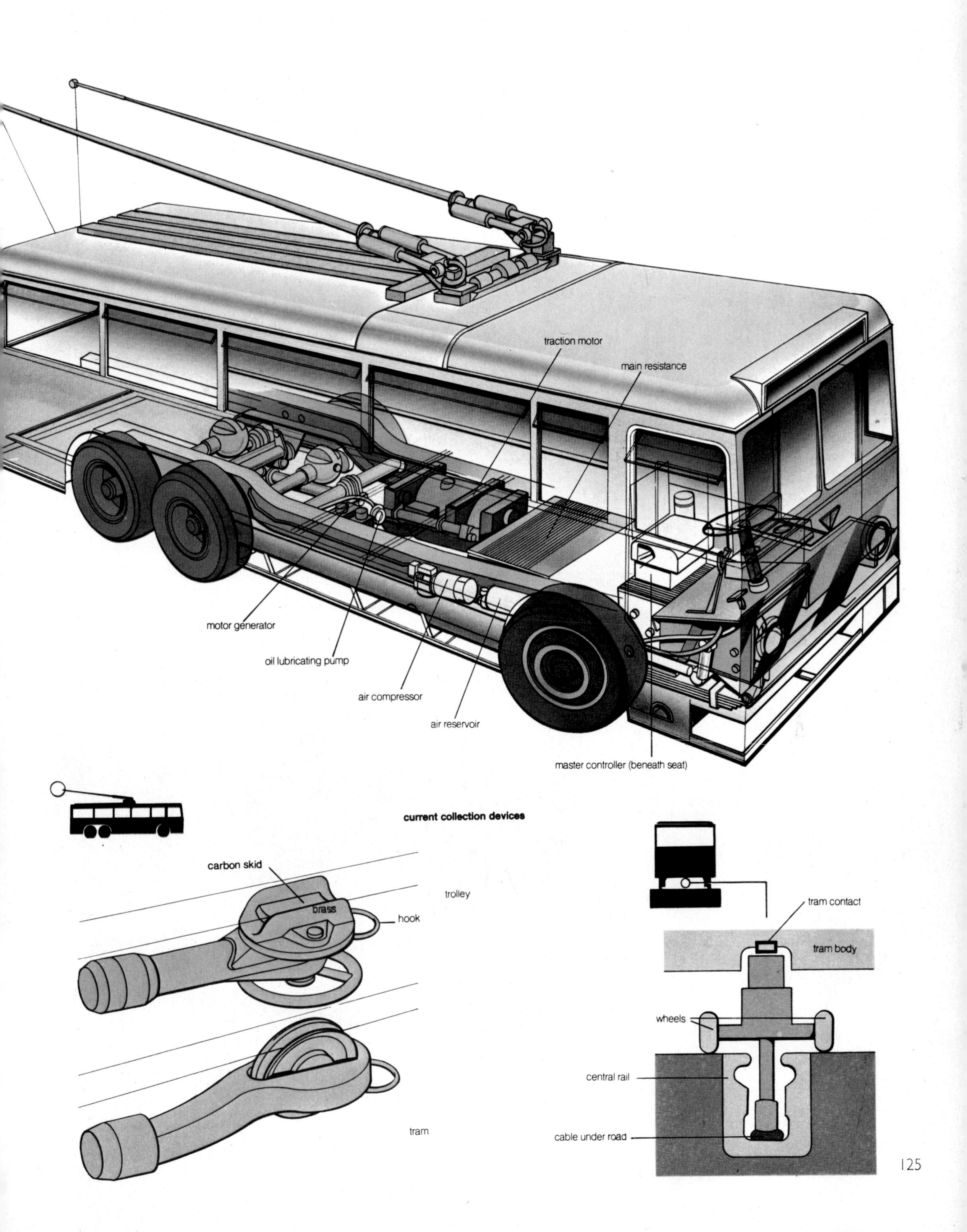
traction motor
main resistance
motor generator
oil lubricating pump
air compressor
air reservoir
master controller (beneath seat)
current collection devices
carbon skid
brass
hook
trolley
tram
tram contact
tram body
wheels
central rail
cable under road

Monorail systems

Monorails are railways with only one rail instead of two. They have been experimentally built for more than a hundred years; there would seem to be an advantage in that one rail and its sleepers [cross-ties] would occupy less space than two, but in practice monorail construction tended to be complicated on account of the necessity of keeping the cars upright. There is also the problem of switching the cars from one line to another.

The first monorails used an elevated rail with the cars hanging down on both sides, like pannier bags [saddle bags] on a pony or a bicycle. A monorail was patented in 1821 by Henry Robinson Palmer, engineer to the London Dock Company, and the first line was built in 1824 to run between the Royal Victualling Yard and the Thames. The elevated wooden rail was a plank on edge bridging strong wooden supports, into which it was set, with an iron bar on top to take the wear from the double-flanged wheels of the cars. A similar line was built to carry bricks to River Lea barges from a brickworks at Cheshunt in 1825. The cars, pulled by a horse and a tow rope, were in two parts, one on each side of the rail, hanging from a framework which carried the wheels.

Later, monorails on this principle were built by a Frenchman, C F M T Lartigue. He put his single rail on top of a series of triangular trestles with their bases on the ground; he also put a guide rail on each side of the trestles on which ran horizontal wheels attached to the cars. The cars thus had both vertical and sideways support and were suitable for higher speeds than the earlier type.

A steam-operated line on this principle was built in Syria in 1869 by J L Hadden. The locomotive had two vertical boilers, one on each side of the pannier-type vehicle.

An electric Lartigue line was opened in central France in 1894, and there were proposals to build a network of them on Long Island in the USA, radiating from Brooklyn. There was a demonstration in London in 1886 on a short line, trains being hauled by a two-boiler Mallet steam locomotive. This had two double-flanged driving wheels running on the raised centre rail and guiding wheels running on tracks on each side of the trestle. Trains were switched from one track to another by moving a whole section of track sideways to line up with another section. In 1888 a line on this principle was laid in Ireland from Listowel to Ballybunion, a distance of $9\frac{1}{2}$ miles; it ran until 1924. There were three locomotives, each with two horizontal boilers hanging one each side of the centre wheels. They were capable of 27 mph ($43\frac{1}{2}$ km/h); the carriages were built with the lower parts in two sections, between which were the wheels.

The Lartigue design was adapted further by F B Behr, who built a three-mile electric line near Brussels in 1897. The monorail itself was again at the top of an 'A' shaped trestle, but there were two balancing and guiding rails on each side, so that although the weight of the car was carried by one rail, there were really five rails in all. The car weighed 55 tons and had two four-wheeled bogies (that is, four wheels in line on each bogie). It was built in England and had motors putting out a total of 600 horsepower. The car ran at 83 mph (134 km/h) and was said to have reached 100 mph (161 km/h) in private trials. It was extensively tested by representatives of the Belgian, French and Russian governments, and Behr came near to success in achieving wide-scale application of his design.

An attempt to build a monorail with one rail laid on the ground in order to save space led to the use of a gyroscope to keep the train upright. A gyroscope is a rapidly spinning flywheel which resists any attempt to alter the angle of the axis on which it spins.

A true monorail, running on a single rail, was built for military purposes by Louis Brennan, an Irishman who also invented a steerable torpedo. Brennan applied for monorail

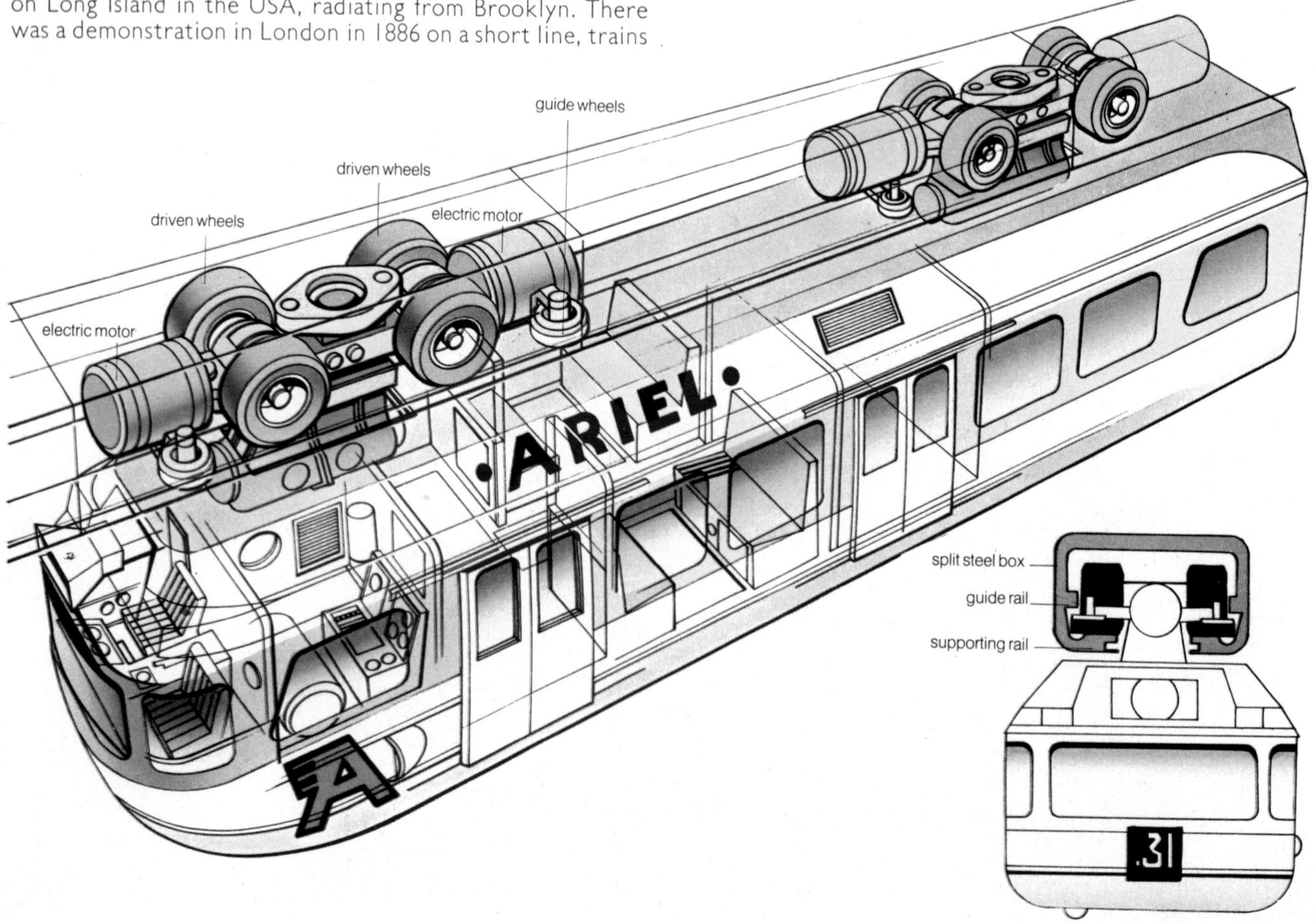

patents in 1903, exhibited a large working model in 1907 and a full-size 22-ton car in 1909–10. It was held upright by two gyroscopes, spinning in opposite directions, and carried 50 people or ten tons of freight.

A similar car carrying only six passengers and a driver was demonstrated in Berlin in 1909 by August Scherl, who had taken out a patent in 1908 and later came to an agreement with Brennan to use his patents also. Both systems allowed the cars to lean over, like bicycles, on curves. Scherl's was an electric car; Brennan's was powered by an internal combustion engine rather than steam so as not to show any tell-tale smoke when used by the military. A steam-driven gyroscopic system was designed by Peter Schilovsky, a Russian nobleman. This reached only the model stage; it was held upright by a single steam-driven gyroscope placed in the tender.

The disadvantage with gyroscopic monorail systems was that they required power to drive the gyroscope to keep the train upright even when it was not moving.

Systems were built which ran on single rails on the ground but used a guide rail at the top to keep the train upright. Wheels on top of the train engaged with the guiding rail. The structural support necessary for the guide rail immediately nullified the economy in land use which was the main argument in favour of monorails.

The best known such system was designed by H H Tunis and built by August Belmont. It was $1\frac{1}{2}$ miles long (2.4 km) and ran between Barton Station on the New York, New Haven & Hartford Railroad and City Island (Marshall's Corner) in $1\frac{1}{2}$ minutes. The overhead guide rail was arranged to make the single car lean over on a curve and the line was designed for high speeds. It ran for four months in 1910, but on 17 July of that year the driver took a curve too slowly, the guidance system failed and the car crashed with 100 people on board. It never ran again.

The most successful modern monorails have been the invention of Dr Axel L Wenner-Gren, an industrialist born in Sweden. Alweg lines use a concrete beam carried on concrete supports; the beam can be high in the air, at ground level or in a tunnel, as required. The cars straddle the beam, supported by rubber-tyred wheels on top of the beam; there are also horizontal wheels in two rows on each side underneath, bearing on the sides of the beam near the top and bottom of it. Thus there are five bearing surfaces, as in the Behr system, but combined to use a single beam instead of a massive steel trestle framework. The carrying wheels come up into the centre line of the cars, suitably enclosed. Electric current is picked up from power lines at the side of the beam. A number of successful lines have been built on the Alweg system, including a line $8\frac{1}{4}$ miles (13.3 km) long between Tokyo and its Haneda airport.

There are several other 'saddle' type systems on the same principle as the Alweg, including a small industrial system used on building sites and for agricultural purposes which can run without a driver. With all these systems, trains are diverted from one track to another by moving pieces of track sideways to bring in another piece of track to form a new link, or by using a flexible section of track to give the same result.

Other systems Another monorail system suspends the car beneath an overhead carrying rail. The wheels must be over the centre line of the car, so the support connected between rail and car is to one side, or offset. This allows the rail to be supported from the other side. Such a system was built between the towns of Barmen and Elberfeld in Germany in 1898–1901 and was extended in 1903 to a length of 8.2 miles (13 km). It has run successfully ever since, with a remarkable safety record. Tests in the river valley between the towns showed that a monorail would be more suitable than a conventional railway in the restricted space available because monorail cars could take sharper curves in comfort. The rail is suspended on a steel structure, mostly over the River Wupper itself. The switches or points on the line are in the form of a switch tongue forming an inclined plane, which is placed over the rail; the car wheels rise on this plane and are thus led to the siding.

An experimental line using the same principle of suspension, but with the car driven by means of an aircraft propeller, was designed by George Bennie and built at Milngavie (Scotland) in 1930. The line was too short for high speeds, but it was claimed that 200 mph (322 km/h) was possible. There was an auxiliary rail below the car on which horizontal wheels ran to control the sway.

A modern system, the SAFEGE developed in France, has suspended cars but with the 'rail' in the form of a steel box section split on the underside to allow the car supports to pass through it. There are two rails inside the box, one on each side of the slot, and the cars are actually suspended from four-wheeled bogies running on the two rails.

Opposite page: the SAFEGE monorail system, recently constructed in France, is called an aerial railway, and is suspended from a steel box. ***Left:*** *the Lartigue system, which ran in Ireland for more than 35 years. There was a pressure-equalizing pipe between the two boilers of the locomotive; the engine ran on three wheels which were about two feet (0.75m) in diameter.*

Underground railways

The first underground railways were those used in mines, with small trucks pushed by hand or, later, drawn by ponies, running on first wooden, then iron, and finally steel rails. Once the steam railway had arrived, however, thoughts soon turned to building passenger railways under the ground in cities to avoid the traffic congestion which was already making itself felt in the streets towards the middle of the 19th century.

The first underground passenger railway was opened in London on 10 January, 1863. This was the Metropolitan Railway, 3.75 miles (6 km) long, which ran from Paddington to Farringdon Street. Its broad gauge (7 ft, 2.13 m) trains, supplied by the Great Western Railway, were soon carrying nearly 27,000 passengers a day. Other underground lines followed in London, and in Budapest, Berlin, Glasgow, Paris and later in the rest of Europe, North and South America, Russia, Japan, China, Spain, Portugal and Scandinavia, and plans and studies for yet more underground railways have already been turned into reality—or soon will be—all over the world. Quite soon every major city able to do so will have its underground railway. The reason is the same as that which inspired the Metropolitan Railway over 100 years ago—traffic congestion.

Above: the Paris Mētro in 1903, at the Place de L'Opēra.
Right: the building of an underground railway by the cut-and-cover method of tunnelling, in Tashkent, capital of Soviet Uzbekistan. It will be the first subway in Central Asia. The city suffered an earthquake in 1966, and the new transport facilities are part of the reconstruction plans.
Opposite page: London had the world's first underground train in 1863. The Victoria Line, opened in 1969, is the most fully automatic.

The first electric tube railway [subway] in the world, the City and South London, was opened in 1890 and all subsequent tube railways have been electrically worked. Sub-surface cut-and-cover lines everywhere are also electrically worked. The early locomotives used on underground railways have given way to multiple-unit trains, with separate motors at various points along the train driving the wheels, but controlled from a single driving cab.

Modern underground railway rolling stock usually has plenty of standing space to cater for peak-hour crowds and a large number of doors, usually opened and closed by the driver or guard, so that passengers can enter and leave the trains quickly at the many, closely spaced stations. Average underground railway speeds are not high—often between 20 and 25 mph (32 to 60km/h) including stops, but the trains are usually much quicker than surface transport in the same area. Where underground trains emerge into the open on the edge of cities, and stations are a greater distance apart, they can often attain well over 60 mph (97 km/h).

The track and electricity supply are usually much the same as that of main-line railways and most underground lines use forms of automatic signalling worked by the trains themselves and similar to that used by orthodox railway systems. The track curcuit is the basic component of automatic signalling of this type on all kinds of railways. Underground railways rely heavily on automatic signalling because of the close headways, the short time intervals between trains.

Some railways have no signals in sight, but the signal 'aspects'—green, yellow and red—are displayed to the driver in the cab of his train. Great advances are being made also with automatic driving, now in use in a number of cities. The Victoria Line system in London, the most fully automatic line now in operation, uses codes in the rails for both safety signalling and automatic driving, the codes being picked up by coils on the train and passed to the driving and monitoring equipment.

Code systems are used on other underground railways but sometimes they feed information to a central computer, which calculates where the train should be at any given time, and instructs the train to slow down, speed up, stop, or take any other action needed.

automatic train control (Victoria line)

Trains are controlled by impulses in the rails: a high frequency at 'command spots' for speed and braking, and a low frequency for acceleration and coasting. Should this signal fail, the train stops automatically. Stopping at stations is automatic.

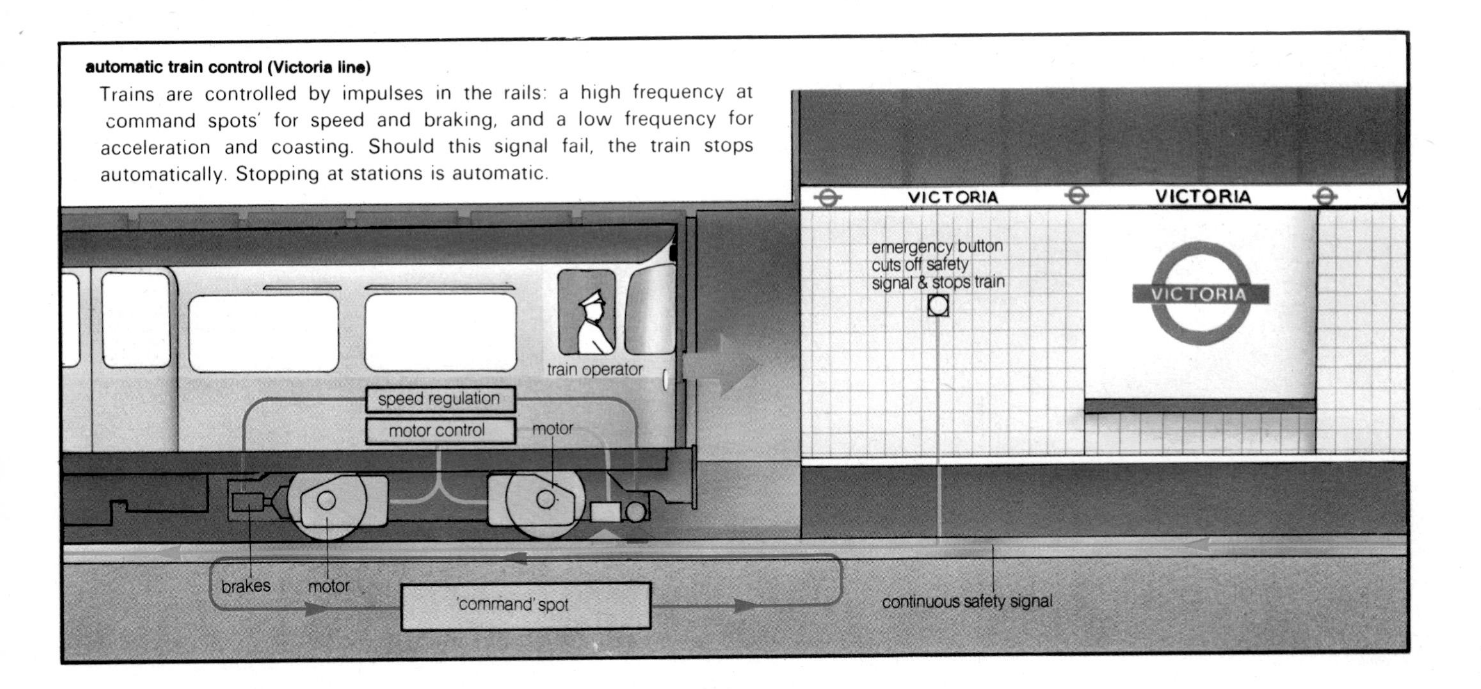

safety signalling system (other underground & tube lines)

Trains are driven manually in obedience to automatic signals, controlled by the trains themselves operating track circuits. The operator is required only to supervise and to start the train by pressing two buttons. Further protection is given by a 'train stop' a short distance past each signal, which stops a train should it pass a signal at 'danger' (red). When the signal shows 'clear' (green), the train stop arm is held down by compressed air.

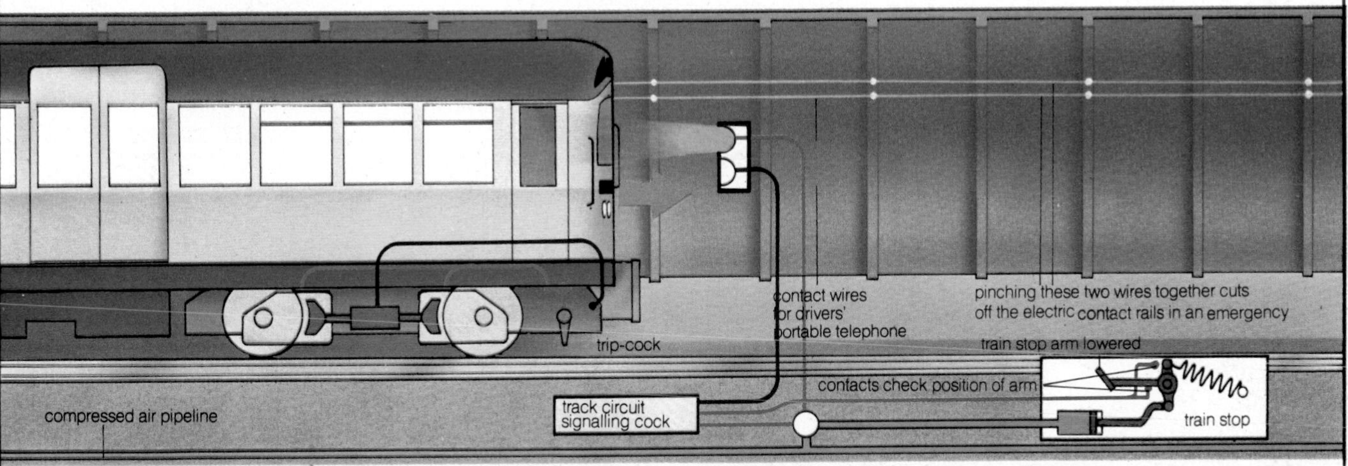

When the signal is at 'danger' the compressed air to the train stop is cut off and the arm is raised by a spring. The train stop arm operates a 'trip cock' on the train, stopping the train in its own length. The train stop is designed to be 'fail safe'.

Should the electricity or the compressed air supply fail, the train stop arm will spring up. Electric contacts check that the arm is in the correct position and that it is not damaged before the signal clears; each light has two bulbs in case one burns out.

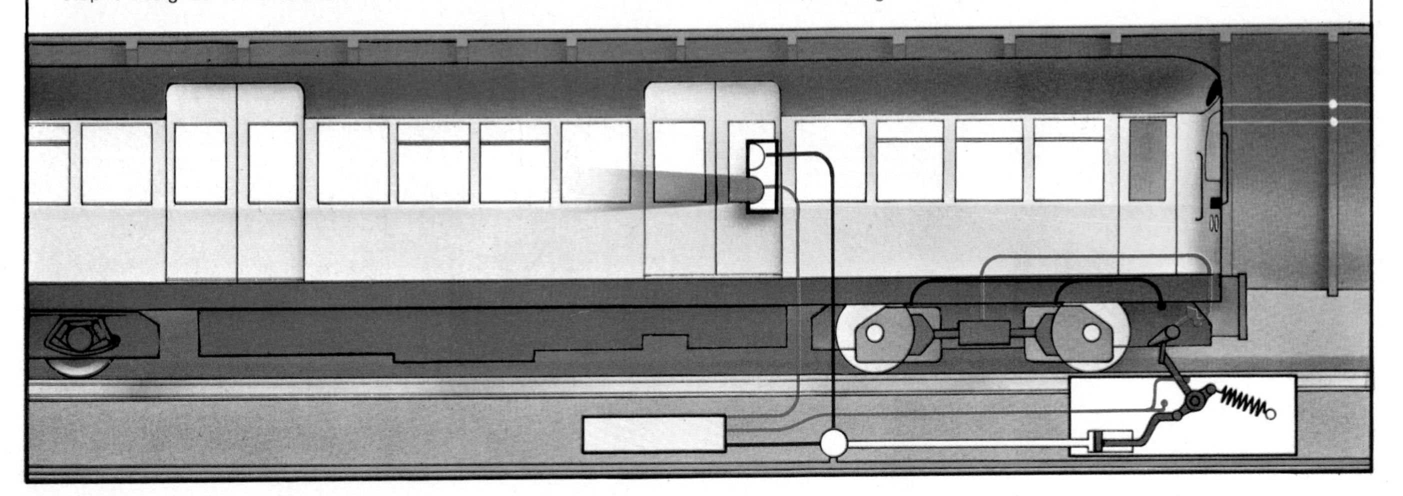

THE AGE OF THE AUTOMOBILE

Rapid technological advance during the nineteenth century made the automobile a reality. Henry Ford's mastery of assembly line construction put America on wheels, ending the isolation of rural communities and setting an example of industrial efficiency for the rest of the world to emulate. But the car is not an unmixed blessing. Cities have been spoiled, their neighbourhoods obliterated as highways smash through; environmental pollution, shortage of natural resources and the survival of public transport are problems caused largely by the car. Yet the car itself is still the object of endless fascination.

The history of the motor car

The first practical cars driven by internal combustion engines appeared in the 1880s, but steam driven carriages had been produced with varying degrees of success for over a hundred years previously. The car was not invented by any one man, but was the outcome of the work of many individuals, working either independently or in collaboration with others.

The first workable four-stroke internal combustion engine was invented in 1876 by Nikolaus Otto in Germany, after many years of experimental work with gas engines. Following Otto's work, Karl Benz and Gottlieb Daimler, working independently, produced the world's first cars with internal combustion engines. The Benz, of 1885, was a three-wheeled vehicle with a tubular steel chassis and an open wooden two-seater body. The single front wheel was steered by a tiller, and the two large rear wheels were driven by chains. The single-cylinder petrol engine operated on the four-stroke principle, with electric ignition and water cooling, and was mounted horizontally over the rear axle. At 250 to 300 revs/min it produced about $\frac{1}{2}$ hp and drove the car at about 8 or 10 mile/h (13 to 16 km/h).

Daimler's car, the first four-wheeled design, was a converted carriage fitted with a vertical single-cylinder engine developing $1\frac{1}{2}$ hp and running at up to 900 revs/min. It was produced in 1886.

By 1900 in most industrial countries various car designs had appeared, and several important innovations were made during this period.

Among the improvements made to car designs at this time were the introduction of the float-type carburettor by Maybach in 1892; the steering wheel (Vacheron) in 1894; the Michelin brothers' pneumatic tyre, and Lanchester's propeller shaft transmission (1895); multi-cylinder engines by Mors (V4) and Daimler (four in line); in 1899 the honeycomb radiator, gate gearchange, and floor-mounted accelerator (Daimler); and the universal joint for shaft drive to sprung rear axles (Renault).

The first motor race was held in France in 1895, over a distance of about 750 miles (1200 km) from Paris to Bordeaux and back. The overall winner was a 4 hp twin-cylinder Panhard and Levassor, driven by Emile Levassor. The race took 48 hours and Levassor averaged 15 mile/h (24 km/h), despite falling asleep once on the return journey to Paris.

In Britain, the law restricting cars to 4 mile/h (6.5 km/h) in the country and 2 mile/h (3.2 km/h) in towns, and insisting that every vehicle should be preceded by a man on foot waving a red flag, was repealed in 1896. This was celebrated by the first Emancipation Run from London to Brighton, about 60 miles (96 km), which is still held annually as a reliability trial for Veteran cars. (Veteran cars are those built before 1918; Vintage cars were built between 1918 and 1930.)

Such sporting events served to stimulate design improvements and increase public interest in motoring.

By the beginning of the 20th century the car had ceased to be a novel 'horseless carriage' and had become established as a modern and efficient means of transport. The work of two Frenchmen, Georges Bouton and Count Albert de Dion, had led to the development of lightweight, high speed engines. Their 1903 'Populaire' produced 8 hp at 1500 revs/min, with a cubic capacity of 846 cm³ (52 in³) and a weight of only 40 lb (18 kg). To handle the requirements of this high speed air cooled engine, Bouton had to design an improved ignition system which bore many similarities to the modern contact breaker ignition.

The improvements in engine design, which led to high road speeds, forced the pace of development in braking and transmission systems. The first brakes were based on those used on horse-drawn vehicles and on bicycles. A solid block of wood, leather or metal was forced against the wheel rims

Above: the Benz car of 1886 had a single-cylinder engine which produced 0.9 horsepower. Maximum speed was 9mph (15kph).
Opposite page: the clumsy steam carriage built by Nicholas Joseph Cugnot in 1770, now in a Paris museum, was the first full-sized mechanically-powered road vehicle. Cugnot was ahead of his time; he could not build a boiler sufficiently strong to provide the kind of pressure needed. The layout also leaves much to be desired; the heavy boiler ahead of, and supported by, the axle of the steered wheel would have made the 'car' somewhat unstable.

by a hand-operated lever, or a contracting band of friction material acted on the propeller shaft in conjunction with externally-contracting brakes fitted to drums on the rear wheels. Asbestos brake linings were patented by Herbert Frood in England in 1908, and were much more effective than the cotton based linings then in use. The disc brake was invented in 1902 by Frederick Lanchester, then the following year the Mercedes company (formerly Daimler) produced a braking system with internally-expanding shoes inside a brake drum. Four-wheel braking was first employed by the Italian company of Isotta-Franchini in 1911.

The early 1900s saw the growth of the American automobile industry, and the stream of innovations that originated in the USA during this period included automatic transmission (1904), coil and distributor ignition (1908), the electric starter, the dynamo, a car telephone (1911), and hydraulic braking (1920). In 1908 the first Model T Ford came off the production line, and by 1911 Ford of America were producing 1000 per day. During its 19-year production run over 15,000,000 were made.

World War I was the first mechanized war and there was a great demand for mass-produced, standardized engines and components. This led not only to improved production techniques but also to more reliable and efficient vehicles.

Between the World Wars a number of very high quality cars were built, and some of these represented such an exceptionally high standard of craftsmanship and durability that owing to changing economic circumstances it is unlikely that cars of comparable quality will ever be built again. Such classics as the Bugatti Royale, Hispano-Suiza, Rolls-Royce Phantom III, Bentley 8 litre, and Delage are unlikely to be excelled.

Mass-production methods were now well established, and this led to the availability of a wide range of cheap, reliable and comfortable cars which found a ready market in the relatively affluent 1920s. The main components of the cars were well-designed and efficient, and a variety of accessories were introduced, such as reversing lights, radios, automatic chokes, windshield wipers, and chrome-plated trim. During the Depression many smaller firms went out of business or were absorbed by larger ones, but new groups were emerging that had better production facilities and more money to spend on sales promotion and market research.

In America the trend was towards power and luxury, while the European manufacturers concentrated on small low-priced cars like the Austin 7 in England, and the Italian Fiat 500. In Germany, the KDF, which was to become better known as the Volkswagen, was designed by Ferdinand Porsche with the backing of Adolf Hitler. KDF stands for 'Kraft durch Freude'—'Strength through Joy'. The basic shape remains today, and over 12 million have been made.

During World War II the production of private cars was severely restricted as raw materials were diverted to military uses and the factories used to make military vehicles, munitions and aircraft components. When car production recommenced the first models were almost the same as the pre-war designs, and it took a few years for the plants to re-tool sufficiently to produce any really new designs.

Many of the new models had powerful high compression engines, independent front suspension, and in styling were longer, lower and more elaborate. Lightweight single-unit

Above, from top: before 1900 the mechanical parts of cars were arranged wherever they would fit around a bare platform frame. Then in the first decade of this century, the layout became standardized: front engine, rear drive. By 1920, cars had become longer and lower in the interest of stability. During the 1930s, closed cars became common, and body lines became more streamlined.

bodies were adopted, and the use of curved glass for windshields and rear windows greatly improved the drivers' visibility. There was extensive, sometimes excessive, use of chromium plating, and styling became one of the major preoccupations of the industry, with new models being introduced annually that were often mechanically identical to those they were meant to replace.

In the USA the tubeless automobile tyre was introduced in 1948 by the Goodrich company. Power steering, air conditioning, twin headlamps, and wrap-around windshields all originated in the States during the early 1950s. Glass fibre reinforced resins, light and corrosion free, were used on the bodywork of the 1953 Chevrolet Corvette, and for the roof panel of the 1955 Citroën DS19.

Advances in fuel technology allowed the use of higher compression ratios. Overhead valve and overhead camshaft designs, with improved fuel systems (including fuel injection) and better ignition system performance contributed to engine power outputs for a given cubic capacity being greatly increased. The resultant increase in power-to-weight ratio that was possible improved the acceleration, speed, road holding and braking of cars of that time.

Disc brakes, less prone to failure from overheating than drum brakes, at last became widely accepted, over half a century after Lanchester's original design was patented. Further improvements in roadholding and braking resulted from the introduction of radial-ply tyres in 1953. Due to their higher cost, these tyres were at first used only on expensive high-performance cars, but they are now widely used on all kinds of car.

The introduction of new plastic materials for the interior trim was a great asset for the stylists, and a wide range of

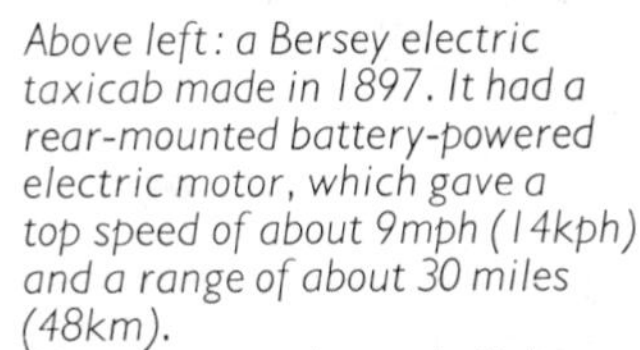

Above left: a Bersey electric taxicab made in 1897. It had a rear-mounted battery-powered electric motor, which gave a top speed of about 9mph (14kph) and a range of about 30 miles (48km).
Above centre: the car built by Siegfried Markus of Austria in 1875. The single-cylinder engine developed about ¾hp and drove the car at about 4mph (6kph).
Above: a 1931 Ford De Luxe Sedan is typical of the cars built in the USA between the wars.
Left: the 1910 Austin 10 had a single-cylinder side-valve engine and a three-speed gearbox; it was a forerunner of the famous Austin 7, which first appeared in 1922.

colour schemes to match the body colours became available. The once universal oil-pressure gauges and ammeters were often replaced by simple warning lights, which are cheaper and less complicated but also less informative.

Car design in the 1960s was greatly influenced by the new interest in safety and pollution control. Mechanical improvements brought higher speeds, better roadholding, braking and acceleration, but many countries began to introduce laws restricting the maximum speed of vehicles. Cars had to be built to comply with the strict new safety and anti-pollution laws of the United States, which were gradually adopted by many other countries.

Above: the Fiat 500 'Topolino', which was produced from 1936 to 1948. 'Topolino' is the Italian equivalent of 'Mickey Mouse'. Produced in large numbers, the car brought motoring within the reach of many people.
Top of page: the 1953 Chevrolet Corvette had bodywork made of glass-reinforced plastic, which combined strength and lightness, giving the car a good power-to-weight ratio.

In addition to the improved performance, cars became more comfortable and easier to drive. Heating and ventilating equipment became standard on even the small cheap cars where it was previously only available, if at all, as an extra. Automatic transmission, power brakes and power steering also gained widespread acceptance.

The electrical system, which had an increasingly heavy load to handle, was improved by the introduction of the alternator to replace the dynamo, and the use of circuit breakers instead of fuses.

The ordinary reciprocating piston engine may well eventually be replaced by the Wankel engine, or possibly by small gas turbine units. Due to dense traffic conditions, pollution problems, and the possibility of inadequate fuel supplies, inner city motoring may be restricted to small electrically powered cars, with rechargeable batteries giving a range of about 50 miles between charges.

Fuel cells may be developed to a useful stage within the next half century, and one car has already been built which is powered by a hybrid fuel cell and battery unit. The fuel cells provide the main driving power, with the batteries being used to give extra power for acceleration. The main problem with fuel cells is that they are large and heavy in relation to their power output.

The modern car

It had long been man's dream to discover some form of mechanized transport, but it was not until the industrial revolution was well under way in the mid-nineteenth century, bringing breakthroughs in engineering techniques, that this dream became a reality. It was with the invention of the internal combustion engine, and Karl Benz's application of it to a self-propelled road vehicle, that the car became practical. Within 50 years the car had developed into a highly developed machine containing over 20,000 parts, all working together to provide a convenient method of transport.

There are many variations in car design—both in exterior styling and in engineering layout—but for most purposes a 'standard' model can be considered with a front-mounted engine, rear-driven wheels and comparatively simple suspension, brake and steering systems.

Most modern cars consist of a rigid steel unitary or monocoque structure which combines bodywork and frame. Some, however, and most older cars, had a separate chassis on which the mechanical parts were mounted.

The engine The heart of any motor car is the engine—the high precision machine which converts the latent energy of petrol [gasoline] into mechanical energy, or movement. Petrol is drawn from the petrol tank by a fuel pump, which can be electrical or mechanical (driven from the engine itself). It is then passed to the carburettor, a device which vaporizes the fuel, providing a highly combustible mixture of fuel and air, the ratio of which changes in proportion with the position of the accelerator pedal. The mixture passes through the inlet manifold and is fed, in turn, to each of the cylinders through the inlet valves. The potentially explosive vapour is compressed in the combustion chamber by the rising piston and fired by a high voltage spark from the spark plug in the top of the chamber. The piston is thus forced down the cylinder, and on its return stroke an exhaust valve is opened to allow the burnt gases to escape through the exhaust manifold, into a silencer [muffler] to quieten the explosive roar of the engine, and out through the exhaust pipe at the rear of the car.

Each piston drives a connecting rod, the other end of which is connected to the crankshaft; this runs along the length of the engine and converts the up-and-down movement of the pistons into a circular or rotary movement. At the end of the crankshaft is a heavy flywheel which provides the momentum to keep the engine running smoothly, and because of the great forces involved, the crankshaft is mounted on precision bearings which (like all moving engine parts) are fed with lubricating oil drawn from the sump by a mechanically operated oil pump.

For the engine to run efficiently, it is necessary that the inlet and exhaust valves are opened or closed at exactly the right moment in relation to the position of the piston. The crankshaft therefore drives a camshaft (via a driving chain) which has cams (eccentric lobes) along its length; as the camshaft rotates, the lobes operate pushrods to move rocker arms which in turn open and close spring-loaded valves in synchronization with the movement of the pistons. In some high-performance cars, the camshaft is mounted over the valve assembly—this eliminates both pushrods and rocker gear and increases efficiency by reducing friction.

In the same way that the valves must open and close at the correct moment, the spark at the spark plug must also be accurately timed. This is done by the ignition system. The current for the spark originates from the battery, usually at 12 volts DC; this is passed through the contact points in the distributor, which provides bursts of current at the exact moment required for ignition at each cylinder. In order to provide a good spark, the 12 V current is passed through an induction coil (after the contact break points) and the voltage increased to about 30,000 V. This is then fed to the spark plugs, where the current jumps the gap and provides the igniting spark.

Naturally a machine which contains a series of almost continuous explosions becomes very hot. Most engines are cooled by water which is pumped around the cylinder block by a water pump and through a radiator which can efficiently dispose of unwanted heat. Behind the radiator is a fan which draws air through the cooling vents and increases the efficiency of the cooling system. Both the fan and the water pump are driven by a belt from the crankshaft; this belt also drives a dynamo or alternator which, operating through a voltage regulator, continuously charges the battery.

A more detailed description of the internal combustion engine and its accessories will be found in the *Methods of Propulsion* section of this book.

A schematic diagram showing the layout of the typical car. Many of the various components are described in the text of this book. Other possible layouts include rear-engine design, as in the Volkswagon, and front-wheel drive, as in the Mini, which also has a transverse engine.

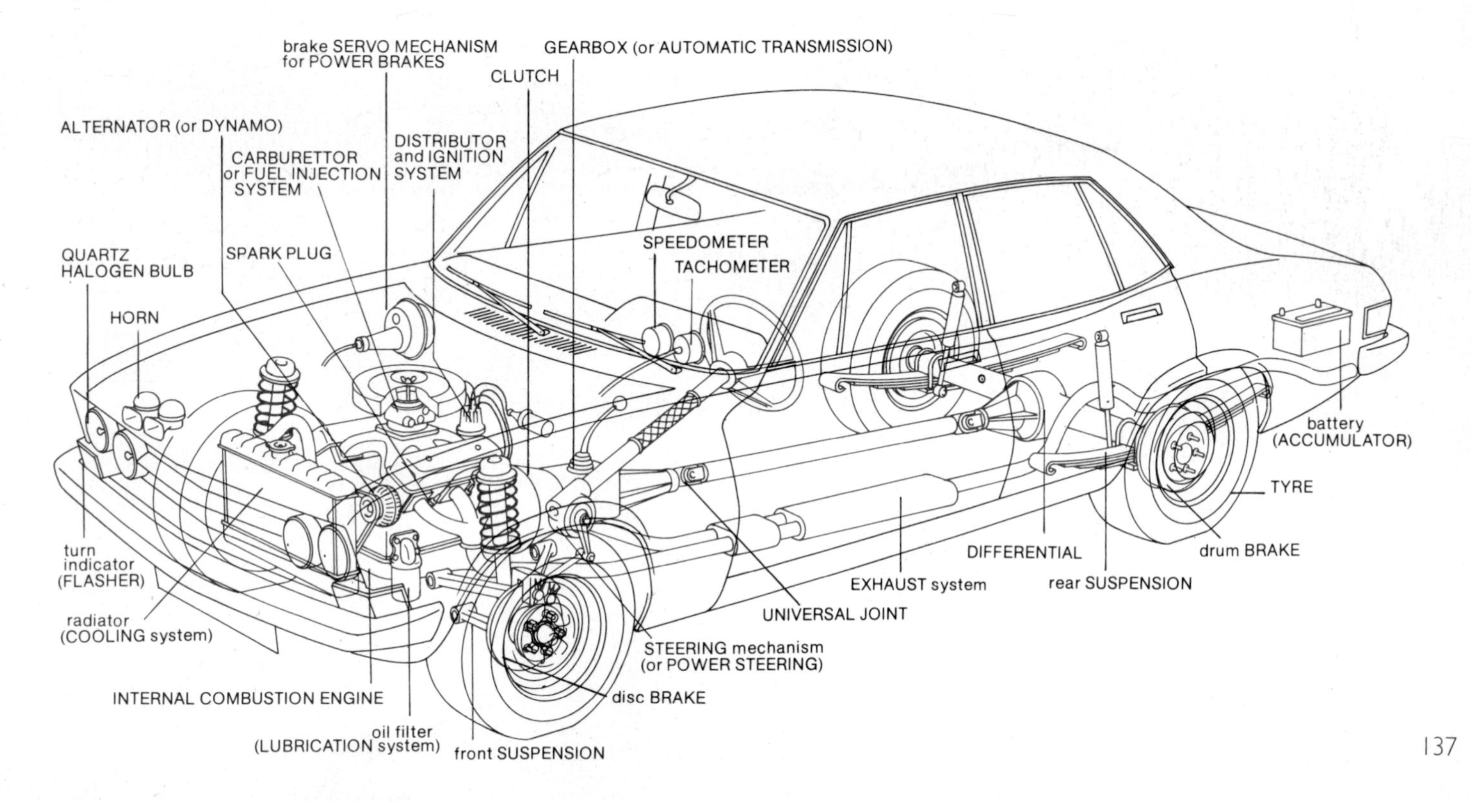

Below: a Morris Mini-Cooper, photographed in 1961. It has a transverse-mounted four-cylinder engine, front-wheel drive and front disc brakes. Bottom of page: the Rolls-Royce Camargue, which costs £35,772 in England.

Below: the British Leyland Range Rover, with four-wheel drive, is a sophisticated version of the Land Rover, one of the world's most versatile vehicles.
Left: a 1961 'E' type Jaguar, which can do 150 mph.

The 1955 Ford (left) featured a 'wrap-around windshield'. In the USA, car manufacturers increasingly concentrated on bodywork design as a means of selling their product. The 'fins' on the rear deck of the 1959 Cadillac (below) were expensive to manufacture and difficult to repair. In the meantime, the distribution of cars and the finance of their purchase had become such a morass of near-fraudulant practices that the business was investigated by Congress; one result was the required posting of the manufacturer's price sticker on each car, giving a breakdown of costs.

The gearbox A gearbox is a set of gears with a shifting lever which provides a selection of gear ratios between two components of a machine, such as a machine tool or an automobile.

The most familiar application of the gearbox is in the car. The gearbox, or transmission, is located in the drive train between the engine and the drive wheels, and is a necessary part of the car because an internal combustion engine does not have much power at low speeds. In order to move the car from a standing start, the gear ratio between the engine and the driving wheels must be such that the crankshaft of the engine is turning over at a relatively high speed. A simple three-speed gearbox allows the crankshaft to turn at roughly four, eight or twelve times for each revolution of the wheels, depending on the gear selected.

A three-speed gearbox consists of a clutch shaft with a clutch gear on it which is turning when the clutch is engaged; a layshaft [countershaft], which has several gears on it, one of which is always meshed with the clutch gear and is turned by it; and a transmission shaft, which transmits the power to the propeller shaft [drive shaft], and which has two gears on it, one larger than the other, splined so that they can slide on the shaft. Each of the two gears on the transmission shaft is fitted with a shifting yoke, a bracket for pushing it back and forth on the shaft. The shifting yokes are selected and shifted by the driver by means of the shifting lever, which pivots between the driver's compartment and the top of the gearbox case.

When the driver selects first gear, or low gear, the larger of the two gears on the transmission shaft is pushed along the shaft until it meshes with the smallest gear on the layshaft. Then the clutch is engaged, allowing power to be transmitted from the engine through the gearbox to the wheels. The gear ratio between the transmission shaft and the layshaft is 3 : 1, and the effective ratio between the crankshaft and the wheels is about 12 : 1 (because of further reduction gearing in the differential, another set of gears which transmits power between the propeller shaft and the drive wheels).

When the car is moving fast enough, about ten mile/h (16 km/h), the driver shifts to second gear, engaging the small transmission gear with the large layshaft gear. The gear ratio is now 2 : 1 and the ratio between the crankshaft and the wheels about 8 : 1. For cruising speed the driver shifts to third gear (high gear), forcing the smaller transmission gear axially (lengthwise) against the clutch gear. These have teeth on the sides of them which engage, and the gear ratio is now 1 : 1; that is, the transmission shaft and the clutch shaft (hence the crankshaft) are turning at the same speed. The ratio between the crankshaft and the wheels is now about 4 : 1. Thus the speed of the engine is always within the range of efficiency for the engine while the car moves from a standing start to speed.

The reverse gear is at the back end of the layshaft, and turns a small idler gear which meshes with the large transmission gear when the driver selects reverse. (When two gears mesh, they turn in opposite directions; the inclusion of an idler gear between them means that they turn in the same direction, so that the car reverses its direction.)

In a gear system, speed reduction means an increase in torque. (Torque is a twisting force, such as the effort needed to loosen a tight cap on a jar.) Thus the gearbox, when first gear is selected, transmits less speed from the small layshaft gear to the large transmission gear (because when two gears turn together the larger gear turns slower), but transmits more torque from the crankshaft, to 'twist' the propeller shaft and overcome the inertia of the car to get it moving.

American cars with large engines have usually had three-speed gearboxes, but smaller European cars usually have four or five forward gears, because the usable range of speed of the engine is smaller. The gearbox on a large lorry may have sixteen forward gears or more, because of the torque required to get the weight of a loaded lorry moving.

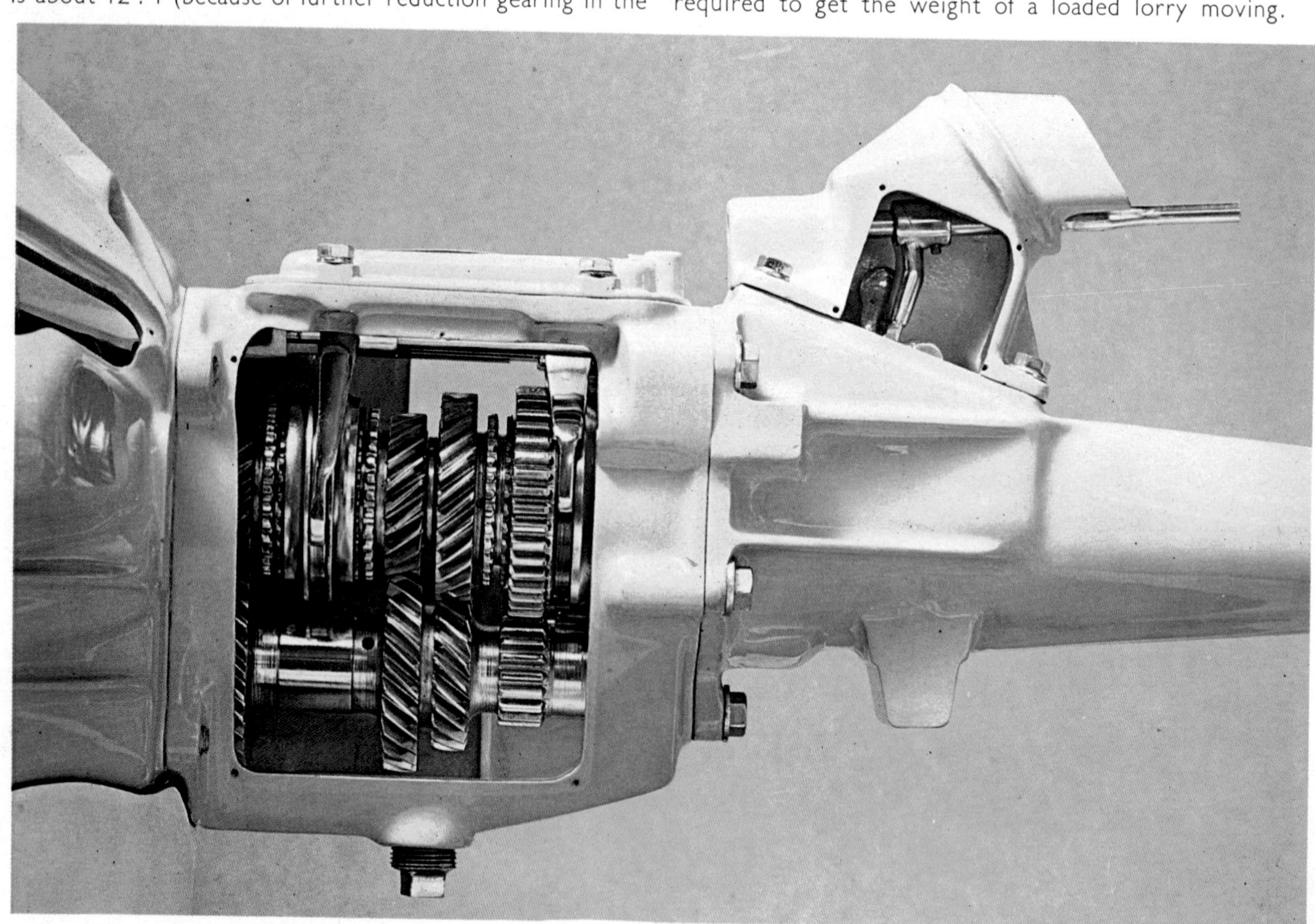

In the early days of motor cars, gearboxes were simple devices such as described above, and it took some muscle power and skill to shift the gears smoothly. The edges of the gear teeth were chamfered (rounded) so that they would mesh as smoothly as possible. Down-shifting was particularly complicated, requiring double-clutching (disengaging the clutch, revving the engine to a higher speed, and re-engaging the clutch after shifting gears).

As more and more people took up driving, it became necessary to make shifting easier, and gearbox design became more complicated. Syncromesh gears were designed, in which the gears are made to run at the correct speed before they mesh so that they can do so without grinding. One way of doing this is to provide conical sections on the sides of the gears which fit into one another; friction starts the gear turning before the teeth actually mesh. A balking provision is made in some designs to prevent the gears from meshing until they are running at speed. Some synchromesh gearboxes may have all the gears meshing all the time; power is not transmitted until a sliding 'dog' axially engages the appropriate gear. At first most gearboxes provided syncromesh only on the upper gears, but most models nowadays provide it on all gears.

Nowadays nearly all American cars, and many others as well, have automatic transmissions, but some drivers like the feeling of control over the vehicle that they get from operating a manual gearbox. People who drive in competitions prefer manual gearboxes because skilful gear-changing can mean split-second advantages in races.

Opposite page: a cutaway view of a British Ford gearbox designed for use with the V-4 engine. The shifting lever protrudes on the upper right. Below: a Ford Cortina gearbox, which has four forward speeds.

All forward gears are permanently engaged, but free-wheel until locked to the transmission shaft. They are locked gradually and smoothly by the syncromesh, a splined ring which slides and engages a clutch to synchronize gear and shaft, then moves on to bridge two sets of teeth. The baulk ring gives an intermediate semi-locked stage.

The gear lever moves in two planes, moving the linkage rod back and forth or twisting it. Twisting it makes the actuating tooth engage with links for gears. Unused links are locked.

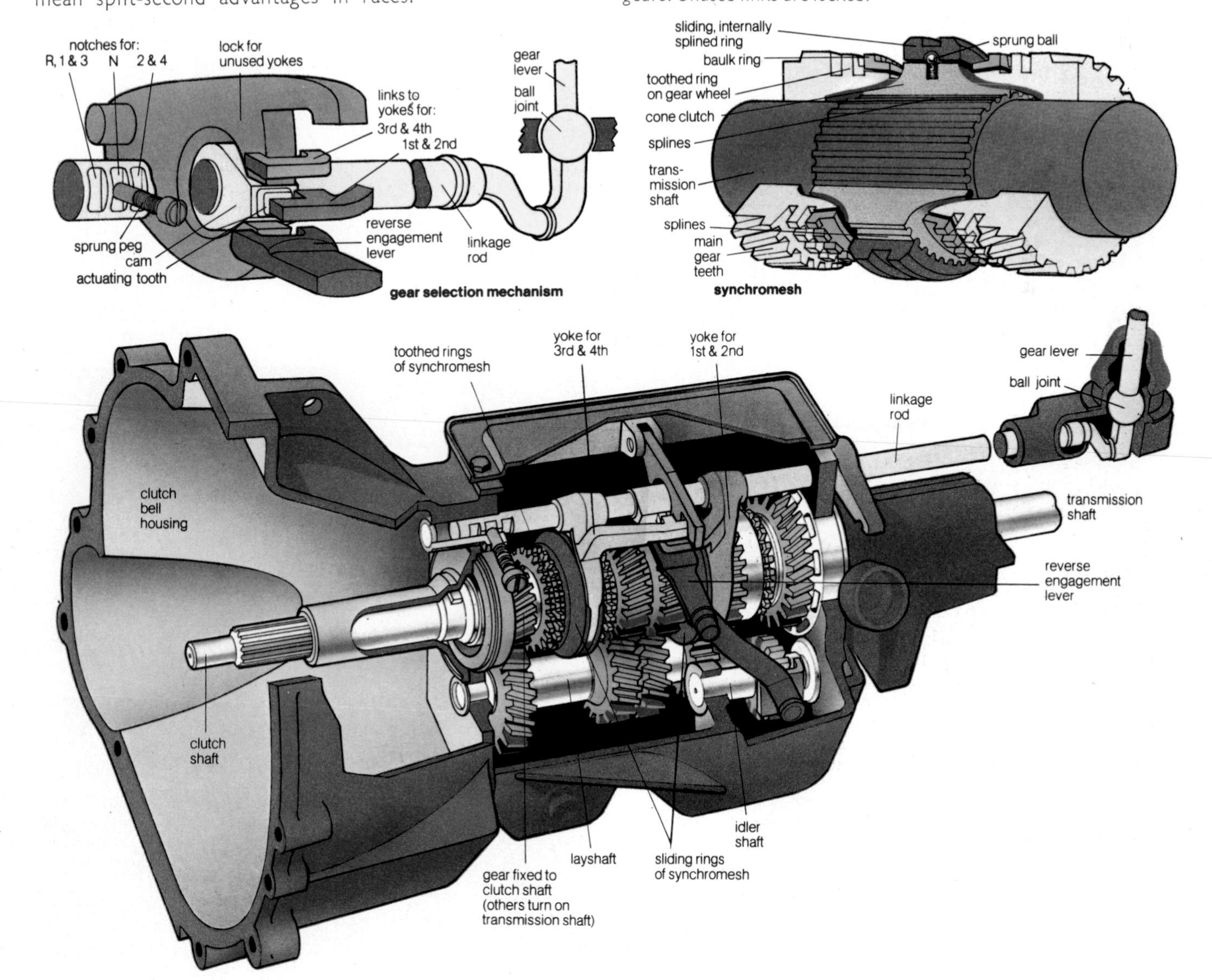

Automatic transmission With automatic transmission the most appropriate gear is selected without any action from the driver. There is no clutch pedal for the driver to control, and selection and engagement of gears are performed automatically. All the driver does is operate a control lever for a number of guiding positions: P for park, R for reverse, N for neutral, D for forward drive, and L to hold the lower gear. There are stopping devices to prevent the accidental selection of reverse and park while the vehicle is in motion, and another safety feature ensures that the engine only starts in neutral or park. Fully automatic gear changes, both up and down, occur with the lever in the D position.

(Automatic transmission should not be confused with semi-automatic transmission, where the driver changes gears manually but is assisted by an automatic clutch. With the self-changing or preselector gearbox, the driver predetermines the choice of gear and controls the clutch, but the gearchange is performed automatically.)

Automatic transmission must provide smooth acceleration from rest without jerking. The engine must therefore be allowed to slip with respect to the wheels, so that it can turn over while the vehicle is stationary, then by increasing the engine speed progressively transmit more and more power to the wheels. In addition, automatic transmission should provide a sufficient range of gear ratios to accommodate all possible vehicle loads and speeds. The gear ratio is the ratio of the

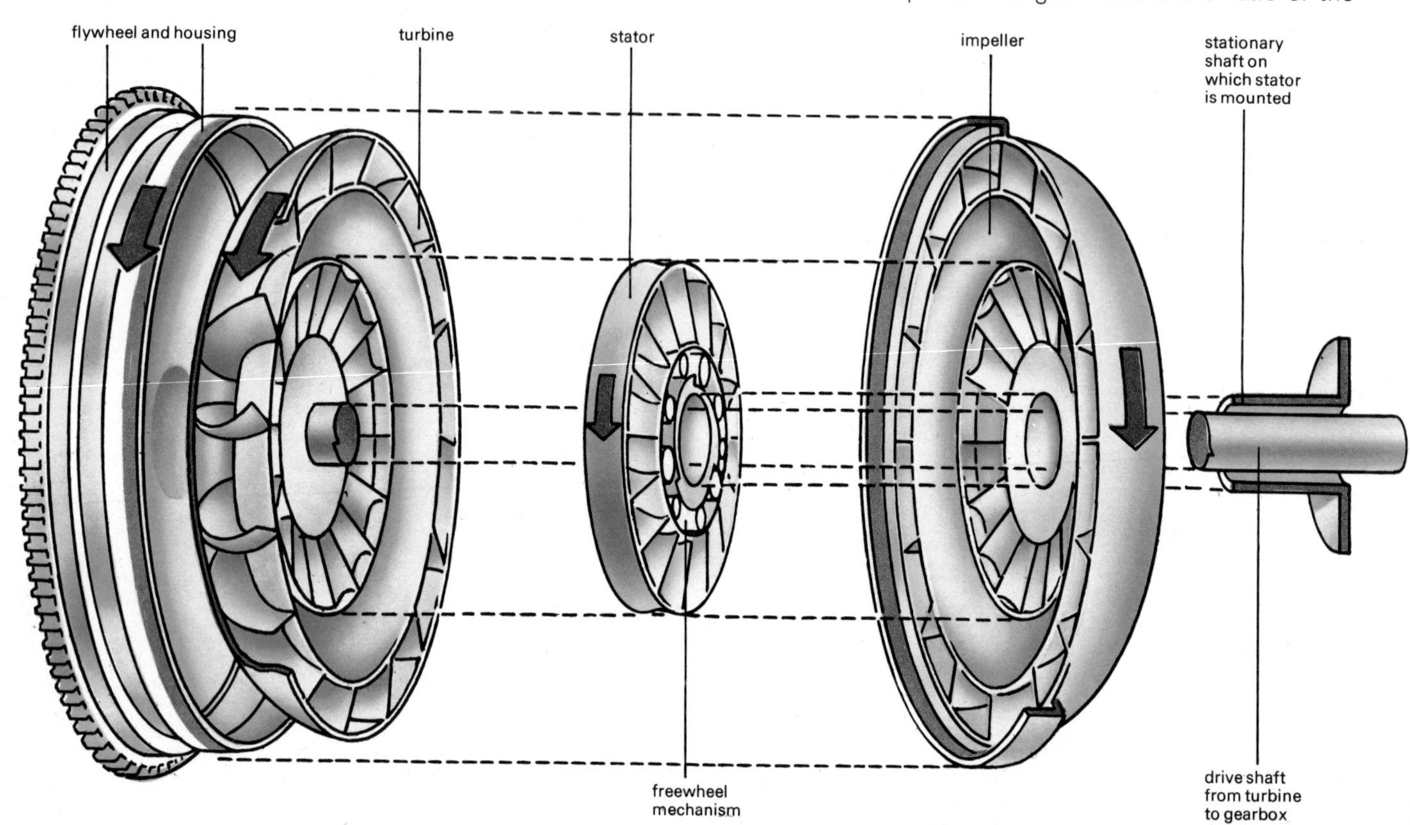

Above: a torque converter works like two fans facing each other: spinning one makes the other revolve. The casing is full of oil, which is thrown outside by the impeller and returns through the turbine, carrying it around. At low speeds the turbine slips, but the stator, turning on a one-way freewheel, angles the oil flow to minimize the slippage.
Right: a Borg-Warner model 45 automatic transmission, which is designed for smaller cars and has four ratios, achieved with three sets of planetary gears and five multi-plate clutches. The cutaway torque converter is at the left.
Far right: the Variamatic transmission, which transmits power by means of V-belts.

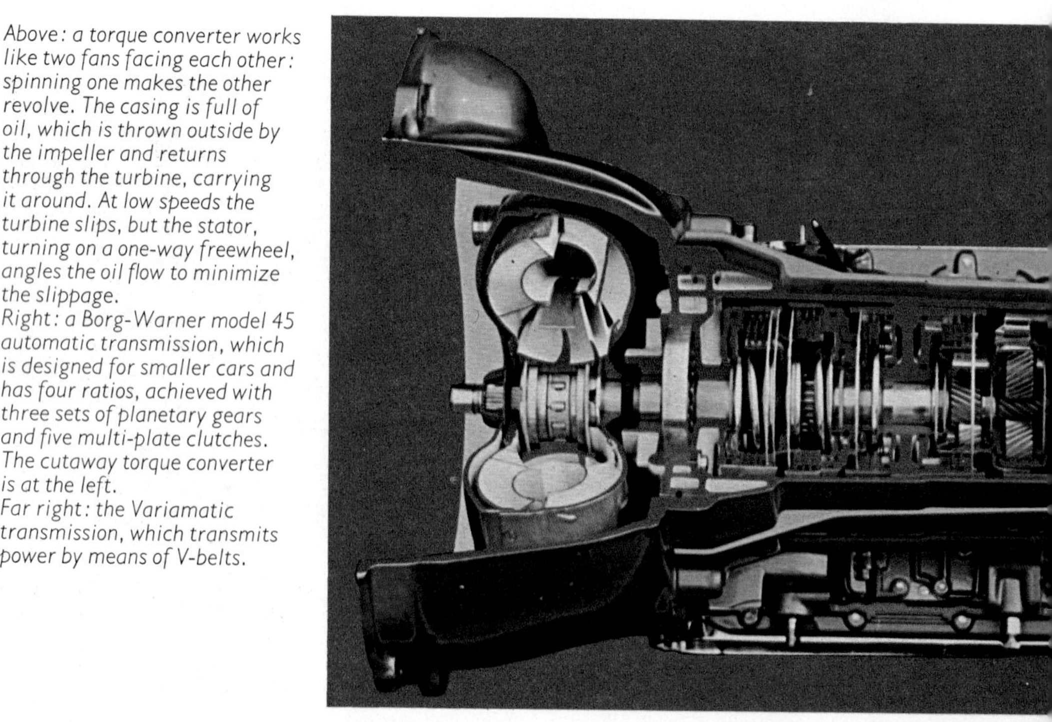

engine speed (measured in revolutions per minute) to the speed of the output shaft from the automatic transmission (measured in the same units).

Normally this range is provided by a number of different ratios (usually three or four) but one system has been developed which provides a continuously variable gear ratio. This can be achieved by a system of friction belts running over pulleys of varying diameter, provided that the amount of power to be transmitted is not great. Such schemes have been tried at various times since the beginning of the century, but the Variomatic is the only one now in use, in the Daf car, built in Holland. In this, a centrifugal friction clutch forms the coupling mechanism and takes up the drive when the engine is accelerated. The final drive is provided by belts of wedge section passing over pulleys that expand or contract to alter their effective diameters.

The pulleys are controlled by centrifugal governors, built into the driving pulleys and sensitive to engine speed, and by vacuum servomechanisms which respond to the degree of vacuum in the engine inlet manifold. These two devices are balanced against each other so as to achieve gearing which is a reasonable compromise between performance and economy. When maximum acceleration is desired, the accelerator pedal is fully depressed. This floods the inlet manifold with a mixture of fuel and air, reducing the vacuum (increasing the pressure) and cancelling the vacuum servomechanism. The transmission therefore responds to engine speed alone and assumes a gear ratio which maintains the engine at the speed which produces maximum torque. When eventually the highest available gear ratio is reached, further acceleration is accompanied by a rise in engine speed until the car reaches maximum speed.

Such a mechanism is limited in its capacity, and is at present suitable only for cars of fairly low power. In such applications its high mechanical efficiency is important. Mechanical efficiency is the ratio of the work output of a machine to the work input and for the Variomatic is in excess of 90% over most of the working range, reaching a maximum of 94%.

Vehicles with larger engines use an automatic gearbox offering a series of gear ratios which are selected and engaged automatically by clutches energized hydraulically or (less commonly) electrically. The controls for these are still governed by the inlet manifold vacuum and engine speed, with additional overriding controls linked to the accelerator pedal and the gearbox output shaft, which gives a measure of the vehicle's speed.

The gears themselves are almost invariably planetary or epicyclic sets in constant mesh. The term 'planetary' is sometimes used for epicyclic gears because the system is made up of an inner sun gearwheel which is in mesh with two small planet gears. These planet gears rotate about the sun wheel, and are themselves surrounded by a toothed outer ring. The various gear ratios are produced by locking one or other of the gear elements, for example the planet carrier, by means of a clutch. In an automatic gearbox there are several of these systems interconnected to provide the necessary gear combinations.

Epicyclic gearboxes of this kind are invariably associated with some form of fluid drive or hydrokinetic coupling—a form of clutch in which a fluid (oil) transmits torque from the input member (called impeller or pump) driven by the engine, to the output member or turbine which drives the gearbox.

The simplest kind, first used in 1929 by Daimler in conjunction with a preselector gearbox and featured in Rolls-Royce and Mercedes-Benz cars until quite recently, is the fluid coupling. Once known as the fluid flywheel, it consists essentially of two radially-vaned saucers face to face, separated slightly to avoid direct contact. Oil is flung centrifugally from the pump to the turbine, which makes it rotate. Thus torque is transmitted without any rigid mechanical connection between the driving and the driven members, which can therefore slip relative to each other. (Slip refers to any difference in the rotational speeds between the pump and the turbine). The important feature of the fluid coupling is that the output and input torques are always equal. Any difference between input and output speeds is a measure of the slip taking place and of the efficiency of the device. At best, a fluid coupling reaches about 97% efficiency; but until it has reached coupling point (that is, until the slip has been reduced to the minimal 3%) it is very inefficient. Careful matching of the coupling to the engine is essential: if it is to carry the full engine torque at high efficiency and low slip, it must have a torque capacity much greater than that which the engine can deliver. Otherwise, slip would be severe and excessive heat would be generated, eventually boiling the fluid.

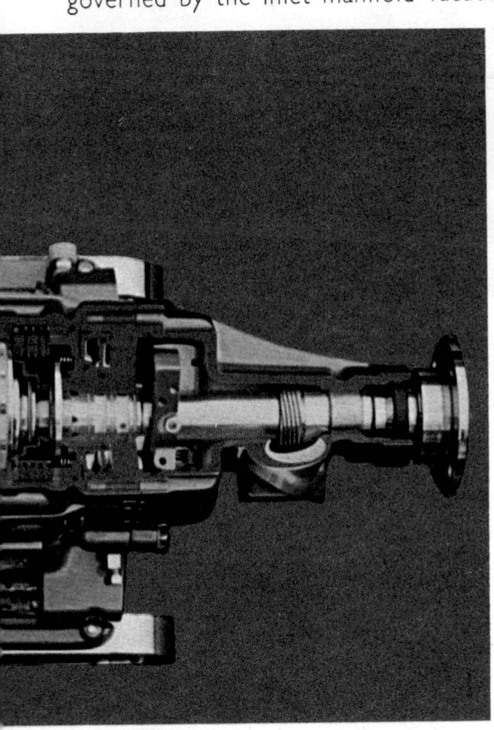

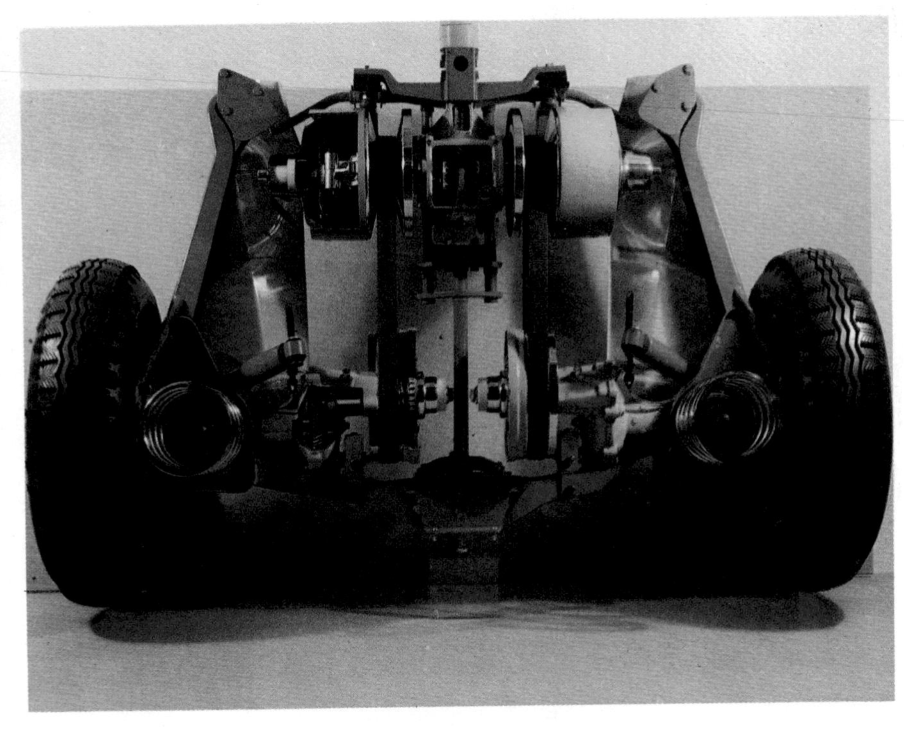

A more complex form of hydrokinetic coupling is the fluid converter, so called because it converts or multiplies the input torque. To achieve this, a third element is added between the pump and the turbine, and the blade forms are curved instead of straight. The third element, the reactor or stator, is fixed so that fluid flow from the pump is diverted to strike the turbine blading at a more effective angle, thus amplifying the torque applied to the turbine. Actually the torque on the turbine equals the sum of the torques on pump and reactor, which means that the torque applied to the gearbox can be several times higher than that produced by the engine.

This sort of converter is used in construction machinery, but is unsuitable for cars: it is most efficient at about 40% slip, and is grossly inefficient at the extreme ends of the slip range, where all the engine power is dissipated in heat. The answer to this problem is to mount the reactor on a free-wheel or over-running clutch; then, when coupling point is reached, the reactor is free to move in the same direction as the turbine but is always locked against rotation in the opposite direction. This system is used in most cars with automatic transmission.

At coupling point, therefore, the reactor automatically removes itself from the circuit, in which it can no longer play an effective part. The mechanism then acts purely as a fluid coupling, with no torque multiplication or conversion. Hence it is called a converter coupling in the strict terminology, though the motor industry casually refers to it as a torque converter. In effect, the device combines the best characteristics of fluid converter and fluid coupling, but at some cost of efficiency to each. The transition from converter to coupling after reaching coupling point is slow, since the curved blading necessary for converter operation is not entirely suitable for coupling operation. Furthermore, the coupling point is normally reached when input and output torques are equal, but at that point the output speed will still be about 10% less than the input speed.

The full load efficiency of a converter coupling is usually below 90% in the converter range, rising to about 95% in the coupling range. This efficiency is directly related to the product of the torque multiplication ratio and the slip ratio. It follows that when driven gently at low speeds, such a device is of poor efficiency; engine efficiency, however, is low when the load is light and the input speed is high (a situation which favours coupling efficiency), so great care has to be taken in setting the central mechanism to select the most appropriate gear ratio for the circumstances. Careful matching of the converter to the engine is also essential.

The combination of a converter coupling (which slips enough to allow the engine to idle while the car is stationary, and grips enough to transmit most of the engine's power at high speeds) with an epicyclic automatic gearbox is almost

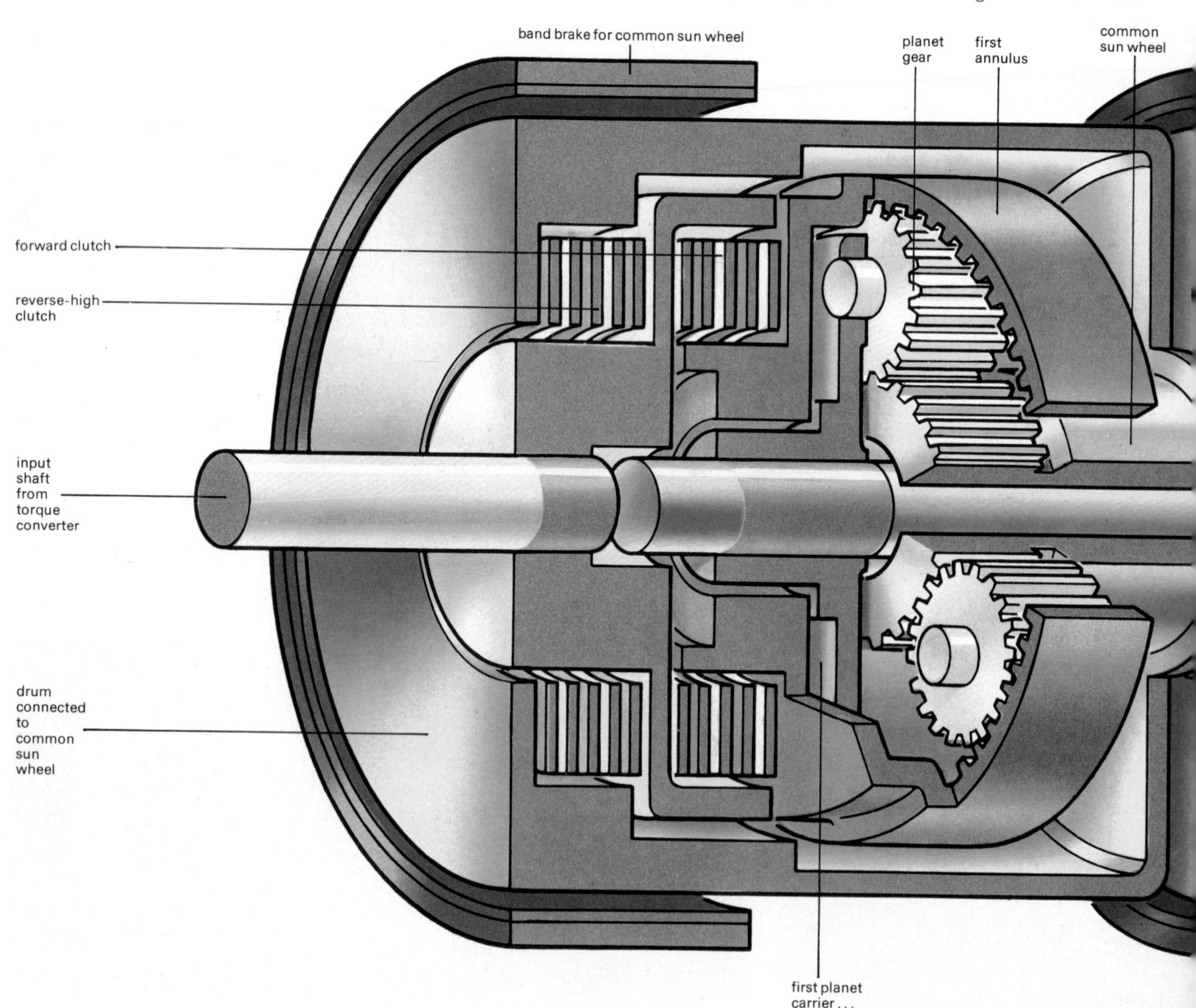

universal in the many cars and other vehicles now equipped with automatic transmission. Most developments have been concerned with simplifying (and cheapening) the manufacture of the transmission, together with a number of refinements in the control system. Of the latter, the most important are those which give the driver some means of inhibiting the gearchanges that would otherwise take place automatically, so that he can hold the transmission in a low gear to improve the car's acceleration. Much attention is also being given to making gearchanges as smooth as possible.

There have been a number of variations on this basic arrangement. The General Motors Hydramatic transmission has appeared in several versions: some had two fluid couplings, one always being full and operational while the other was drained or filled with fluid according to the gear ratio engaged. Another version had a single coupling with a free-running stator built in and anchored to the gearbox output shaft rather than to the casing: this gave some torque multiplication, about 30%, but minimized drag losses. Varying the angle of the stator blades has also been tried with some success in the GM Super Turbine 300, and in an earlier Buick transmission featuring a five-element, twin-turbine converter. These refinements, and others such as a lock-up clutch to eliminate slip when in the coupling position, have not been found to justify their cost. Others, such as the substitution of bevel-gear differentials for the conventional

Below left: schematic diagram of a three-speed automatic transmission. It has two sets of planetary gears, two multi-plate clutches and two band brakes. Parts which are linked and turn together are shown in the same colour.
Directly below: a planetary gear set has three main parts: a central sun wheel, an outer annulus (inward facing) gear, and planet wheels mounted in a carrier. In the diagrams, moving clockwise from the top, (1) if the carrier is held and the sun wheel turned, the annulus travels slowly in the opposite direction; (2) if the sun wheel is locked to the annulus, the unit revolves as one piece; (3) if the sun wheel is locked and the carrier turned, the annulus revolves slowly in the same direction.

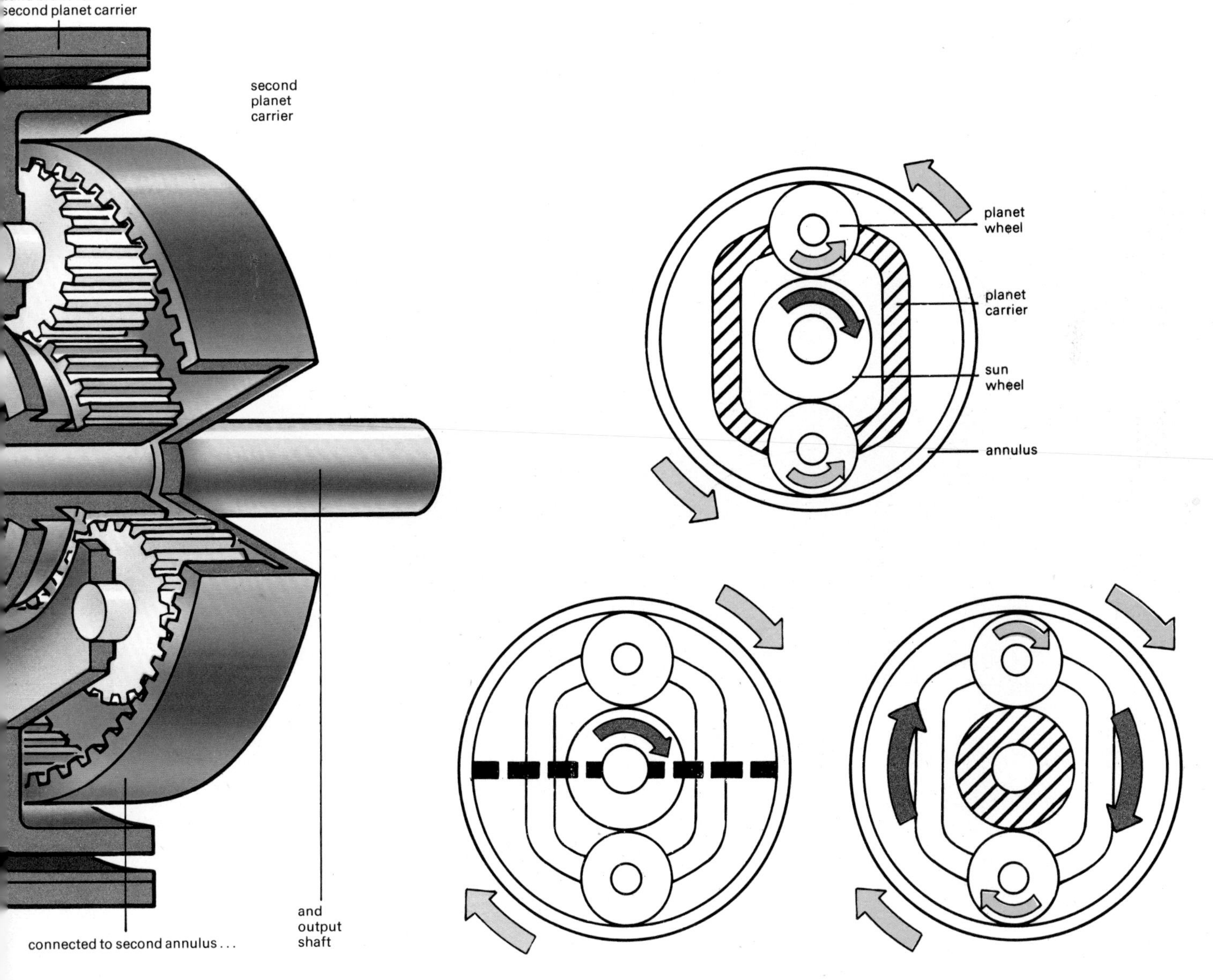

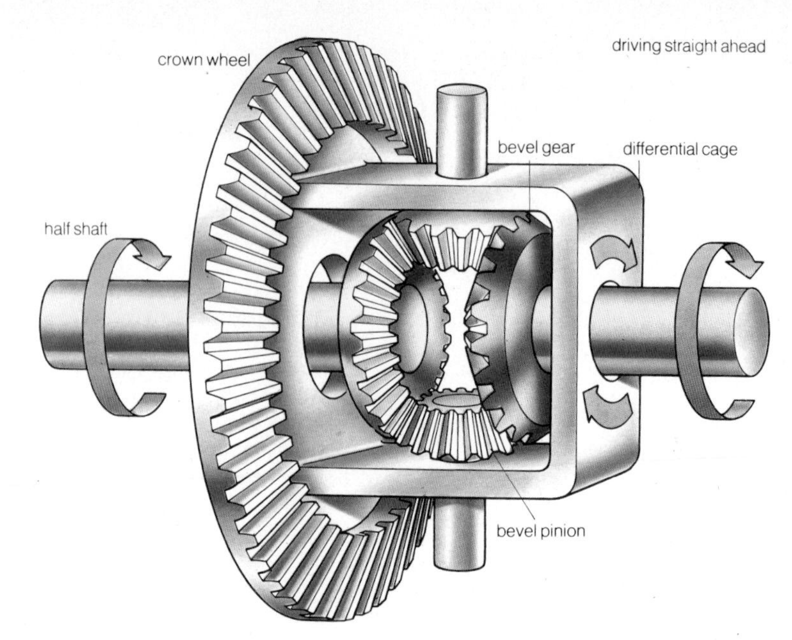

Below: a cutaway view of a truck axle. The large gear at the right is the crown gear of the differential; one of the driven wheels of the truck is mounted on the flange with the holes in it. Power from the engine is transmitted through the transmission shaft, which protrudes from the unit above the crown gear and has half a universal joint fitted to it, to the unit in the green casing at the rear, which in turn drives the pinion in the differential. The unit in the green casing is an electric two-speed gearbox, which doubles the gear ratios available.

epicyclic gear trains in the British gearbox made by Automotive Products, have been more successful.

The differential The differential is a gear assembly in a motor vehicle which allows the propeller shaft, or drive shaft, to turn the driven wheels at different speeds when the vehicle is going around a curve. When a vehicle goes around a curve, the wheel on the inside of the curve travels less distance than the other, and so must turn more slowly, for safety in handling and to keep tyre wear to a minimum. A four-wheel-drive vehicle, such as a Jeep or a Land-Rover, has two differentials. For maximum traction, a four-wheel-drive vehicle has been designed with three differentials, separating the front wheels, the rear wheels and the front from the rear, allowing each wheel to turn at its own speed under power. (The only car which does not have a differential is the Daf car, built in Holland, which has a belt drive system allowing slippage of the belt on the pulleys.)

The differential is encased in a housing, which is located on most cars (having rear wheel drive) in the middle of the rear axles between the wheels. (It is sometimes called the 'cabbage head' because of its bulbous appearance.) The drive shaft enters the housing in the front and one axle enters at each side. A pinion gear, which is splined into the end of the drive shaft, turns a bevelled crown–gear which is fastened onto the end of one of the axles. An assembly of four small bevelled gears (two pinions and two star gears) is bolted to the crown gear and turns with it. The other axle is driven by the small pinion gear opposite the crown gear. The assembly drives both axles at the same speed when the vehicle is being driven in a straight line, but allows the axle opposite the

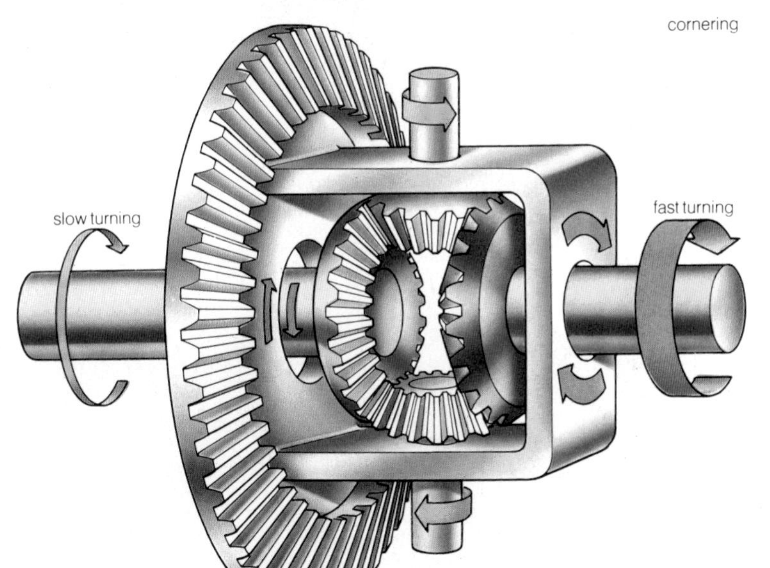

Far left: on a straight road, the main shafts rotate at the same speed, the bevel pinions turning with the cage but remaining stationary about their own axes. Near left: in cornering, one shaft turns faster and one slower than the cage (yellow arrows). The bevel pinions allow this by turning on their own axes inside the cage.

crown gear to turn slower or faster, as required.

Some units are designed to give a limited-slip or slip-lock differential, to equalize power between the wheels on a slippery or a soft road surface, providing safe handling and minimizing the likelihood of getting stuck in snow or soft earth.

The gear ratio (ratio of the number of teeth on one gear to the number of teeth on the other) between the crown gear and the pinion gear is one of the factors that determines the performance characteristics of the car, such as acceleration and top speed.

Early cars had pinion and crown gears with straight teeth on them, which resulted in noisy operation of the differential and allowed play in the gear teeth, causing undue wear. Today the pinion and crown gears are helical gears, which means that the toothed surfaces are bevelled and the teeth themselves are curved. This design eliminates play between the teeth, because as the gears spin together one tooth is in full contact before the previous tooth leaves. A properly constructed differential should last the life of the car without any maintenance at all.

In order to produce a particularly quiet differential, the pinion and crown gears are lapped together in a lapping machine which duplicates the operating conditions of the completed differential. After lapping, the two gears are kept together as a set. They are inspected together in a machine in a quiet room which determines the exact thickness of shims (sheet metal discs used to ensure a close fit) required in the assembly to ensure quiet operation; then they go to the differential assembly line. All the gears in the system are installed against roller bearings, the proper shimming is installed; then the unit is test-run, filled with a heavy oil and sealed. Quiet operation of the differential is essential in a vehicle with unit-body construction, as opposed to a separate body bolted to a frame, because noise from the differential will be transmitted by the body itself.

Suspension systems Modern car suspension systems have to do considerably more than just cushion the occupants from irregularities in the road surface: they have to ensure that the wheels stay in contact with the road to give adequate grip for accelerating, braking and cornering. The ancient system used on carts of a solid beam axle rigidly attached to the chassis with a wheel at each end would not only be uncomfortable but unstable at anything more than a walking pace.

The first cars drew heavily on carriage practice, and were thus simply sprung on leaf springs, with narrow tyres, and an Ackermann type steering gear. The sophisticated suspension systems of today are the result of development which still continues.

The tyre alone would not provide much comfort; between the wheel and the body it is necessary to have springs. Some carriages had the body suspended by straps from the chassis extremities, but the semi-elliptical multi-leaf spring was an early development. Leaf springs are still widely used on cars, particularly on the rear axles. One of the advantages is that the inter-leaf friction provides a bit of damping. The leaf spring also performs the function, on a rear axle, of locating the axle, holding it in a set position except for allowing an up-and-down motion.

The other most common type of spring is the coil spring, a helically-wound rod of spring steel, which operates in torsion rather than bending. Closely allied to the coil spring is the torsion bar, which is little more than a straightened coil: one end is clamped to the chassis and the other to the suspension, so that as the latter rotates about its fittings it twists the rod, whose tension then helps to stabilize the suspension.

If a car had only springs without any damping, a road shock would set it bouncing, and theoretically it could bounce for a very long time. This would not only be disturbing to the occupants but would be unsafe, because it might result in the wheels leaving the road, so dampers or shock absorbers are necessary. Early 'shocks' were often of the simple friction type, consisting of a pivoted arm attached to the axle so that its movement turned friction discs like a clutch, whose resistance provided the damping. Nowadays shock absorbers are usually hydraulic and telescopic, consisting of a piston inside a sealed cylinder, one attached to the chassis and the other to the axle. Holes in the piston allow fluid (oil) in the cylinder to leak from one side of the piston to the other, absorbing energy in the process. Some designs use a compressible gas rather than a hydraulic fluid, allowing the shock absorber to perform a spring function as well as absorbing shock.

There are also springs which take advantage of the elasticity of other substances than spring steel. Rubber and air (or some other gas) have their advantages. For example, rubber can be varied in shape and composition to allow it to work in shear as well as compression (shear is a strain or stress in which parallel planes of a substance remain parallel but are allowed to move parallel to each other). Also, owing to its hysteresis properties, rubber can be effectively self-damping. Hysteresis is the retardation of an effect beyond the cause of it: high hysteresis rubber is rubber with less bounce. British Leyland Minis use a simple and elegant form of rubber spring.

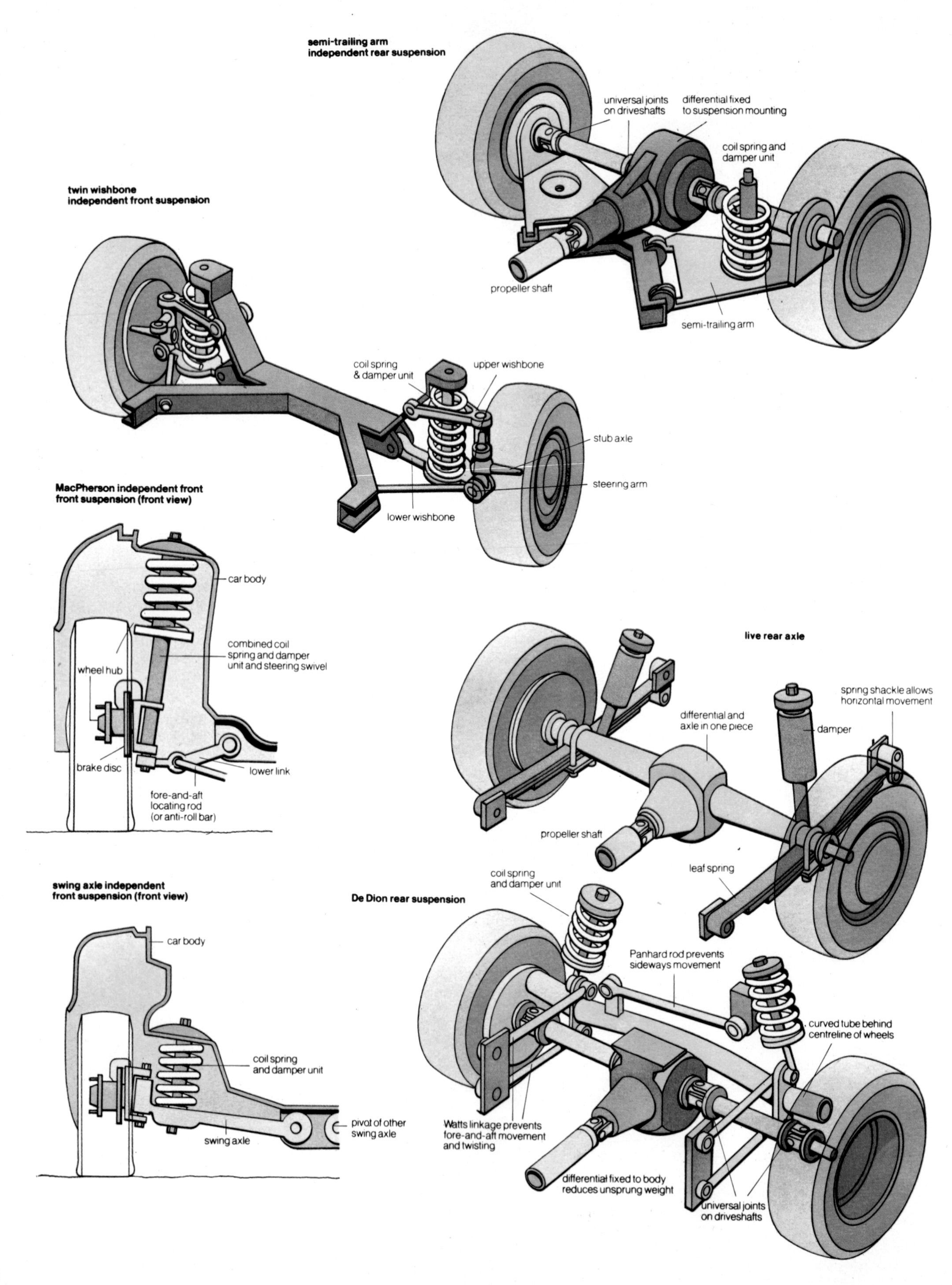

semi-trailing arm
independent rear suspension
universal joints
on driveshafts
differential fixed
to suspension mounting
coil spring and
damper unit
propeller shaft
semi-trailing arm
twin wishbone
independent front suspension
coil spring
& damper unit
upper wishbone
stub axle
steering arm
lower wishbone
MacPherson independent front
front suspension (front view)
car body
combined coil
spring and damper
unit and steering swivel
wheel hub
brake disc
lower link
fore-and-aft
locating rod
(or anti-roll bar)
live rear axle
spring shackle allows
horizontal movement
differential and
axle in one piece
damper
propeller shaft
leaf spring
swing axle independent
front suspension (front view)
car body
coil spring
and damper unit
swing axle
pivot of other
swing axle
De Dion rear suspension
coil spring
and damper unit
Panhard rod prevents
sideways movement
curved tube behind
centreline of wheels
Watts linkage prevents
fore-and-aft movement
and twisting
differential fixed to body
reduces unsprung weight
universal joints
on driveshafts

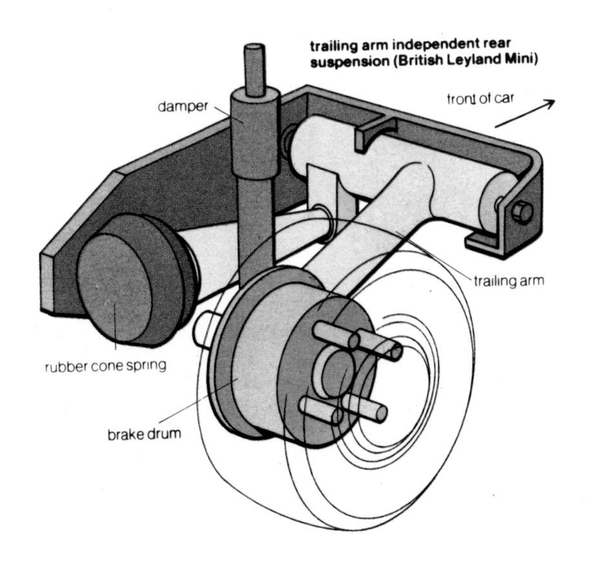

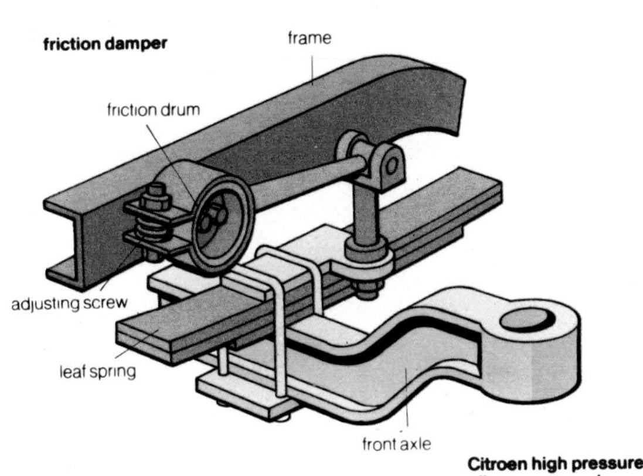

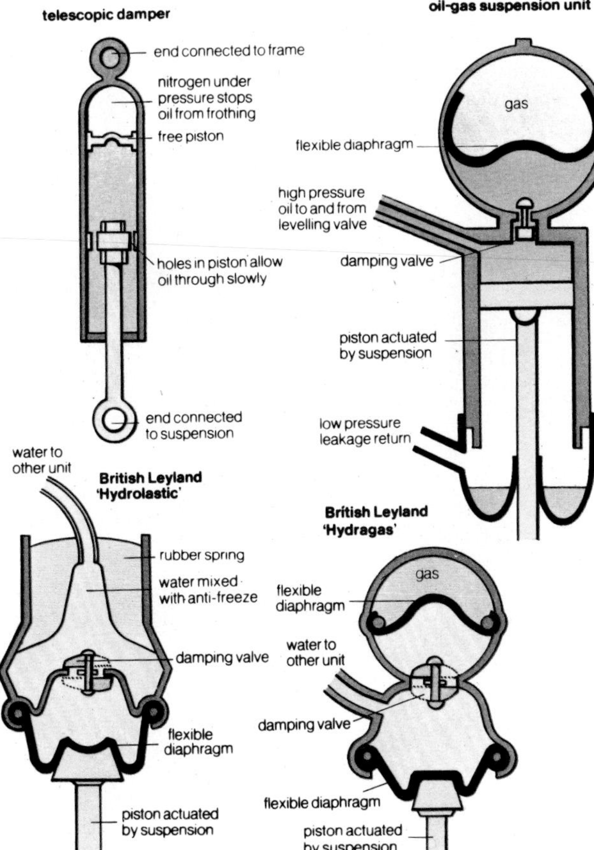

Left: suspension types. The ideals are to keep the wheels upright (best done by twin wishbone and MacPherson front, and live and De Dion rear systems), low unsprung weight (twin wishbone front; trailing, semi-trailing and De Dion rear) and simplicity (swing axle front; live, trailing and semi-trailing rear).
On this page, lower: gas pressure damper, and fluid 'spring' units.

Air springs come in two basic types: high or low pressure. They resemble balloons in principle, which means that in general they act simply as springs and cannot be used to locate any part of the suspension system. Their stiffness comes from three basic parameters: the internal pressure, the load-carrying area, and the volume. A low-pressure spring will have perhaps 70 psi (48 bar) and a high-pressure device up to ten times as much, which means that it can be much more compact.

An example of the high-pressure spring is that fitted to Citroëns. It is a metal sphere, divided in half by a flexible diaphragm, on one side of which is a gas, and on the other oil. A piston and connecting rod assembly, attached to the wheel by a system of levers and arms, acts on the oil and thus compresses the gas: as the wheel oscillates the piston moves up and down, alternately compressing the gas and allowing it to expand, providing the springing medium. By including a source of high-pressure oil it is possible to provide a ride of a given height, pumping or bleeding oil from the system as necessary. The British Leyland 'Hydragas' system is similar, but the fluid used is a mixture of water and anti-freeze, the piston takes the form of a flexible diaphragm, and the hydraulics are connected front-to-rear instead of to a central reservoir. This system is cheaper to build than the Citroën, requiring no extra pump and circuitry, but does not give self-levelling.

Air springs in general give a soft ride, do not require any extra damping, and by means of the fore-and-aft interconnection can be made quite stiff in roll and pitch. The 'Hydrolastic' system is basically a rubber spring as in the early Minis, but with front-to-rear fluid interconnection.

The tyres, springs and so on provide the carrying and comfort function of the suspension system; the axles and other linkage, however, have to do with handling and road-holding, and thus with safety, for they control the movement of the body with respect to the wheel and the road (the suspension 'geometry'). All these functions—comfort, capacity, understeer and oversteer, steering response and so forth—are the result of compromise, and the ultimate handling characteristic of the car depends upon subjective decisions at the development stage.

The ancient solid-beam axle is still in wide use, both in front (common on trucks) and in the rear. In the former case, stub axles swivelling on kingpins allow steering to take place. Leaf springs in conjunction with a beam axle at the rear give fore-and-aft and sideways location, and this is still standard on many mass-produced cars. Further location (essential if coil springs are used) can be provided by trailing arms, A-brackets, Panhard rods, Watts linkage and so forth; the more precise the location the more precise the handling of the car. Disadvantages of the beam axle are that a bump on

one wheel tends to affect the opposite wheel, and that they tend to be heavy, so increasing the ratio of sprung to unsprung weight, affecting comfort detrimentally, a sort of 'tail wagging the dog' effect. On the other hand, the beam axle keeps the wheels rigidly upright, providing better cornering.

The swing axle is essentially a beam axle divided in half. VW used it on the Beetle for years, and the British division of Chrysler still use it at the front of the Imp. It has an unfortunate 'jacking-up' effect: when cornering, the body rolls around the heavily laden outside wheel, which then adopts the wrong angle to the road. Thus cornering power is lost and the swing axle can be tricky at high speeds.

The most common independent suspension is the twin wishbone layout, which because of its configuration is compact, allows large vertical wheel movements, gives good camber characteristics and a tight steering lock (small turning radius). Similar characteristics apply to the MacPherson strut type of suspension, but this does away with the top wishbone, using the damper and coil unit as a triangulating locating member.

Some cars (including VW) use trailing-arm front suspension, but this means that the front wheels adopt the same angle to the road as the body, which gives undesirable camber change.

Independent rear suspensions follow similar principles to those at the front; the most common is the semi-trailing arm type, which is geometrically a combination of swing axles and trailing arms: such layouts, if well-designed, give acceptable camber change, are compact and are lighter than a beam axle. The De Dion rear suspension system uses a relatively light hollow tube to connect the wheels, mounting the differential directly on to the chassis, so that the desirable geometric characteristics of the beam axle are kept without the weight.

The brakes In the internally-expanding type of drum brake, two brake shoes on a fixed mounting are pushed against the inside of a rotating drum to create the required braking force. In some cases the shoes are operated mechanically but

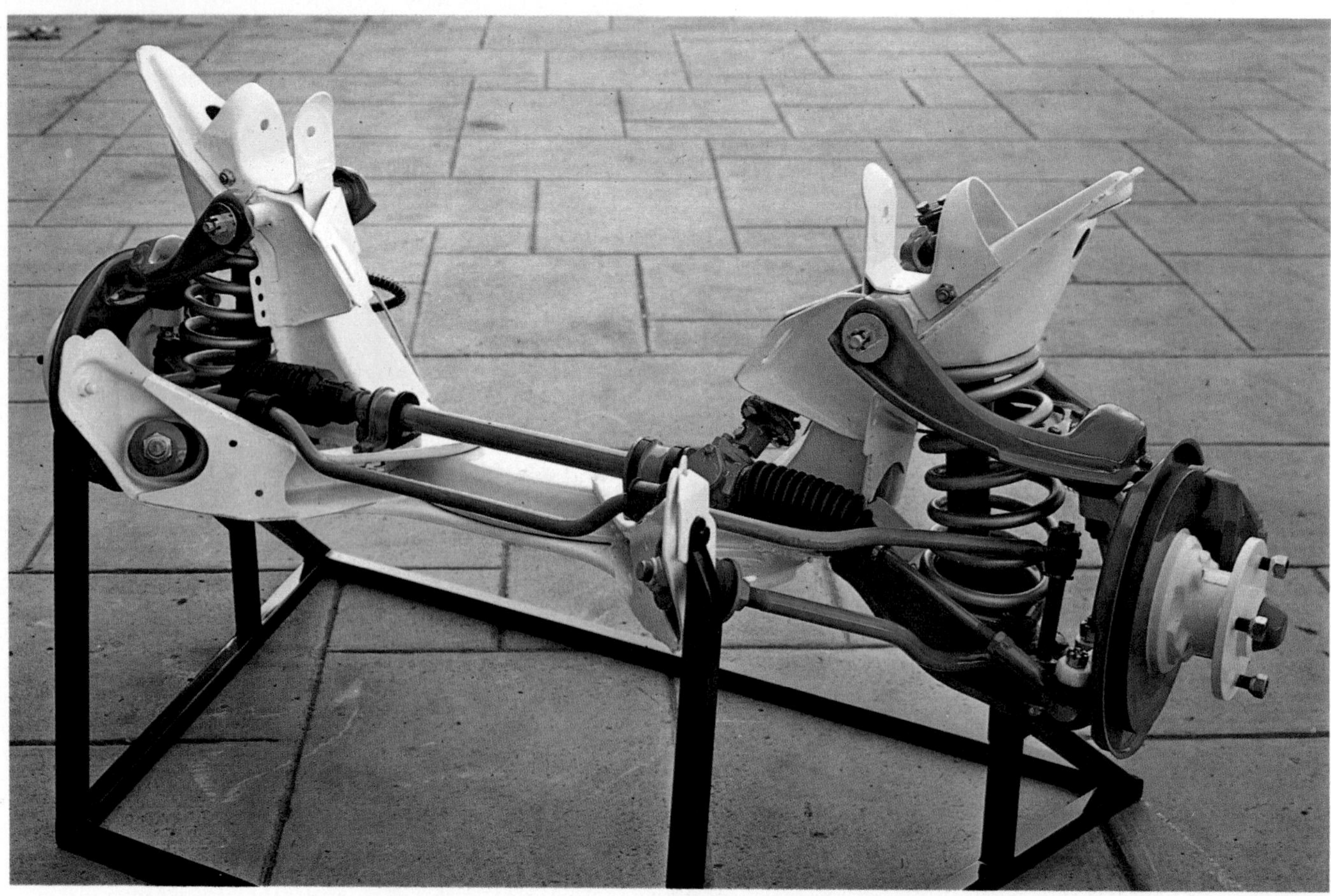

nowadays hydraulic or compressed air systems are often used for applying the brakes. Brakes of this type are widely used on trucks, buses and motor cycles; on modern cars they are usually found at the rear only.

The metal brake drum dissipates heat quite well, particularly if there are fins on its outer surface to increase the area for heat dissipation, but the internal forces created by the shoes can cause distortion which leads to cracking and failure if temperatures become excessive. Self-energizing types give powerful braking but their effectiveness falls off rapidly at high operating temperatures through what is called 'fade', where the higher temperature reduces the lining friction level. Drum brakes are also particularly liable to lose their effectiveness for a short period after immersion in water—after driving through a flood or a ford it is necessary to proceed cautiously and use the brakes frequently but lightly until they dry out.

On a drum brake of the type shown, one shoe is self-energizing and the other is not. If the drum rotation is clockwise, it tends to pull the shoe on the right harder against the inner surface, so that the braking effect is increased; this shoe is called the leading shoe. The other shoe is pushed off by the drum and its braking effect is reduced; this is called the trailing shoe. The leading shoe will wear faster because it does more work than the trailing shoe.

To gain particular advantages of power or wear resistance, drum brakes have also been designed with two leading shoes, with two trailing shoes and with three or even four shoes. The duo-servo type uses two leading shoes linked together in such a way that a very high output indeed is obtained.

The power of car brakes must always be matched to the load carried by each axle and there is also the effect of weight transfer from rear axle to front axle during normal braking. Cars usually, therefore, have more powerful brakes at the front, and, because more energy has to be dissipated, these are often of disc type rather than drum.

The disc brake consists essentially of a revolving disc which can be gripped between two brake pads. The caliper

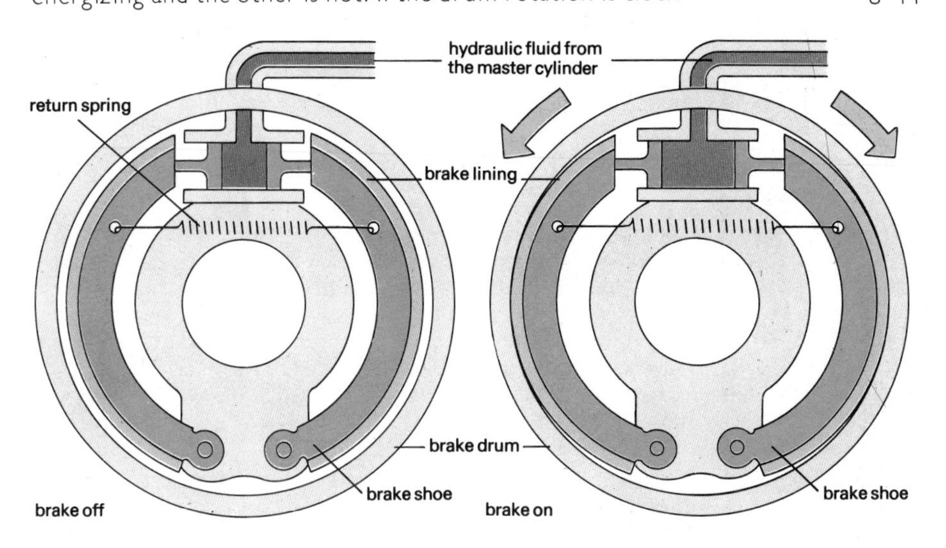

Opposite page, top: a test of four-wheel brakes on a Rickenbacker car in Los Angeles in the 1920s. A car dealer stopped the car halfway down the steps.
Opposite page, below: twin wishbone front suspension of the Ford Cortina; this is a left-hand drive model. The rack and pinion steering and anti-roll bar can be seen.
On this page: an internally expanding brake drum system. When the brake pedal is depressed the piston in the master cylinder transmits increased pressure throughout the entire system.

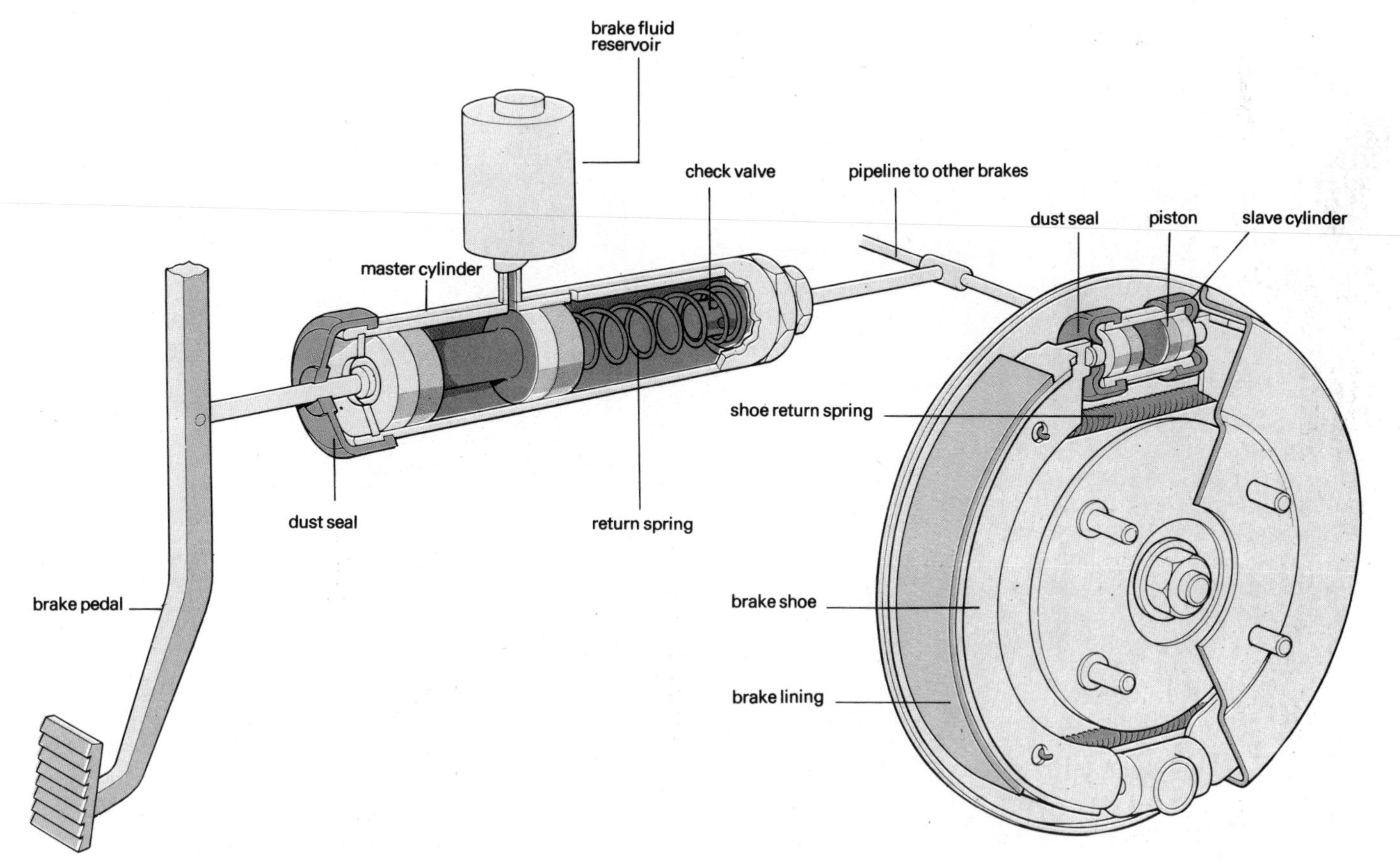

type brake commonly found on bicycles is a familiar example, but disc brakes for cars are mostly hydraulically operated and the pads cover between one sixth and one ninth of the swept area of the disc. Because the disc is less subject to distortion than a brake drum it can run at much higher temperatures without ill effect; the disc brake can therefore be used for much heavier duty than a drum brake provided an appropriate grade of lining is used.

Disc brakes are now used very widely on the front axle of cars where, within a given size of wheel, they dissipate about twice as much energy as the rear brakes. High performance cars sometimes have disc brakes on both axles. A rather different design of disc brake, having linings equal in size to the full swept area of the discs, is sometimes found on tractors.

All cars need a parking brake which is mechanically operated. Although this commonly acts at the rear, in some cases it acts at the front wheels. When the parking brake is on a drum braked axle it usually operates the same shoes that are controlled by the foot brake system; when disc brakes are involved, the parking brake often has its own pads in an attachment to the main brakes.

The materials of which the rubbing surface of brakes are made must give a good level of friction or excessive force would be necessary. At the same time, the life of these parts must be acceptable for the class of application. Experience has shown that the best results are obtained when the moving part is of metal and the stationary parts are lined with a composition friction material which wears much faster than the metal but can readily be replaced.

For general use, such as on vehicle brakes, cast iron has been found to be the metal surface most suitable from all points of view at reasonable cost for both drum and disc brakes. Drums can be made of a cast iron lining with a light alloy body for improved heat conductivity but these are expensive; discs can be made with internal air passages.

Brake linings for general use are made of a carefully chosen mixture of asbestos fibre, metallic particles and non-metallic ingredients bonded together with a temperature resistant synthetic resin.

Power brakes The ordinary hydraulic brake system of a car consists of a master cylinder actuated by the brake pedal, and linked by tubing to slave cylinders at each wheel brake. In the case of conventional drum brakes, there is a total of eight pistons actuated by the single piston in the master cylinder.

The force that must be applied to the brake shoes to stop the car is considerably greater than that exerted by the driver's foot, and this is catered for by building mechanical advantage into the system. Suppose that the brake pedal is pivoted so that a certain force applied by the foot exerts three times that force on the piston in the master cylinder (the foot, of course, has to move three times as far as the piston, by the laws that apply to any lever). If the master piston is $\frac{3}{4}$ inch in diameter, each of the front brake pistons is $1\frac{1}{2}$ inch in diameter, and each of the rear ones 1 inch (front brakes are always larger, since braking throws the weight of the car forward), then the force from the driver's foot will be amplified 48 times at the front and 21 times at the rear, a total of 69. The foot has to move 69 times as far as any of the pistons, but since the brake shoes are already more or less touching the drums, this gives an acceptable pedal travel of a few inches.

Disc brakes, however, are not self-energizing, and need considerably more force to apply them. They also have larger pistons than drum brakes, and may have four per brake instead of two. This would result in an unacceptably long pedal travel, or a very high foot pressure, so on nearly all cars they are power assisted. Large drum brakes, for example on buses and trucks, also need assistance.

Power assisted brakes were originally designed for aircraft, and were gradually introduced on road vehicles from the 1940s on. Nowadays systems on vehicles of all types are

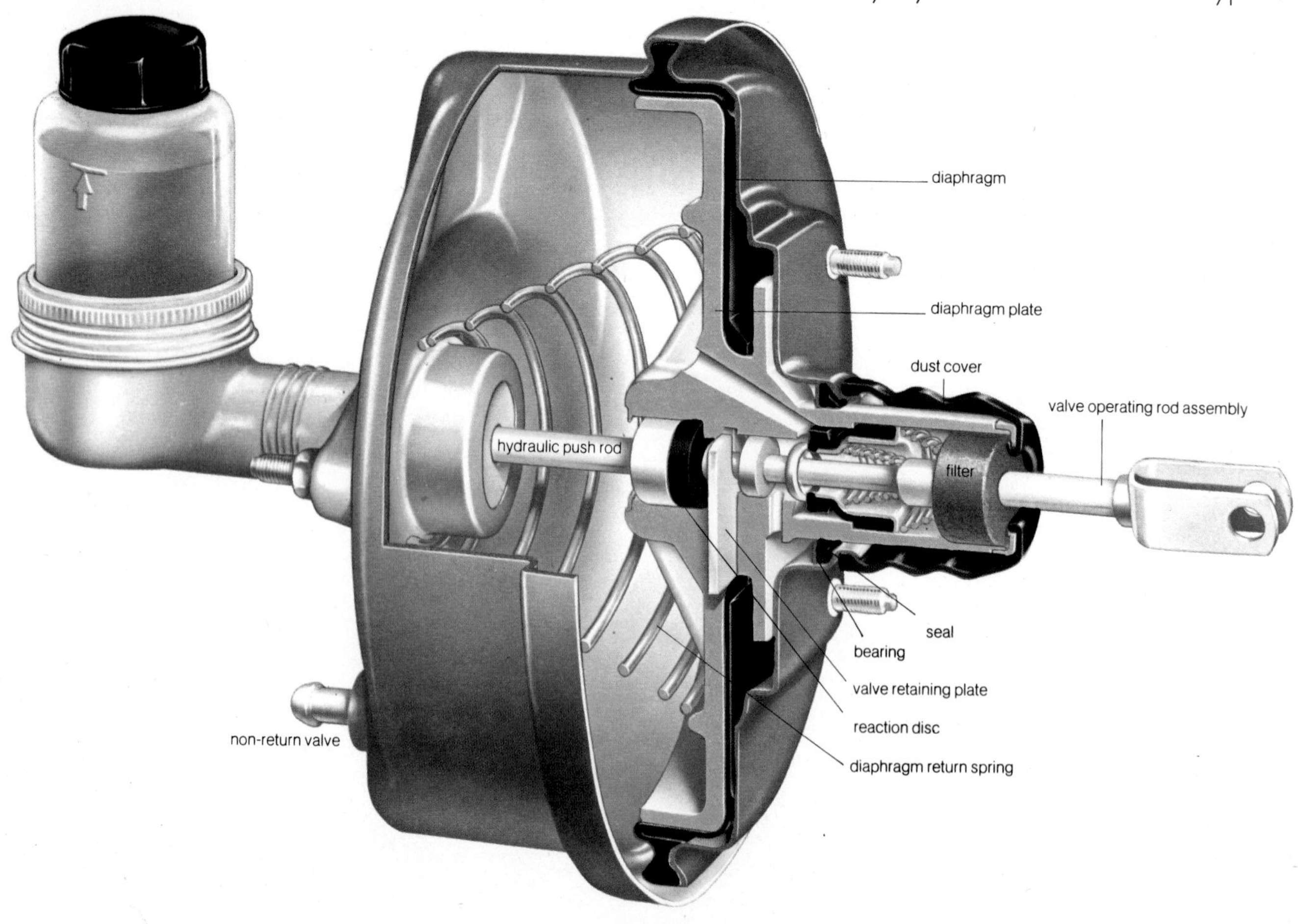

designed so that with power assistance, no more pressure on the brake pedal is required to stop a bus than a small car.

Most power brakes on private cars are vacuum-servo assisted. A normal petrol [gasoline] engine has a partial vacuum, about 10 psi (0.7 bar) below air pressure, in the inlet manifold while it is running. This can be tapped and used as a source of power; it is particularly useful for brakes because the vacuum is at its highest when the driver suddenly removes his foot from the accelerator, as he would when braking. On diesel engines, this vacuum is not available, so a separate vacuum pump driven from the engine is used.

The servo contains a large rubber diaphragm in a chamber connected to the inlet manifold on both sides. One side is directly connected, and the other through an air control valve opened and shut by hydraulic pressure from the master cylinder. The diaphram is pressed towards this side by a large spring.

Hydraulic fluid from the master cylinder enters a slave cylinder with a slave piston in it. There is a hole in the slave piston, and if the servo were not working (for example, if the engine stalled) pressure from the master cylinder would pass through this hole direct to the brakes.

A passage leads off the slave cylinder 'upstream' of the slave piston to a small auxiliary cylinder containing a piston linked to the air control valve. Pressure from the master cylinder moves this piston, thus opening the valve, which admits air from the outside to one side of the diaphragm. Since this air is at a higher pressure than the inlet manifold vacuum, it forces the diaphragm across its chamber, against the spring. A rod connected to the diaphragm moves against the slave piston, closing the hole in it and forcing it down the slave cylinder so that it applies the brakes. (If the hole were not closed, the high pressure between the slave piston and the brake would be transferred back to the master cylinder, forcing the brake pedal violently up).

As soon as the driver feels he has applied enough pressure to the brakes, he stops pressing the pedal down and the air control valve partly shuts, so that just enough air is admitted to hold the brakes as far on as they were. Because the amount of air admitted is proportional to the pressure on the brake pedal, the driver can control the amount of air entering, and thus the pressure on the brakes, exactly. When the brake pedal is released completely, the air control valve shuts completely and the equalized vacuum on either side of the diaphragm allows the return spring to push it back to its original position.

Below: in this type of vacuum assisted power brake unit, the vacuum assists the driver's foot on the pedal rather than assisting the cylinder. The pushrods, with a diaphragm in between assisted by the vacuum, are a direct mechanical link in case of vacuum failure. There will be no vacuum, for example, when the engine is not running.

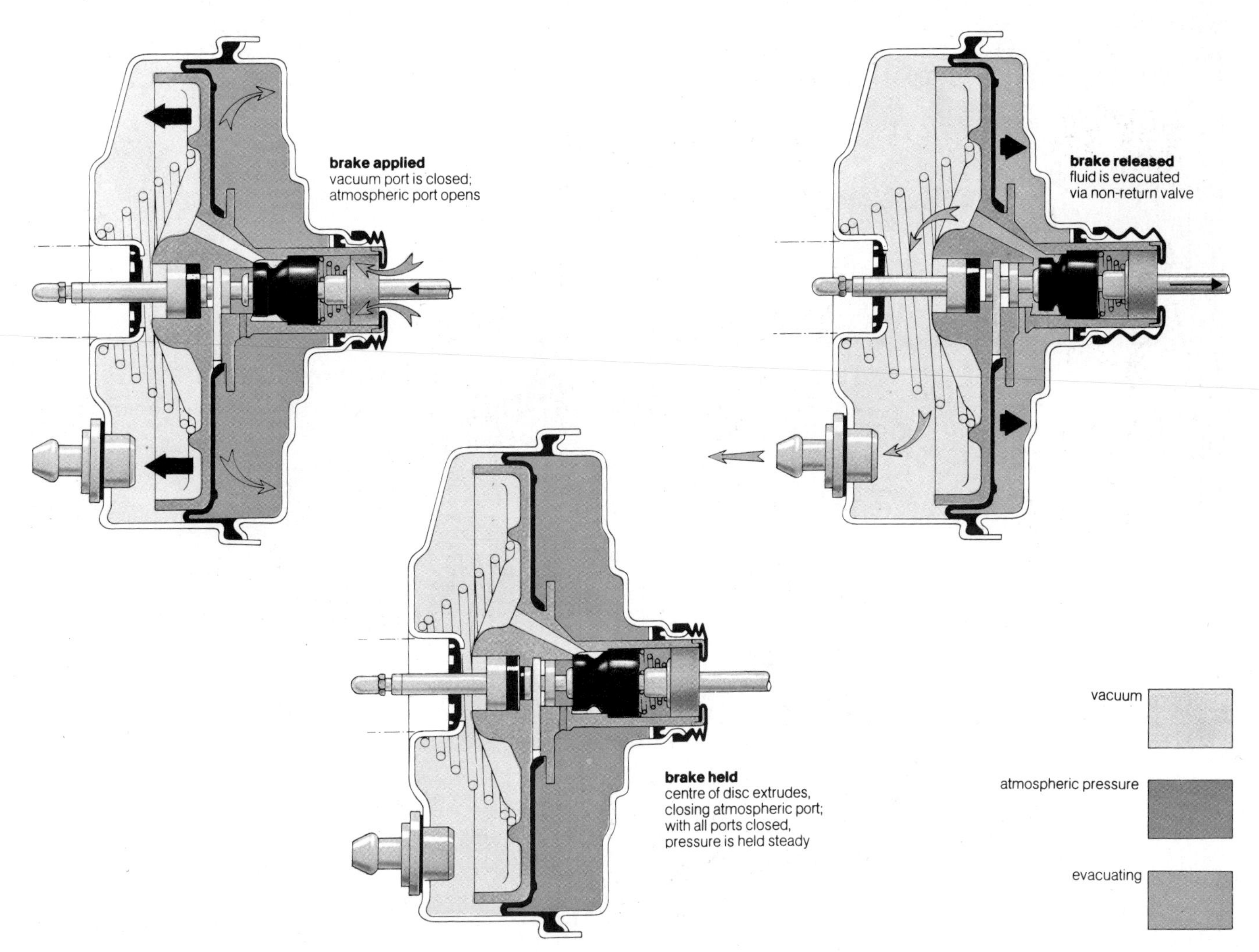

The system is 'fail safe' because a leak in the air system, or failure of the manifold vacuum for any reason, leaves the diaphragm in the 'off' position, so the hole in the slave piston is unblocked and the master cylinder works the brakes directly, without assistance. The extra pressure the driver feels when operating without assistance warns him of the fault.

Large commercial vehicles generally have air pressure brakes. These work in much the same way as hydraulic brakes, but by air instead of oil pressure. This is supplied by an engine-driven compressor rather than a hydraulic pump. There is a low pressure warning device, as well as an air pressure gauge, in the cab, and duplicate systems for trailer brakes and as a standby.

The steering system The simplest form of steering is that which was used for centuries on horse-drawn vehicles: it consists of a beam axle, pivoted in the middle and having a wheel on each end. It works quite effectively at walking pace, but as many a cowboy movie shows, when cornering at higher speeds it is unstable, because of the difference in track (distance between front and rear wheels) and is liable to overturn easily.

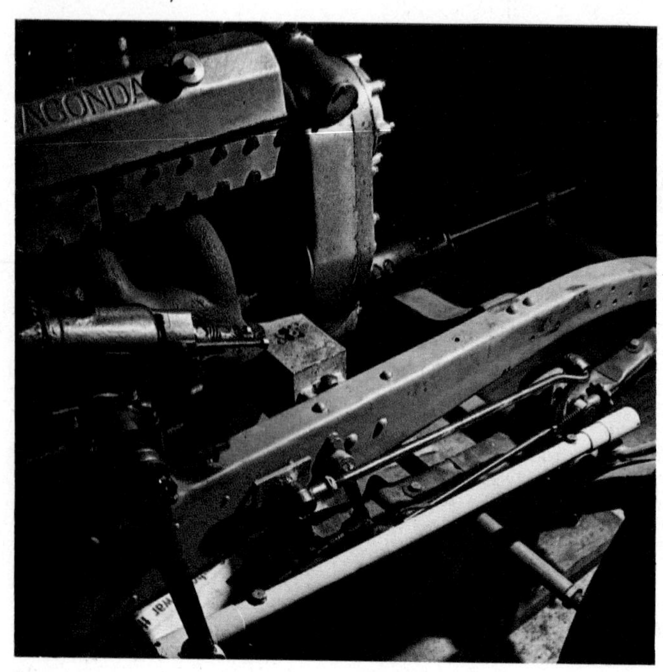

Above: a 1929 Lagonda, in the process of restoration. The photo shows the Marles steering box, drop arm, drag link and track rod. Also visible are the gaitered leaf spring, brake rod, and anti-tramp rod, which stops the axle from twisting.
Right: the diagram shows the Ackermann steering principle, in which the front wheels adopt different angles to travel on a circular course. The rack and pinion steering gear is the simplest and most positive and is fitted to many small cars.

In 1818 a German called Lankensperger took out a British patent on a system of steering in which the steering wheels are separately pivoted at the ends of the shaft. All four wheels follow a true radius about a single point on a line which includes the rear axle. The geometry was named after the patent agent, whose name was Ackermann; in fact few modern cars operate on pure Ackermann principles, but the name and the idea have stuck.

The steering wheel in a modern car is made of a steel skeleton covered with hard plastic or leather. The hub is splined to the steering column (splines are longitudinal grooves similar to gear teeth). The steering column extends from the interior of the car down to the front end underneath, where the steering gear is located. Nowadays both the hub of the steering wheel and the column itself are made to be energy-absorbing, or collapsible, for safety reasons, to reduce chest damage in case of accident. At the lower end of the steering column is the steering box.

The purpose of the steering box is to provide leverage, or mechanical advantage, since the effort required to turn the front wheels is much greater than that which could be exerted by the driver through a direct mechanism. There are four types of box.

The worm and peg (or worm and nut) system consists of a coarse thread (the worm) on the end of the column and a peg mounted on a cross rocker shaft, or a meshing nut surrounding the worm and fastened to the rocker shaft. As the column is rotated the peg or nut rides up and down in the worm, thus turning the shaft. The cam and roller system is a variant of this in which the peg is replaced by a cam (a sort of V-shaped roller) which meshes with the worm. The recirculating ball gear is a more elaborate variant. A bearing casing takes the place of a nut, and instead of threads, ball-bearings ride in the coarse groove of the worm. The bearing casing can have one or two 'threads' which are connected end-to-end by a tube (or tubes) to allow the balls to recirculate back to

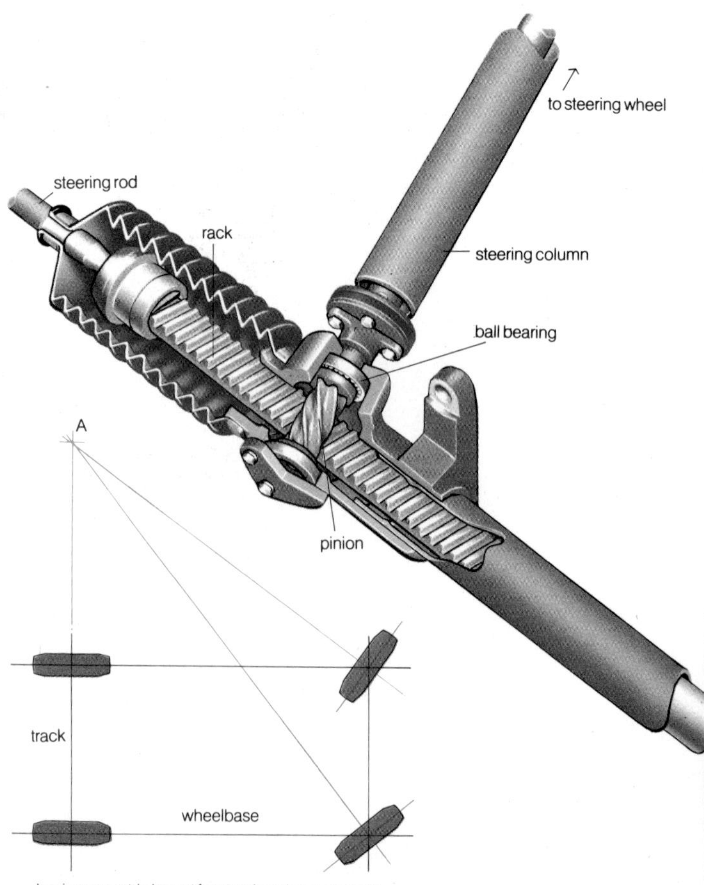

basic geometric layout for steering about a point A

the end of the worm thread when they have in effect rolled off the edge. The rack and pinion system consists simply of a pinion gear on the end of the column meshing with a rack (a flat section of gear teeth).

All but the last type of box require a system of linkages to take the movement created by the drop arm (attached to the steering box rocker shaft) to the steering arms on the wheels. Each manufacturer designs his own system so as to give the correct linkage geometry, to allow the links to follow vertical wheel movement (for example when the wheels hit bumps in the road) and to fit into the available space. This means a combination of several pieces of linkage: operating arm, idler arm, and track rods [tie rods]. In this respect the rack and pinion system scores highly, for its linkage is simple: a track rod from each end of the rack connected to the steering arms on the wheels. Its main disadvantage is that the rack bridges the centre of the car, getting in the way of a front-mounted engine.

The attitude or angle of the wheel and axle to the road are important in steering. Castor means that the contact patch of the tyre on the road about which the wheel swivels is slightly behind the axis of the king-pin or swivel-pin, so that there is a self-aligning torque which keeps the wheels pointing straight ahead when travelling in a straight line. King-pin inclination has the effect that, as the wheel is turned, it tends to lift the car slightly (like a door on rising hinges) so that in addition to the turning torque required an extra load has to be applied; steering effort rises and is thus partially related to cornering forces, and steering becomes self-centring. Camber is the inclination of the wheel to the vertical; the angle is called camber angle. Modern cars have camber angles of less than one degree, and camber is usually positive (the tops of the wheels are tilted outwards). The purpose of it in modern cars is to prevent the wheels from tilting too much inward because of heavy loads or wear resulting in play in the king-pins or wheel bearings. The camber must be carefully adjusted; if the camber of the two steered wheels is not equal, the steering of the car will have a tendency to 'wander' in the direction of the greatest camber and tyre wear will be uneven.

Power steering The car of today is larger and heavier than earlier cars; the tyres are wider, further apart and inflated to lower pressures. In addition, the trend of development has been to place more than half the weight on the front wheels, especially the weight of the engine, which itself is larger and heavier than in the early days.

To make cars easier to steer, the gear ratio in the steering box at the end of the steering column was changed so that turning the wheel required less torque (see dynamics), but this increased the number of turns of the steering wheel required on modern cars without power steering compared to $2\frac{1}{2}$ or 3 turns for cars built before 1940. Modern cars with power steering only require about three turns.

Power-assisted steering was first developed in the 1920s; one of the first devices was developed by an engineer at Pierce Arrow, an American maker of luxury cars. The Cadillac division of General Motors was going to offer power steering as optional equipment on some models in the early 1930s, but the Depression interfered with development. During World War II power steering was fitted to military vehicles; in 1952 Chrysler began offering it, and it is now standard equipment on many of the biggest American cars.

Electric devices were tried, but power steering today is always hydraulic, with oil pressure of perhaps 1000 psi (70 kg/cm^2) maintained by a pump driven by the engine of the car. The system is a servomechanism, or servoloop, which makes a correction to compensate for the torque applied to the steering wheel by the driver. It consists of an actuator and a control valve. The actuator is a hydraulic cylinder with

Above: a Lotus Elan, showing rack and pinion steering system, as well as double wishbone suspension.
Left: the cam and peg steering system. The cam is essentially a worm gear; in one variant, the peg is replaced by recirculating ball bearings.

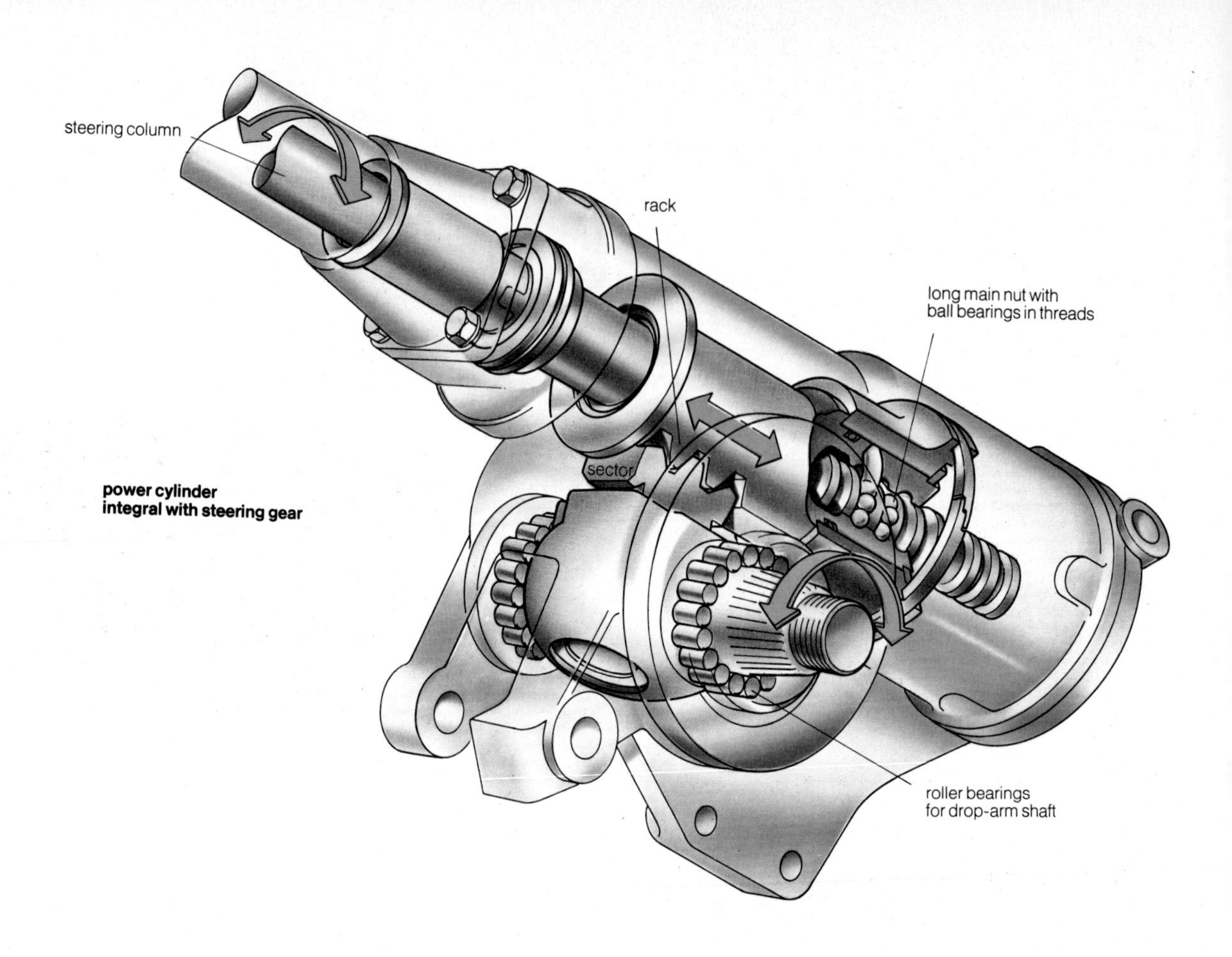

This power steering design uses rotary valves within the steering box. When there is no torque on the steering wheel, that is, when the car is travelling in a straight line, all valves are open and there is equal fluid pressure at the front and back of the rack member. When the steering wheel is turned in one direction or the other, appropriate valves close and the travel of the rack member on the long main nut is assisted by the appropriate pressure. If the pressure fails for any reason, the mechanical operation is harder but still fully controlled. The steering linkage itself is connected to the drop-arm shaft and ratios are such that a short travel of the rack equals a large degree of turning of the steered wheels of the car.

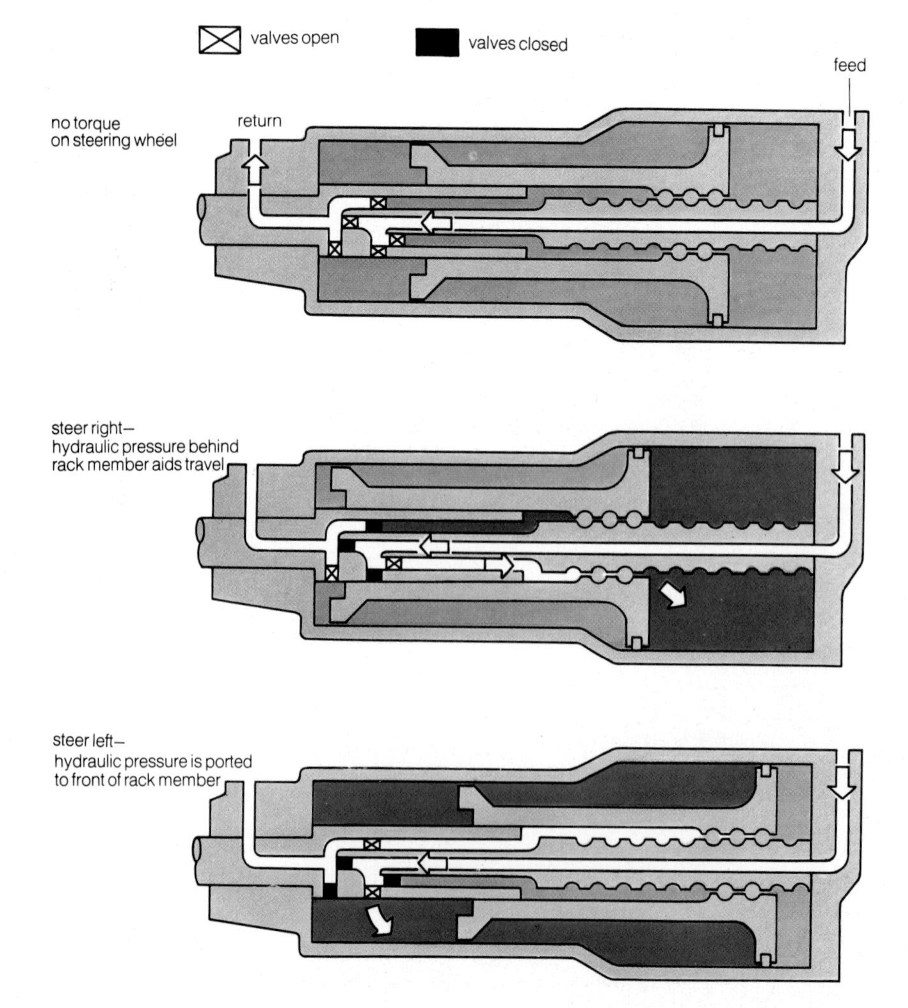

a piston, or ram, which is free to travel in either direction from the centre. The function of the control valve is to respond to the torque from the steering wheel by actuating smaller valves at each end of the cylinder. The system is designed to assist the steering linkage, rather than to replace it, and it does not do all of the work of steering, but leaves some of it for the driver. Thus if the hydraulics fail the car can still be steered, though with greater effort, and at all times the feel of the road is mechanically transmitted from the front wheels to the hands of the driver on the steering wheel, an essential element of safe driving. The power steering makes a positive contribution to safe driving in that if the driver hits a small obstacle in the road or has a flat tyre at speed, the power unit makes it easier to keep the car under control. Many large cars fitted with wide, stiff radial ply tyres would be nearly impossible to steer at parking speeds without power steering.

Hydrostatic systems, designed for off-the-road vehicles, are exceptions to some of this, because they dispense with the steering column and the steering box, and the steering wheel and the steered wheels are connected only by hydraulic tubes or hoses.

The power steering system includes a reservoir to hold the oil. Oil pressure is always provided when the engine is running, but when the system is at rest, that is, when the steering wheel is not being turned, equal pressure is available to each side of the piston in the actuator, so that it does not move.

There are basically two types of power steering systems: those which have the control valve located within the steering box, in which case it is usually a rotary valve, and those in which the valve is integral with the actuator, when it is an axial spool valve.

In a rotary valve system, the valve is integral with the steering column and operated directly by rotation of the steering wheel. In some systems the actuator is part of the steering linkage; in the Adwest rack-and-pinion system, the actuator is mounted on the rack itself. In others, such as the Marles Varamatic, the actuator as well as the valve is integral with the steering box, and operates the Pitman arm, which connects the end of the steering column to the steering linkage between the front wheels. A rotary valve is an input shaft inside a valve sleeve, with longitudinal slots machined on the shaft and on the inside of the sleeve. When the steering wheel is not being turned, the slots are lined up so that oil flows with equal pressure into ports in both directions; when the slots are misaligned by torque on the steering wheel, the oil flows all in one direction. This system can be designed so that the more the shaft is turned the more power assistance is given, with the least assistance near the straight-ahead position.

An axial spool valve, which reacts laterally, is usually integral with the actuator in the steering linkage, particularly on commercial vehicles. The axial load fed into the steering linkage by turning the steering wheel actuates the valve. In these systems the actuator is often connected at one end to the cross piece of the steering linkage with the piston end connected to the frame of the car, so that the actuator in effect pushes or pulls against the frame when it is activated.

The speedometer The speedometer is the device on the dashboard of the car which indicates two things: the speed of the vehicle and total distance travelled. It may also have a 'trip' facility whereby the distance travelled on a journey may be shown.

The type of speedometer fitted to cars, motorcycles, and the majority of commercial vehicles since World War II is the magnetic speedometer, originally developed in the 1920s.

The speedometer is driven from a point on the vehicle transmission which is rotating at a speed directly proportional to road speed; there is often a specially provided point on the gearbox tail shaft. Drive transmission is by means of a flexible shaft of multi-strand wire rotating inside a flexible tube. This assembly is called the flexible drive, or more commonly, the speedometer cable.

The flexible drive is connected to the main spindle of the speedometer, which carries a magnet. Close to this magnet, and pivoted on the same axis, is an aluminium disc or cup called the drag cup. This is connected directly to the indicator or pointer. On the other side of the drag cup from the magnet is a steel stator; when the vehicle moves, the magnet rotates, creating a magnetic field, and the stator is rotated.

Above, top: a typical speedometer. The counting device is the odometer, or mileage counter; the numbers show through a slot in the dial. Above: a ribbon speedometer has a coloured line indicator instead of a pointer. The advancing coloured line is provided by a tape which is stored, like a typewriter ribbon, on the drums at each end of the device. As speed increases, the instrument winds the tape on to the right hand drum against spring pressure, allowing the red line to show through a slot.

Although the aluminium drag cup is non-magnetic, it is conductive; the magnetic field causes an eddy current in the cup thus creating a force field. This causes the cup to try to follow the rotation of the magnet, but it is restrained by a hairspring. As speed increases, the torque on the drag cup also increases and overcomes the hairspring reaction, causing the pointer to move around the scale on the dial.

The most common type of speedometer uses a pointer on a circular or arc scale, but sometimes the design calls for a coloured line moving along a horizontal or vertical straight scale. This effect can be achieved in two ways with the magnetic speedometer. In one instance the drag cup is linked directly to a tube on which is printed a coloured helix. This is positioned behind a slot in the dial so that when it is turned by the drag cup the helix advances along the scale, giving the appearance of a moving strip of colour. An alternative means of providing the same effect is the ribbon speedometer, which is more common in Britain. In this instrument the drag cup is linked by nylon cords to two vertically pivoted drums between which is wound a tape, printed half black and half red. In the static condition the red portion of the tape is wound around the left-hand drum and held there by hairspring tension. As speed increases,the drag cup winds the tape on to the right-hand drum, so that the leading edge of the red portion appears in the slot on the dial.

The device which acts as a distance recorder is called the odometer. It comprises a series of adjacent rotating drums or counters each numbered 0 to 9, positioned behind a slot in a dial so that only one number on each counter is showing at any one time. Hidden between the counters are tiny double-sided plastic transfer pinions, which are mounted eccentrically to the main counter axis on thin carrier plates, which in turn are anchored so that they cannot turn with the counters.

The transfer pinion has two sets of teeth engaging, one on each side, with the internal gearing on the counters. The internal gearing of the counter has twenty teeth on one side and just two 'knockover' teeth on the other. Each time a counter comes up to the '9' position, a knockover tooth engages with the transfer pinion to the right, and drives it far enough to rotate it one tenth of a turn, before disengaging.

Where a 'trip' facility is provided, the transfer pinions are mounted externally to the counters on a spring-loaded spindle, which allows them to disengage when the counters are reset to zero.

Drive to the counters can be provided by a system of worm-geared shafts driven from the main spindle of the speedometer, if the manufacturer does not wish to vary the ratio between distance travelled and turns of the speedometer cable. The ratio is usually 1000 turns to a mile. In some cases, however, car makers want to be able to alter the axle ratio or the tyre size without changing the pinion in the gearbox which drives the speedometer. To accommodate this need, the counters are driven through an eccentric spindle geared to the main spindle and carrying a pawl which engages with a ratchet wheel. By varying the number of teeth and the throw of the eccentric, variations in gear ratio can be provided.

Calibration of the speed indication part of the instrument to suit each vehicle manufacturer's requirements on gearing and accuracy is carried out by fully magnetizing the magnet and then demagnetizing it until the desired readings are obtained.

The tachometer The tachometer or revolution indicator is an instrument for measuring engine speed, and is calibrated to indicate the number of revolutions the engine crankshaft makes in any one minute. The number of revolutions per minute of the engine as a vehicle is shifted through the several gears is an indication of engine performance. The tachometer is useful to racing car drivers, drivers of large commercial vehicles which have many gear ratios available, and drivers of ordinary cars who try to get the best performance possible from the engine, or better fuel mileage.

Right: the speedometer cable turns the shaft, which has a short worm on it, operating the odometer by means of a ratchet. The internal teeth in the odometer are arranged so that one revolution of a counter turns the next counter one-tenth of a revolution.
Opposite page: two types of tachometer. The blocking oscillator type (below) has the primary loop fitted directly on to the instrument. The impulse tachometer (above) is used by mechanics and engineers; placed on any solid part of the engine, it indicates the revolutions on a 'broken-back' scale.

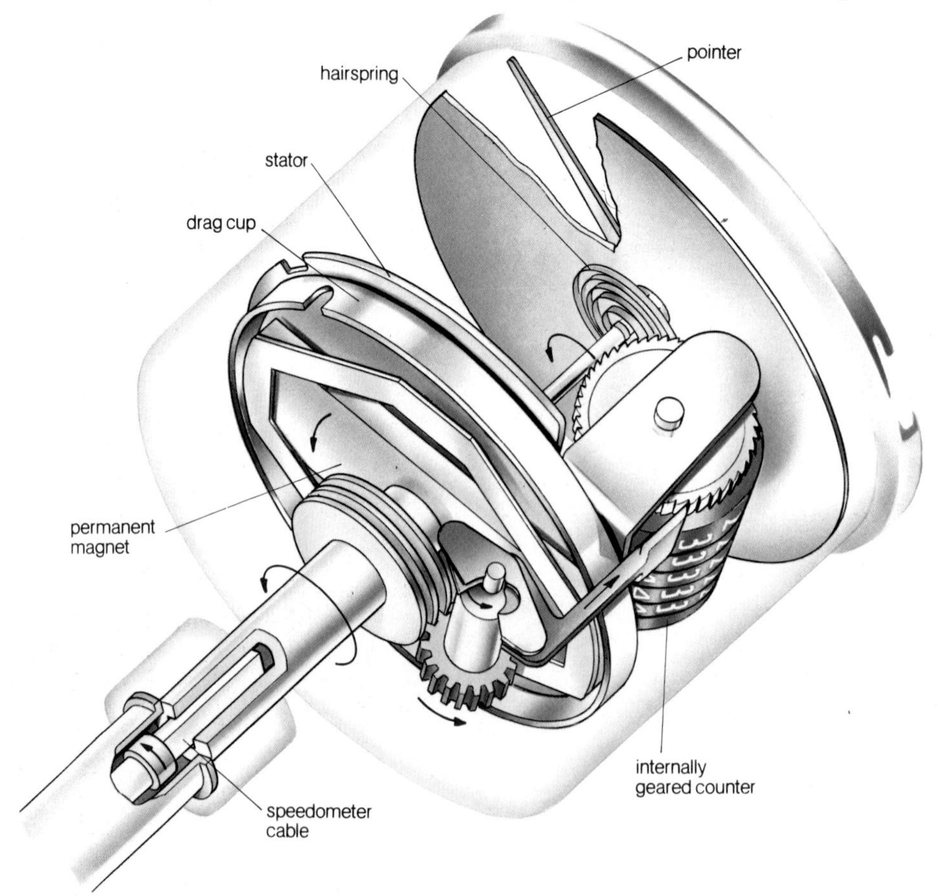

Originally, this instrument was driven mechanically, using a magnetic movement similar to that of a speedometer. Later, however, with developments in electronic circuitry, it was possible to design electronic tachometers operated by pulses from the ignition system.

There are several variations of tachometer design and operation; only four designs are described here which will cover the main forms of triggering used for tachometers.

The coil ignition or original impulse tachometer is used by car manufacturers and is triggered by the current pulses produced in the low tension circuit by the contact breaker, which is operated by a cam on the distributor drive. The change of current in the low tension (LT) side of the coil generates the high tension (HT) voltage supplied through the distributor to the spark plugs. On the internal combustion engine this pulsing is directly proportional to engine speed and is therefore suitable for triggering the tachometer. This is done by the transformer principle of primary and secondary coils, the primary loop being formed from the LT lead. On a typical instrument, the loop is located on the rear of the instrument case by a metal strap which slots, either side, into the other half of the transformer core on which is wound the fine wire bobbin forming the secondary coil. This senses the induction pulse, which is amplified and fed to the triggering circuit designed into the circuit board. Since the pulse obtained from the contact breaker is of a particular shape, the trigger circuit is constructed to respond to this particular input signal. The circuitry interprets the frequency of the pulses as a voltage, which is fed to a moving coil voltmeter—this is the indicating dial of the instrument and is therefore calibrated in revs/min rather than volts. The more cylinders an engine has, the more pulses it generates per revolution, so tachometers have to be designed for engines with a particular number of cylinders.

The tachometer requires a power supply, which is obtained from the vehicle's battery via the ignition switch. Because the voltage fluctuates slightly, it is stabilized within the tachometer. Final calibration of the movement is completed by adjusting the resistance across the moving coil meter.

A later design (known as a blocking oscillator type), used by vehicle manufacturers fitting tachometers as original equipment, incorporates the primary loop within the tachometer case, which permits easier installation on production lines.

With the advent of transistorized and capacitor-discharge ignition systems to improve ignition characteristics and engine performance, the LT current was reduced to a few milliamps, thus changing the pulse shape, which made it ineffective for a current-triggered tachometer. This entailed a change to a different circuit design measuring the voltage pulse at the contact breaker terminal on the coil. The voltage triggered tachometer is now generally being used by vehicle manufacturers in preference to the current type.

The induction tachometer was perfected for use on the diesel engine, which does not have an electric ignition system. It senses rotational or peripheral speed by means of a magnetic perception head. When a ferrous object moves in close proximity past the sensor there is a variation of the reluctance in the sensor's magnetic circuit. The resultant electromotive force (EMF) or pulses generated in the perception head coil are then transmitted via the two terminals on

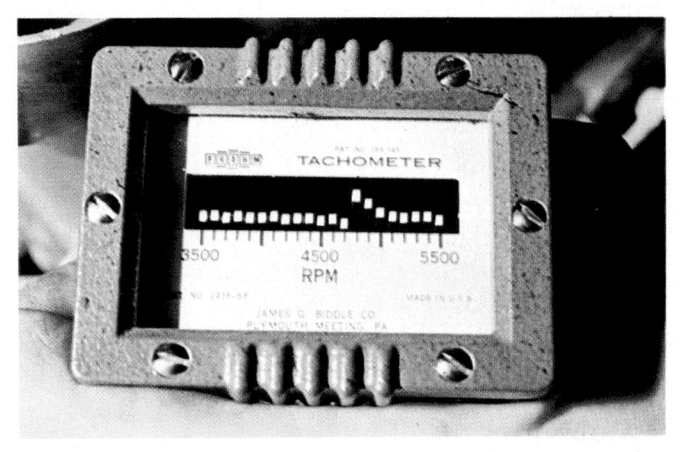

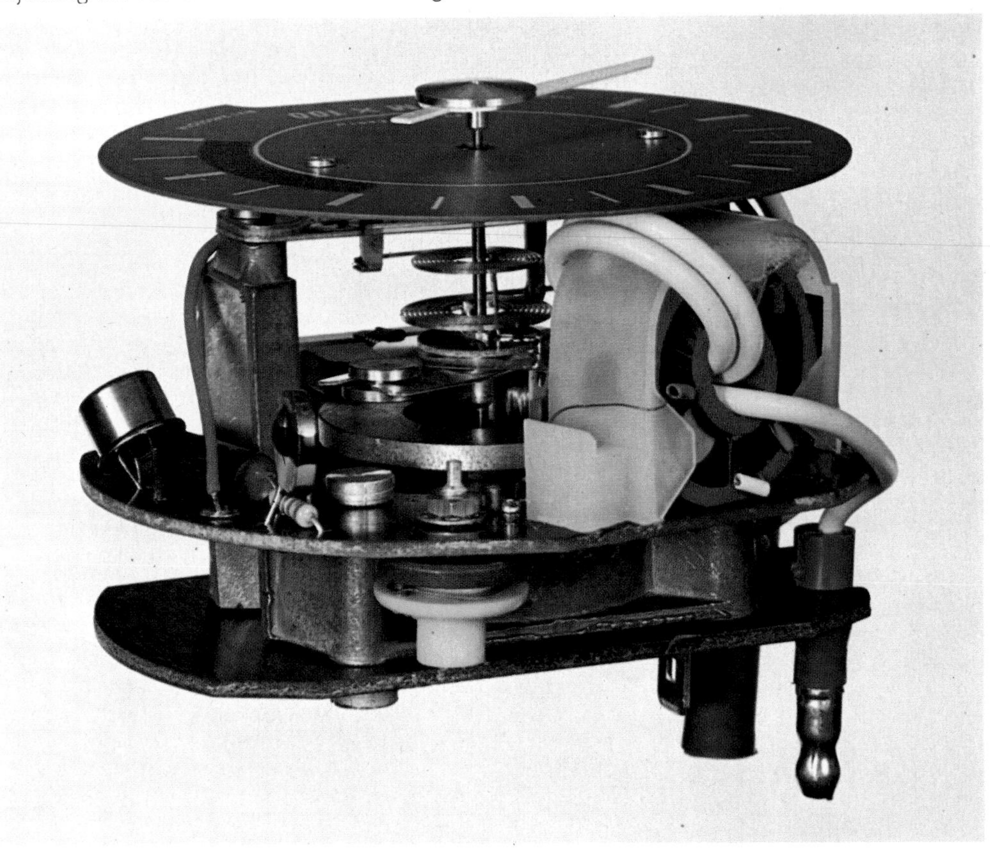

the sensor to the tachometer head. The instrument is also connected to the vehicle's 12 or 24 volt DC supply. The number of equi-spaced ferrous lobes permitted to pass the sensor per revolution varies depending on the ratio of the take-off shaft to crankshaft.

The self-generating tachometer was supplied to certain manufacturers for original equipment and has since been superseded. The installation used a generator which was connected directly to the instrument. The generator was mechanically driven from a suitable ratio take-off, producing an alternating current (AC) voltage proportional to the speed of rotation of the generator shaft. By rectifying this voltage, a direct relationship between voltage and engine speed was obtained on the tachometer scale.

The high-tension tachometer was specifically designed for marine engines and operates off the HT plug leads. It is connected to a 12 volt supply and senses the HT voltage via a screened trigger cable which is twisted around each plug lead. This enables the tachometer to operate effectively irrespective of the type of ignition system. To permit use on a wide range of engines with different numbers of cylinders, and thus pulses per revolution, a changeover switch is incorporated.

Horns and klaxons As the speed of motor vehicles increased around the end of the 19th century, some means of audible warning became necessary. The first type of horn to gain general acceptance was the bulb horn, a trumpet-like instrument operated by squeezing a rubber bulb. When the bulb was squeezed, the blast of air produced caused a reed to vibrate, and the sound waves this produced were amplified by the shape of the horn.

A more efficient warning device was the mechanical klaxon, which used a vibrating diaphragm (similar in principle to a loudspeaker cone) to produce the sound. The klaxon was operated by a spring loaded plunger, which had a corrugated surface within the body of the klaxon that ran down the back of the diaphragm causing it to vibrate rapidly. The sound produced was amplified by a horn in front of the diaphragm.

The plunger-operated klaxon continued in use up to the 1920s despite the introduction of the electric klaxon in 1912. This had a DC electric motor which drove a corrugated disc that vibrated the diaphragm, producing a higher, stronger note than the manual klaxon. Another form of manual klaxon also had a corrugated disc, which was turned by a handle.

A typical modern electric horn consists of an electric coil and a metal diaphragm, with a set of contacts between them. When the horn button is pressed, current flows in the coil, setting up a magnetic field which attracts the diaphragm towards the coil. As the diaphragm moves it opens the contacts, cutting off the supply of current to the coil, in a similar manner to the contact arrangement on an electric bell. When the current stops flowing in the coil, the magnetic field collapses and the diaphragm springs forward away from the coil. This allows the contacts to close again, completing the circuit to the coil, and the cycle is repeated as long as the horn button remains depressed. The vibrations of the diaphragm are amplified by a small horn, and there is often an adjusting screw provided, which alters the amount of movement of the diaphragm by altering the opening and closing positions of the contacts, thus altering the pitch of the sound.

The latest types of horn are driven by compressed air and produce powerful high single or multiple notes. The compressed air is supplied by a compressor driven by a DC electric motor, and there are usually two, three or five horns, which are of different sizes and so produce notes of different frequencies.

The compressor is basically of the rotary vane type, and in some horns there is a gear mechanism which drives a disc valve that distributes the air to each horn in turn, giving an alternating note. The disc valve may be controlled by an electromagnet so that the horns can either be operated in sequence (consecutive operation) or all together (concurrent operation), the mode of operation being selected by means of a switch.

To prevent an arc across the horn button contacts caused by the heavy current needed to drive the compressor motor, the horn button operates a relay which has heavy duty contacts that complete the circuit to the motor.

Tyres The pneumatic tyre [tire] was invented by a Scot, R W Thomson, and first patented by him in 1845. A set of tyres made according to Thomson's design were fitted to a horse drawn carriage and covered more than 1000 miles (1600 km) before they needed replacing. It was not until nearly 50 years later, however, that the modern tyre industry was founded by J B Dunlop, an Irishman from Belfast.

A modern vehicle tyre consists of an inner layer of fabric plies which are wrapped around bead wires at their inner edges. The bead wires hold the tyre in position on the wheel rim. The fabric plies are coated with rubber which is moulded to form the sidewalls and the tread of the tyre. Behind the tread is a reinforcing band, or breaker, usually made of steel, rayon or glass fibre. The radial ply tyres fitted to most modern cars differ from cross ply tyres in that they are constructed with very flexible sidewalls and have breakers which are almost inextensible. These properties are achieved by altering the disposition of the fabric plies in the tyre 'carcass'; in particular, the cords in the fabric plies run from one bead wire to the other making an angle of 90° with the 'crown' of the tyre (the circumferential line around the middle of the tread). Radial tyres have better cornering and wear characteristics at high speed than cross ply tyres.

The main materials in tyre construction are steel wire for the inextensible beads, textile fabric or steel for the casing and reinforcing breakers, and of course rubber mixed with various additives to give the required strength and resistance to wear and fatigue. For tyre casings, rayon, nylon and polyester are the most commonly used materials, although thin steel cable is often found in truck tyres. The breaker was originally made of steel, and still is in truck tyres, but nowadays rayon and other materials such as glass fibre are more common in car tyres. The rubber mixes are made from modern synthetic rubbers for car tyres: heavier truck tyres tend to be made of natural rubber because it is cooler running than synthetic rubbers.

The first operation is to prepare the necessary rubber mixes, working the rubber into a plastic state, in an internal mixer, and milling into it sulphur for vulcanization and other ingredients for the different types of rubber mix needed in the various parts of the tyre. In the tread, a relatively high loading of finely powdered carbon black is used to give resistance to tread wear. The casing rubbers are mixed to have strength in the thin layers which bond the casing cords together, and to have resistance to fatigue under repeated flexings and continuous tension. For use on the bead wires, a hard mix with a high sulphur content is prepared, which will set into a solid mass on moulding.

The bead wires themselves are prepared from high tensile steel wire, assembled as a ribbon of five or six strands, side by side, and enclosed in the hard rubber mix to form a tape. A number of runs of this tape are wound up on a former of the correct size and, on vulcanization, the bead is set in a solid and virtually unstretchable form.

Cotton, which was originally used, has been replaced in tyre casings by modern synthetic textiles, which, as they are made of long continuous filaments, have a much greater strength for a given thickness than the old cotton thread spun from a collection of short hair-like fibres.

The material for the plies and for the breakers consists of a practically weftless fabric, with the strength all in the warp, and with the sheet held together only by a system of fine, widely spaced weft threads. This construction is used to eliminate the 'knuckles' which occur in a normal cross-woven fabric, where warp and weft threads cross, and to reduce the sawing and chafing which takes place when such a fabric is flexed under load. The sheet of 'weftless' fabric is made into a sandwich, between two films of rubber, in a calendering operation (pressing between rollers). The rubber-coated fabric is then cut into strips, with the threads running in an appropriate direction, for use in building up the casings of the different types of tyre.

Both radial and cross ply tyres are built up on collapsible steel formers. The layers of fabric plies are secured by turning their edges round the coil of bead wire enclosed in suitable reinforcing and packing strips.

Opposite page: an old brass trumpet-type car horn.
Below: the diagrams show the construction of a cross-ply tyre and a radial. The radial has a flexible wall and a stiff, braced tread, which improve its grip and wear resistance. The 'Denovo' tyre is an extra-strong radial which stays in place on the rim when it is deflated. The lubricant canister contains a gel and a liquid which mix to seal and partly reflate the tyre, so it can be driven, minimizing the danger of 'blow-outs'.

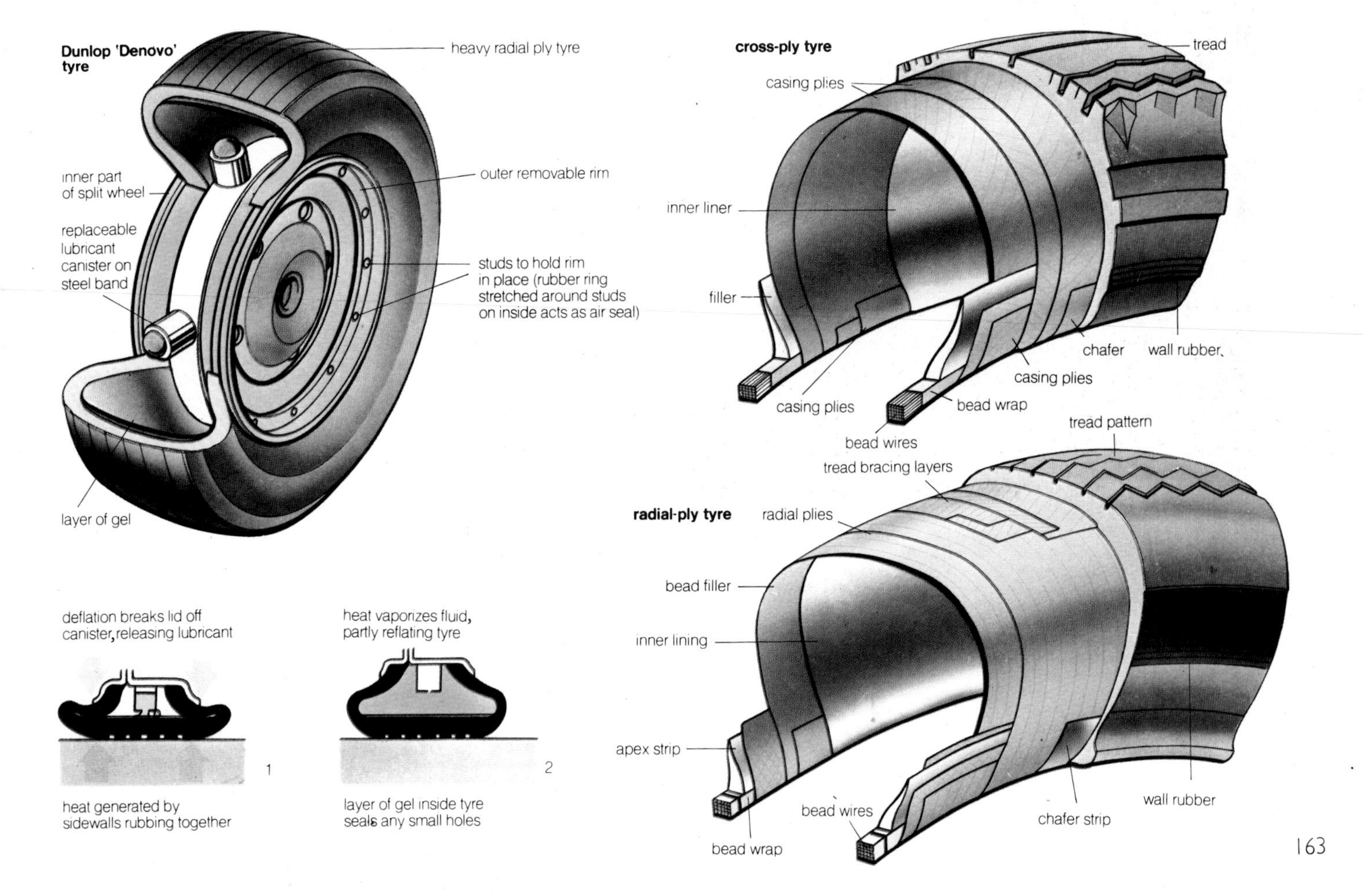

The process of adding the smooth strip of extruded rubber compound which will form the tread differs in the two types of tyre. The cross ply receives its tread while still on the almost flat building drum. Afterwards the complete cylindrical tyre is shaped up to the usual doughnut form as it is introduced into the mould. The radial ply tyre, from its nature, demands different treatment. Here the casing must be taken from the nearly flat cylindrical building drum, and shaped up to the required toroidal form before the unstretchable rigid breaker bands are added. On top of these bands the tread is then fitted.

The final stage in tyre manufacture consists of moulding the built-up raw tyre in a suitably designed steel shell mould. This is either engraved with the tread and wall pattern or has die-castings riveted inside it, to make up the complicated pattern of ribs, blocks, grooves and fine slots, which give the final tyre its road grip and even wear characteristics.

The mould is situated in a press containing a cylindrical rubber diaphragm which is inflated under high pressure inside the tyre as the mould closes. This operation forces the raw plastic tyre into the pattern on the inside of the mould. Heat is then applied, in the form of steam, both from cavities in the press through the outer mould, and by supplying steam to the inside of the diaphragm within the tyre. This heating causes the chemical combination of the rubber with the sulphur which has been included in the various rubber mixes used in the components of the tyre. The result, when the moulding is finished, is a tyre of permanent shape and of suitable physical properties for its intended use.

Racing cars

The early motor cars were fragile, temperamental machines; they were built by men who were discovering design requirements as they went along. The first competitions were endurance contests rather than races, but beginning in 1895 motor racing became extremely popular and has been so ever since. Town-to-town races were the most common type; transcontinental races have been run, including one from Paris to Peking.

Racing has always been dangerous, and racing competitions on public roads, common in the early days, are now strictly limited. The worst accident in the history of racing happened in 1955 at the Le Mans sports car event; 83 spectators were killed, and for a while it was thought that the entire sport would be abandoned. The famous annual 500-mile event at Indianapolis, Indiana has averaged about one death a year, and is considered to be relatively safe.

There are several types of racing, each regulated by its own professional organization in various countries.

Grand Prix racing From the very beginning, racing contributed to the development of the automobile. An American newspaper publisher who was interested in this aspect of racing established the Gordon Bennett Cup, and six races were held from 1900 to 1906. The rule was that each country could enter three cars and that the cars had to be entirely constructed by the industry of the entering nation. In 1906, the French established the Grand Prix, which had no restriction on the number of cars which could be entered, and the Bennett Cup was abandoned. Since World War I, each European nation has held its own Grand Prix. Nations in North and South America also take part, and Japan and Australia are considering having their own Grand Prix.

A divergence in international racing took place before World War I when it became apparent that American racing was more for entertainment than engineering. Wooden board tracks were used for a while; the most famous American track is the Indianapolis Motor Speedway, a $2\frac{1}{2}$ mile (4 km) banked dirt track with identical corners which has been lapped at 200 miles an hour (331 km/h) by the type of car bred for this type of racing. For some time the annual Indianapolis 500 race qualified as a Grand Prix in order to make the sport truly international, but now it no longer qualifies.

In Europe, Mercedes had competed sporadically; when Hitler came to power in 1933 he recognized the prestige in winning races and gave government support to Mercedes. At the same time the German Auto-Union was commissioned to build their own car. Mercedes, with its enormous engineering skill, was able to build a powerful (400 bhp) car within the rules, which then limited the weight of the car to 750 Kg (1650 lb). The battle between Mercedes' classic front-engined car and the Auto-Union revolutionary rear engine design resulted in a 'golden age' of racing from 1934 to 1939. The cars, despite high power and narrow tyres, were surprisingly safe, although Richard Seaman, the only Englishman to join the Mercedes team, was killed in the Belgian Grand Prix in 1939.

After World War II, the racing nations banded together to set rules and to make the sport international. There had been European Champions before, but now the international series of races became a series which resulted in a World Champion Driver on the basis of points. The first World Champion was the Italian Giuseppe Farina in 1950. Although attempts were made to build fully enclosed Grand Prix cars, the rules now effectively require that the cars be single-seaters with exposed wheels. Since the war, the Grand Prix car has been known as the Formula 1. At first the races were run using left-over cars from before the war, and superchargers on engines were common, but the rules formulated by the Fédération Internationale de l'Automobile (FIA; founded in 1904) which specify engine displacements have nullified the advantage of supercharging; the last supercharged car to win a championship was the Italian Alfetta in 1951.

Mercedes built an exceptionally advanced car which driven by Juan Fangio of Argentina won almost every race for two years. The company pulled out of racing in 1955, and after that the competition was mainly between Italy and Britain. The BRM (British Racing Motor) and the Vanwall (a notably aerodynamic car built by bearing manufacturer Tony Vandervell) could sometimes beat the Italian cars, but British engineering superiority was established by a series of light cars powered by simple engines. Much of the weight saving came from adopting rear engine design, eliminating the heavy propeller shaft and enabling the driver to lie semi-supine, reducing frontal area and wind drag. The first such car to win a world championship was the Cooper, which enjoyed success during 1959–60.

The firm of Lotus was started by Colin Chapman, an engineer and businessman whose racing success built up demand for his sport cars. In order to reduce weight as the Grand Prix rules kept changing, Chapman introduced the monocoque chassis, using sheet metal to build the frame and outer skin of the car in one unit, in 1962. Another of his innovations was wider tyres, to enable as much power as possible to be transmitted to the ground. He was aided in this by tyre makers Goodyear and Firestone, who wanted to enter a field where Dunlop had had a virtual monopoly.

A modern racing tyre is tubeless and light in weight with comparatively low pressure, in order to avoid explosive blow-outs. It is purpose-built in different versions for wet and dry conditions, the 'dry' tyre having no treads in order to obtain maximum contact with the road. (Similar 'racing slicks' are also used on dragsters.)

Honda of Japan produced a complicated twelve-cylinder car which won the 1965 Mexican Grand Prix. This was the last of the $1\frac{1}{2}$ litre (1500 cc) Formula 1; the rule was changed

Above: this sports car race shows the different types of cars which can qualify under the same set of rules. The car on the left, number 2, is a Chevrolet Corvette, which is essentially a production car; number 3 is a Porsche 917, a specially designed car not produced in great numbers.
Left: Luigi Fagioli training in 1934. The car is a Mercedes.
Opposite page: by contrast, the Auto-Union in this picture, taken in Berlin in 1937, would not look out of place in a race today. Aerodynamic design principles in racing cars were already being established.

in 1966 to 3 litre (3000 cc), and the Japanese had further limited success before dropping out in 1968. (The figures refer to the total volume swept by the pistons in the cylinders.)

Races are held on closed circuits, three or four miles long, with the total length of a race varying from 150 to 400 miles. Since 1967 drivers' performances count in 9 out of 11 events; points are counted on the five best of the first six races of the season and the four best of the last five.

In technical respects there has been some stagnation in the design of Formula 1 cars since the 3 litre rule. They tend to be built by small specialist firms instead of major firms with unlimited research funds. Complaints are heard that they all look the same and even sound the same, since they usually use the 480 horsepower Cosworth V-8 engine, introduced in 1967 and developed with a major contribution by British Ford. But the fact that the cars are closely matched means that the result of a race depends on the skill of the driver, and the Grand Prix are not losing their popularity. It is significant that in the Grand Prix the award goes to the driver, while in sports car racing it goes to the car maker.

A dramatic development since the war has been the adoption of aerofoil devices to improve the grip of the car on the track. It has been known ever since aircraft first flew that wings might be used to create reverse lift; the first car to make use of the idea was the Ferrari in 1967. The most effective place to mount them was found to be directly on the rear suspension, leading to the adoption of parasol-type wings high in the air above the car; a series of breakages led to hasty controls, and since 1969 their size and placement has been strictly limited. A good driver must decide how best to position the wings, balancing the improved cornering force against the wind drag on the straight.

Racing formulae After World War II at the same time as Grand Prix racing went international, the fragmentation of racing formulae also began. Midget racing in the USA, with the cars limited to a wheelbase of 66 to 76 inches, got under way in the 1940s; in England, tiny cars with 500 cc motorcycle engines became popular because of their low cost and the close racing which they provided. The new formulae have become training grounds for young drivers.

In 1958 the Italians applied to have their national 'formula junior' recognized internationally. Ingenious British makers such as Lotus, Cooper (who built many of the English 500 cc machines mentioned above) and Brabham not only eclipsed the Italian cars but sent Formula 3 (as it has been called since 1965) into a cost spiral so that a fully-equipped Formula 3 car is out of the reach of many young drivers.

Consequently a variety of national promotional single-seat racing formulae have sprung up. The first of these was Formula Vee, introduced in 1961. Powered by a tuned Volkswagen 'Beetle' engine, a Vee has many other standard elements in its chassis, and has been useful for training drivers in Germany and the USA, as well as Brazil, where twice World Champion Emerson Fittipaldi raced them before coming to Europe.

More powerful new formulae are Britain's Formula Ford and France's Formule Renault. The essence of each is a tubular chassis carrying a mildly tuned 4-cylinder Ford or Renault pushrod engine. Narrow tyres of the road-going type are required and aerodynamic aids are forbidden. Formula Ford has been popular in Britain, Scandinavia and Austria, and has been 'exported' to the USA, Australia and parts of Latin America. Originally conceived as a £1000 car, it now costs closer to three times that. Formule Renault is a little faster, and popular in France, Belgium and Germany. It is heavily sponsored by French interests, and the successful Formule Renault driver is virtually guaranteed a sponsored driving career.

As a means of teaching racecraft to a driver, these formulae are ideal, being outwardly similar to Grand Prix cars and having a minimum of bodywork to be damaged and being easy to work on. For example, a Formula Ford has a gearbox with easily altered gear ratios, just like some of the Grand Prix cars.

New formulae have been introduced as an intermediate step above the promotional formulae. Formula Super-Vee, Formule Super-Renault and Formula Ford 2000 are some of them. It is increasingly difficult for a young driver to decide which class to enter. The next two expensive classes are the ones on which the aspiring driver fixes his sights: the Formula

5000, powered by highly tuned 5 litre American V-8 engines, and Formula 2, in which the most popular unit is the 4-cylinder BMW. The Formula 5000, with 500 bhp, can be faster than a Grand Prix car, but its greater weight usually cancels the difference. A Formula 2, with 300 bhp, is nimbler than either. There are European championships for both, but the F5000 is mainly based in Britain; the F2 is more popular because it gives drivers a chance to race on more foreign circuits.

Sports car racing The sports car (or GT, Gran Turismo) is usually a two seater, sometimes four, designed for nimble handling rather than speed or power. It is, theoretically, a production car intended for private ownership, and must be 'homologated'—that is, a certain number must have been built. There are also special 'prototype' classes. Some builders of Grand Prix cars also make sports cars, such as Ferrari and Lotus. Some of the most famous car races are sports car events, such as the 24-hour Le Mans race, which is really an endurance contest; they are sometimes called Grand Prix events, which means that they are important national events rather than Formula 1 events, although some sports cars are not much different in terms of speed and power from Grand Prix cars.

Hot rod racing Several types of racing are peculiarly American, having arisen because of widespread car ownership in a large country without much public transport, and because of the availability of old cars. Hot rod racing is as much a hobby as a sport. The owner of a car usually does his own modifying of it, in order to compete with others against time or distance. The sport had bad publicity on account of illegal street racing, but is controlled in the USA by the National Hot Rod Association (NHRA). The cars may be ordinary street models, modified for performance, or they may be modified to an extent that makes them unsuitable for street use. For example, a high-performance camshaft may be intended for high-speed use, and may be damaged by stop-and-go city driving, which includes a lot of idling of the engine.

Drag racing Drag racing is an outgrowth of hot rod racing in which the cars compete against the clock on a track called a drag strip, usually $\frac{1}{4}$ mile (393 m) long. The cars are specially built from scratch and are unsuitable for anything except the acceleration contest.

Stock car racing It is estimated that ten million people a year attend stock car racing in the USA. Cars which are not racing cars or sports cars are called stock cars or production cars; the car manufacturers sponsor cars and teams of drivers for the publicity. There are several categories of stock car racing, depending on the type of car being raced and the degree of modification allowed. There are also demolition derbies, in which the drivers purposely damage each other's cars, while trying to avoid crippling damage to their own; the last car still able to move under its own power is the winner. In stock car racing, all glass is removed from the car, the drivers wear crash helmets and safety harnesses, and roll bars are installed inside the car to strengthen it in case it rolls over. Like virtually all American racing, stock car racing is done on oval-shaped dirt tracks. British stock car racing is called 'banger' racing; it is unsponsored and old cars are used.

Rallying Rallying is a sport in which privately owned sports cars or stock cars compete against the clock over ordinary roads. The object is to get from one control point to another; the driver takes a passenger with him to navigate and the ability to choose the best route between control points is as important as the skill of the driver. The object may be to travel at an average legal speed; in a stages rally there may even be penalties for completing a stage at too high an average speed. Safe, legal driving is the essence. In other types of rallies the drivers are allowed to go as fast as they can. The Safari rally, held in East Africa, is an endurance contest over some of the worst roads in the world.

Above: last-minute adjustments to a dragster.
Top of page: a 'banger' which has already seen some service.
Opposite page: Germany, 1974: a Surtees Fina TS 16 formula car.

Special purpose vehicles

Commercial vehicles, carrying freight or passengers by road, play a vital role in modern industrial society. Through its ability to collect goods from the factory or warehouse loading dock and deliver them direct to the customer's door, the truck or van eliminates the double handling implied by rail or waterborne transport.

In the late 1890s a large number of small engineering concerns embarked on the production of truck chassis. Most of them had steam engines, fired by either coal or oil. Load capacities of 2 to 3 tons were the norm, although the fore-runner of the modern articulated vehicle was pioneered in 1898 by Thornycroft. It could haul payloads of up to 5 tons.

Not all the early commercial vehicles were steam powered. In France, De Dion-Bouton concentrated on a single-cylinder internal combustion engine for its light commercials and cars, while in the United States, electric traction was in vogue for goods vehicles; by 1905 Studebaker had created a division to build light electric vans and trucks.

One of the most significant innovations in truck design came in the 1920s, with the inception of the compression-ignition or diesel engine as a power unit for trucks. Its greater economy and reliability—through the elimination of any form of electrical ignition system—proved attractive to commercial vehicle users in Europe, where petrol [gasoline] prices were close to those of diesel fuel.

Unlike the private car, the truck has changed little in basic design since the 1920s. When cars had separate chassis frames, trucks were essentially larger-scale replicas, with a ladder-type frame consisting of two major longitudinal members linked by a series of crossmembers.

Commercial vehicles. Today's long-haul commercial vehicle follows the traditional configuration, although technological advances in both metallurgy and the understanding of structural dynamics have resulted in a considerable improvement in the ratio of payload capacity to unladen weight. The channel-section chassis rails are now of higher grade steel and computer analysis of a truck's road-holding characteristics enables the suspension to be 'tuned' to the bending and flexural behaviour of the main frame.

Integral construction, as used today for most cars, combining the chassis with the body, has now been adopted in the United States and in Germany for buses, where body configuration is fairly standardized for high volume production. Van trailers without chassis are also growing in popularity, the sidewalls providing their structural strength.

Continual advances are being made in truck design aimed at improving driver comfort, convenience and efficiency. Steel leaf springs are the most popular suspension medium, but through the use of sliding—or slipper—ends in place of wrap-around spring eyes and shackles, the rate (or softness) of the spring is able to vary according to the load imposed on it. Thick tapered spring leaves, with no inter-leaf contact, are also helping to create a better standard of ride for the

Above, top: the steam powered Foden truck of 1927 originally would have had solid tyres.
Above, centre: this 20 ton dump truck is intended for off the road use in places such as building sites and quarries.
Above: the drawbar trailer is widely used on the mainland of Europe. This can carry the maximum allowed weight in any European country, typically about 40 tons gross.
Right: in this 1927 steam truck, a vertical boiler was used to save overall length and increase load area, resulting in a vehicle which resembles a conventional truck.

driver, by eliminating the built-in friction stiffness of an orthodox multi-leaf spring.

Most trucks of up to about 24 tons weight are classed as 'rigid' vehicles, constructed as a one-piece chassis running on two or three axles. Four-axled rigid trucks—of up to 30 tons maximum (including load) weight—are operated in some countries, notably Britain and Italy, where the operating laws give them a weight-carrying advantage.

Heavier commercial vehicles are normally operated as tractor-trailer combinations, which can be either of articulated or rigid-plus-drawbar-trailer formation. The articulated 'tractor' is a short-wheelbase chassis with no load platform of its own, but instead carrying a 'fifth-wheel' turntable which engages with a king-pin protruding below the mating turntable coupler on the underside of the semi-trailer. The primary advantage of an articulated vehicle is its facility of keeping the tractor at work during the loading and unloading operations.

The same principle of minimizing chassis idle time is now achieved on rigid chassis by using quick-release interchangeable bodies, known as 'demountables'. In Britain and Germany the demountable body is used increasingly. When parked at the loading point the body is supported on four legs. To pick up the body the truck chassis is reversed beneath it, and lifting jacks, hydraulically or pneumatically powered, raise it sufficiently for the legs to be folded away or removed and stowed.

A rigid vehicle hauling a drawbar trailer, where each half of the combination carries its load on a demountable body, can give great transport flexibility. Having travelled for several hours from its base, the combination can be split in two for easy delivery in city conditions. The rigid vehicle, having uncoupled its trailer, hauls first one and then the other body on a multi-drop delivery run.

Apart from notable exceptions like the under-chassis-engined Büssing models from Germany, all modern heavy and middleweight trucks are powered by vertical in-line or vee-form diesel engines mounted at the front of the chassis. On-highway operation, where overall length is restricted by law, has brought about the popularity of the forward-control —or 'cab-over'—layout, where the driver sits above and to one side of the engine. The flat-fronted forward-control cab can occupy as little as 4 ft (1.22 m) of the vehicle's length. It can also be made to tilt forward bodily, raised by a counter-balancing torsion bar or a small hydraulic pump, enabling maintenance personnel to gain access to the engine and auxiliaries.

For more traditional markets, most manufacturers continue

Top left: this 30 ton gross 8 wheeler is driven by the two rearmost axles, giving it enough traction for moderately rough ground. The crane is powered by a hydraulic pump, which in turn is driven by the truck's engine. Above, top: the main body on this 26 ton waste disposal vehicle can be removed, leaving a platform onto which several of the smaller containers can be loaded by the hydraulic arm. These are then dropped off, filled and reloaded separately. Above: the rear of the top deck of this 32 ton articulated car transporter is lowered hydraulically to allow the cars to drive on to it.

6
7
8
8a
15
16
2
5
11
26
4
12
24
25
10
9
1
3
14
13

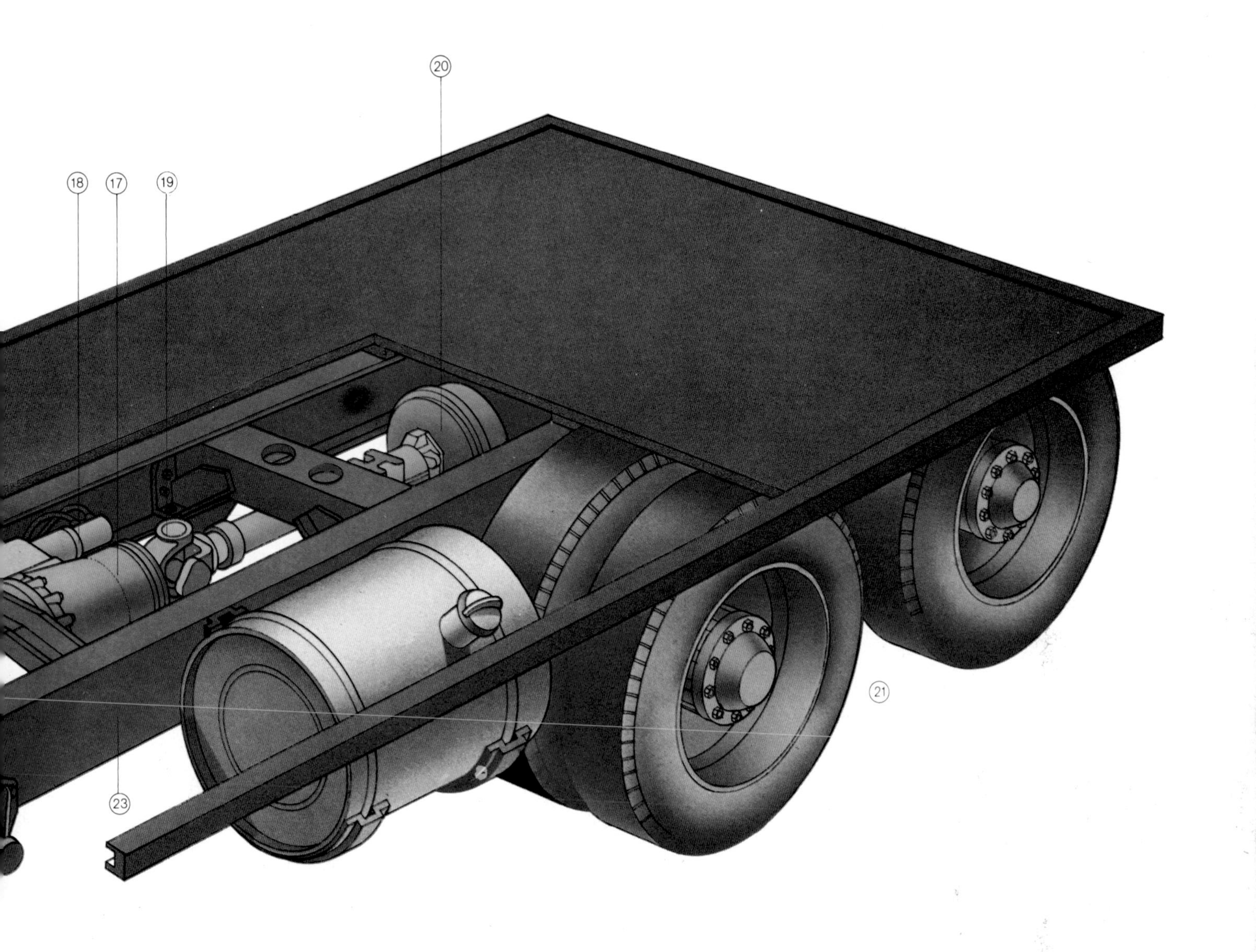

1 radiator grille
2 indicator/side light
3 headlamp
4 reinforced girder
5 radiator
6 passenger seat
7 engine cover
8 six cylinder diesel engine
8a exhaust manifold
9 steering column
10 steering box
11 steering servo
12 steering links
13 16" brake drum/12 stud hub
14 front axle
15 clutch housing (17" single plate)
16 gearbox (9 speed)
17 overdrive unit
18 overdrive selector
19 universal joint
20 differential (one to each rear axle)
21 double driving wheels
22 fuel tank
23 box girder chassis
24 semi elliptical spring
25 chassis/spring mountings
26 axle mounting

A selection of cabs and engines can be built onto the basic chassis of the commercial vehicle, depending on its intended use. The vehicle shown here has a general purpose lightweight fibreglass cab and a six cylinder 330 hp engine with a nine-speed gear box. It has power steering and an air pressure braking system for all six road wheels.

Opposite page: one of the most useful of special purpose vehicles is the fork-lift. Some of them are powered by electric motors or by bottled gas.
Below: the device on top of this airport fire truck is for spreading a large amount of foam on a crash scene as quickly as possible.

to offer trucks with engines in front of the driver, so that engine noise and heat reaching the crew is kept to a minimum.

Diesel engines of vee-formation, from companies like General Motors, Deutz and Perkins have helped the truck designer to minimize engine intrusion into the forward-control cab. Being shorter, the vee-engine is also accommodated more readily under a shallow cab.

Of the various non-reciprocating power units suggested for heavy trucks, only the gas turbine shows the promise of achieving operational viability and then only on non-stop motorway or turnpike running, where at constant load and constant speed its fuel economy approaches that of an equivalent diesel engine. The light weight and compactness of the turbine remain its chief attractions.

In Britain and the United States, truck transmissions are normally of the constant-mesh type, where a degree of driver expertise is called for to achieve clean shifting, although the use of a 'clutch-stop' on gearboxes such as the Fuller Roadranger partially synchronizes gear speeds before engagement.

Synchromesh transmissions are now almost universal on trucks built in mainland Europe and Scandinavia; the manufacturers concerned maintain that private car standards of control should be provided for the truck driver. Fully automatic transmissions of the General Motors Allison type are being fitted increasingly in heavy trucks, particularly those which operate occasionally off the highway. The torque converter fluid coupling is an aid to traction on soft ground.

Bus chassis, which in most cases have the same kind of parallel ladder frame as a truck, often use semi-automatic transmissions to make the driver's job easier, and to minimize clutch-wear and-tear on stop-start operating.

As many as 16 forward speeds are built into some of today's manual truck gearboxes. The Volvo SR61 transmission fitted in the Swedish company's 50 ton F88 and F89 chassis is basically a simple four-ratio synchromesh unit. But a step-down range on the output side enables the same four pairs of gears to serve as 1st, 2nd, 3rd and 4th speed and then as 5th, 6th, 7th and 8th respectively. An optional further gear train allows each of the eight ratio steps to be 'split' giving 16 speeds in all.

A truck's driving axle may also double as a gearbox. The two-speed axles produced on both sides of the Atlantic by Eaton provide an inexpensive means of effectively doubling the number of transmission ratios at the driver's disposal.

Power brakes are universal on commercial vehicles, except the smallest car-derived vans [panel trucks]. Trucks of up to about 14 tons total weight have a hydraulically-actuated brake, with vacuum or compressed-air servo assistance. Heavy trucks are braked using air alone, operating diaphragm chambers mounted behind each wheel. Large springs are now built into these chambers to apply the brakes for parking on many heavy trucks.

On articulated vehicles the brake lines are arranged to 'fail-safe', so that in the event of the trailer becoming uncoupled its brakes will be applied.

Anti-skid equipment which moderates the braking force when wheel-lock is imminent is now being fitted on many articulated vehicles where locking of the driving axle wheels can lead to 'jack-knifing'—a highly unstable condition where the tractor pivots out of control around its semi-trailer coupling. Less advanced equipment designed to combat jack-knifing includes the load-sensing brake valve which regulates the air pressure applied to the brake chambers in accordance with the load on the axle.

protective structure
hydraulic jack
hydraulic controls
handbrake
control unit
footbrake
air filter
diesel engine
radiator filler
radiator
hydraulic pump
transmission
fuel tank
steering rear axle

The tractor One of the most common sights on the farm today is the agricultural tractor, these machines having almost completely replaced the draught animal for farm work in the more developed parts of the world.

Experiments in tractor design using a single cylinder engine mounted on the chassis of an old steam engine were carried out by the Charter Gas Engine Company of Chicago in 1889. Several other firms and individuals produced prototype machines, most of which had some success, and by 1900 a number of crude but effective machines were being produced in America.

In Great Britain in 1904, Dan Albone of Biggleswade produced a twin cylinder machine developing 20 hp with two forward gears and a reverse. Weighing a total of 1.5 tons, this small and efficient machine was the culmination of several years' work and was sold in considerable numbers. Farmers in the United States were more receptive than their counterparts in Europe and eagerly bought machines from the sixty manufacturers in business by 1914.

World War I increased the demand for food production and the use of tractors enabled large areas of land to be brought into cultivation and cropped with cereals. Many American machines of various makes were imported into the British Isles, the most popular being products of the International Harvester Company and later the Ford Motor Company. The British Government imported 7000 Ford tractors; these machines were smaller, lighter, and generally easier to handle than their contemporaries and became very popular. By 1925 Ford was producing over half the total of tractors made in the USA, and the tractor had become accepted as a necessity on many farms and had undergone design improvements which made it much more versatile.

The most radical alteration in design was made by Harry Ferguson with the perfection of the hydraulically controlled mounted implement. By the use of three linkage arms the plough or cultivator effectively became part of the tractor, and the resistance of the soil to the implement was transferred through the upper link and became a downward force on to the tractor to give increased grip and stability.

The Ivel portable traction motor, or agricultural motor, as it was called, was first built about 1903. It had a two-stroke engine with one cylinder and a rotary valve, which burnt a mixture of 20 parts paraffin [kerosene] and one part lubricating oil. There is one gear forward and one reverse, with a friction clutch and chain drive. Note the power take-off pulley.

Most farm tractors are four-wheeled, the rear ones being driven, and the machine is constructed on the unit principle; that is, the engine, gearbox and transmission are so designed that a separate chassis is unnecessary.

This principle was first tried out by the Wallis Tractor Company of Ohio in 1913, and used with great success by Ford. The engine and clutch are directly bolted to the gearbox, which in turn is attached to the rest of the transmission. This forms a rigid backbone for the whole machine. Mounting locations for implements and attachments are usually positioned at the sides as well as the rear of the tractor, and a towbar or an automatic hitch is provided which greatly facilitates towing.

A cab for the operator is usually provided on modern machines, which not only gives protection from the weather but is so designed that in the event of the tractor overturning, the structure will prevent the driver from being crushed.

Headlights on the front and floodlighting on the rear of the tractor enable work to be carried out at night when required. A sprung seat and a roomy operating platform are also essential when working under arduous conditions.

The farm tractor varies in power from 20 to 280 hp, depending on its size and use, the smallest being used for horticultural or vineyard work. These machines are low and narrow to allow for operating under trees and often have wheels shrouded to prevent damage to branches and plants.

The rowcrop version has been developed with a high ground clearance to allow it to straddle crops such as maize, cotton, potatoes and vegetables. Implements for inter-row cultivation can be attached in front of, underneath and behind the tractor, and two or more operations are often carried out at the same time, such as hoeing and the application of insecticide. Such tractors may have a tricycle configuration, having one wheel or two wheels close together at the front and large wheels at the rear, the track of which can be easily adjusted to suit the crop row width.

General purpose farm tractors range from 30 to 120 hp, and are very versatile, operating under a wide range of conditions and performing an extensive number of tasks such as ploughing, sowing, cultivation, crop spraying, haulage and working hay and harvesting machinery.

The advantages of better traction and increased efficiency are obvious and today most manufacturers offer a four wheel drive (FWD) machine. These are usually in the higher power range of 75 to 250 hp, except for specialist machines, such as vineyard tractors. Many FWD tractors are conversions of the basic machine, having a drive shaft brought forward from a transfer gearbox to a modified front axle, or in some cases a hydraulic motor on each front wheel.

The larger machines are capable of very high work rates over difficult terrain. Steering arrangements vary, some having simple front wheel steering, or to give extra manoeuvrability all four wheels steer. In some examples the machines may be articulated, giving a smaller turning circle and the advantage of allowing the two halves to adopt different angles, which ensures a greater ground grip.

Tracked tractors find favour under wet soil conditions because of their low ground pressure, and are advantageous on hard dry soil when the wear on rubber tyres would be excessive. They are usually used for ploughing, heavy cultivation and drainage work, and also for drilling cereal crops when several seed drills can be pulled in formation. Operating speed is lower than wheel tractors, but wider implements can be pulled.

The diesel engine is now almost universally used in the

This Massey Ferguson tractor has an enclosed cab for safety and for driver comfort. It is pulling a general purpose cultivator used for stubble cleaning (churning up weeds), preparing seed beds and so forth. The modern tractor with all its attachments provides an entire cultivating system; some of the largest models can do so much work in one pass that no further field work is necessary until the crop is harvested.

agricultural tractor, being economical and having a good torque range; in some areas gasoline or LPG (liquefied petroleum gas) engines are used to suit local circumstances. The engines have from 1 to 8 cylinders, and turbochargers can be used to increase power outputs. In most cases a four-stroke engine is used. Electric starting is now universal. The engine is usually governed, and operating speeds are mostly in the range of 800 to 2800 rpm. A few air-cooled engines are fitted in tractors but watercooling is much more common. Because of the difficult conditions in which they work, very efficient air and oil filters are needed and access for servicing has to be easy for all except major overhauls.

Tractors have a very heavy duty type of clutch, often ventilated for cooling and operating in two stages. The first stage stops the forward motion of the machine and the second the power take-off shaft and hydraulic pump.

Because of the wide range of operating conditions, many of which have to be carried out at a precise speed (for instance potato planting), a large number of forward gears are desirable and to achieve this a three or four speed gearbox is commonly used together with second high-low ratio. This is often augmented by an epicyclic system to double the number of gears available. Synchromesh is sometimes used to facilitate easier gear changes, and hydraulic transmission is available to give an infinitely variable range of speeds. As considerable stresses are incurred, the whole transmission is much more strongly constructed than that of a road vehicle of a similar horse-power. The power take-off (PTO) shaft is a means of transmitting power to a mounted or trailed implement, and was originally at the rear of the tractor, but in recent years additional PTOs have often been located at the side and or front of the tractor. These can be used for mid and front mounted equipment.

The development of the independent PTO in 1947 was a major improvement. This enables the PTO to be used without interfering with the forward motion of the tractor and is extremely useful when carrying out such tasks as hay baling and forage harvesting. Ground speed PTO, the speed of which is proportional to the rate of travel, is used in certain planting operations.

The use of hydraulic power has been a major factor in making the farm tractor such a versatile machine. At first external rams were used to raise or lower trailed or rowcrop equipment, and for tipping trailers and operating manure loaders. With the invention of the Ferguson three-point linkage, the whole concept of the tractor changed from being a replacement for the horse, which pulled the implement, as the implements are now usually mounted on the tractor. Larger capacity pumps allow hydraulic motors to be used on such machines as hedge cutters and sugar beet harvesters, for example, and by a system of valves several auxiliary services can be used at the same time or in sequence.

A differential lock, usually operated by a foot pedal, is a mechanism which prevents the differential from working in the normal way. Under difficult working conditions, this enables full engine power to be transmitted to both rear wheels. This reduces wheel spin and so increases the tractor's ability to travel through wet areas.

Brakes are extremely important on a tractor, and are used not only for stopping the machine but are also operated individually for making tight turns during field work. Disc brakes are commonly used, as these have a powerful action, are fade resistant, have a prolonged life and require little maintenance.

Ackermann steering is nearly always used on two-wheel drive tractors. Power steering is optional on the smaller two wheel drive tractors, but a standard feature on the heavier and FWD machines. Front-wheel bearings are very sturdy to withstand the shock loads imposed. Suspension of the agricultural tractor is difficult because of the interference this would cause with the traction of the machine, so driver comfort is usually left to the pneumatic tyres and the seating arrangements.

The tracked vehicle A tracked vehicle is a cross-country machine which lays its own road, runs along it, and picks it up behind. The track is made either of metal links pinned together by hinged joints, or of continuous rubber belting. It spreads the weight of the vehicle over a wide area, so reducing sinkage on soft ground. Some tracked vehicles are articulated, the main body being split into two parts, linked by some form of joint. Single-bodied vehicles are those such as tanks or bulldozers,which need to be short enough to do pivot turns, and whose function and layout do not allow the body to be split into two.

Articulated vehicles are long in relation to their breadth and so have less rolling resistance on fertile soils such as clays, or on snow, as the track marks are narrower for a given sinkage compared with single vehicles. They range from small snow vehicles, rather like a tracked scooter, to huge Arctic load carriers. In some the two parts are not of equal size, having the relationship of an articulated truck tractor and trailer. The tracks in all parts of the articulated vehicle are driven.

Articulated tracked vehicles are steered by bending the vehicle at the joint, usually by hydraulic rams. Single bodied vehicles steer by slowing down the inside track and speeding up the outside one; this is done by the transmission. For slow-moving vehicles it is sufficient to disconnect the drive to the inside track by a clutch, then apply a brake to hold it. This system is often used as supplementary steering for fast tracklayers. The World War II Tetrarch tank, and the Bren Gun Carrier, had an excellent system in which the tank's wheels moved sideways, and laid the track on the ground in a curve.

The layout of wheels on which the vehicle runs on the track, and any springing, depends on the speed required. Slow moving machines such as crawler tractors and bulldozers, only doing about 6 mph (10 km/h), have no springing. The wheels are of steel, and run on the track in a manner similar to railways. For high speed, about 40 mph (65 km/h), such as achieved by tanks, other armoured fighting vehicles (AFV) and snow vehicles, springing systems are needed, and the wheels have rubber tyres, solid for AFVs and pneumatic for snow vehicles. The most common springing system is steel torsion-bars running across the floor of the vehicle, with the wheels mounted on the ends of trailing suspension arms. Large shock absorbers are needed to stop bounce.

For low rolling resistance and good tyre life, the wheels must be large, and in order to spread the load evenly along the track there must be as many as possible. The best compromise is five or six each side, as large as can be fitted. The German AFVs of World War II often had interleaved wheels to allow many large ones, but these clogged with mud.

Most high-speed tracked vehicles have an idler wheel, not carrying any weight, and adjustable to tension the track. Vehicles with small wheels, due to some design constraint, may need small support rollers to hold the top run of the track. At the opposite end of the vehicles to the idler is the drive sprocket.

The sprocket teeth engage the track to propel the vehicle. The transmission delivers the engine output to the sprocket at suitable speed and torque and provides steering speed differences for the inner and outer tracks. When steering, the power available from slowing the inner track must be 'regenerated', that is, passed across to the outer track. The simplest layout would have the engine drive an electric generator or hydraulic pump, with motors at each sprocket, but this has limited speed range, and needs a complex control mechanism if it is to be regenerative. Normally the main

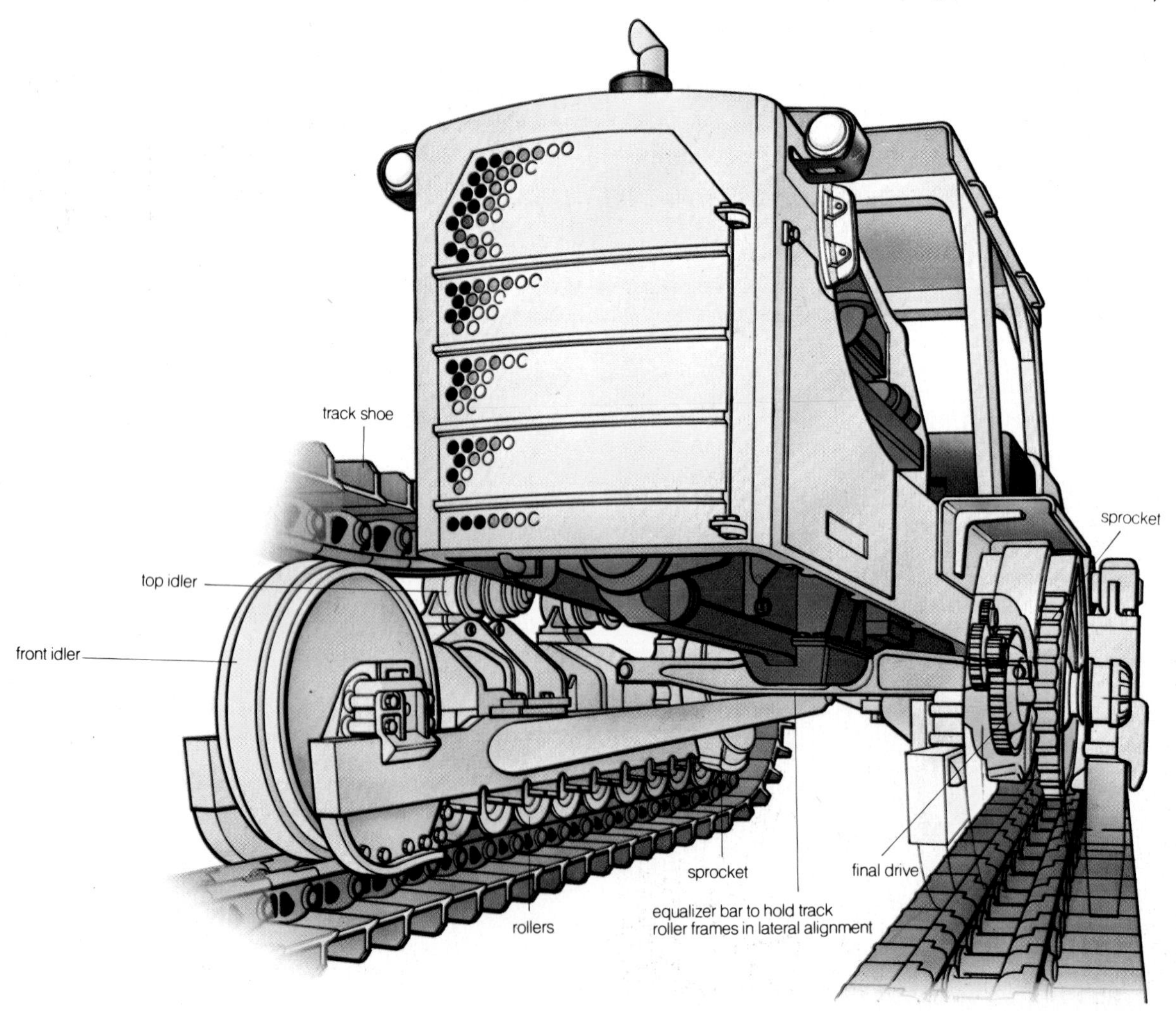

drive is by a conventional gearbox, with this supplemented by the steering drive, the two fed together in differentials. All is usually housed in one unit, using mechanical gears, though recent designs split part of the power through hydraulics. Convenience in the overall layout of the vehicle design is the major influence on choice of front or rear drive.

Slow vehicles such as bulldozers have steel tracks, with steel-pinned joints. The tread has a sharp grouser (projection) for grip. High-speed vehicles have various layouts with rubber-padded soles to limit damage to roads in peacetime, and the pins joining them are sometimes plain steel, but more often rubber-bushed. The links are normally steel, although aluminium can be used where weight-saving is important. For snow and marsh vehicles the tracks must be so wide rubber belting is usually used, for lightness, reinforced by steel cords within the belt, and steel cross-bars as grousers for grip and for the sprocket to engage, and shaped to guide the wheels. These 'band tracks' are, however, vulnerable on stony ground.

There have been many successful vehicles that had wheels at the front and tracks at the rear. Pioneer work was led by Kegresse, with tracks at the rear of a Citroen truck. In the 1930s the Germany Army developed halftracked armoured personnel carriers, followed soon after by the Americans. These vehicles were derived from trucks. The engine was at the front, with a conventional front axle whose wheels provided the steering. The Germans used a steel link track, of high quality but complex design, the pins being sealed and lubricated. The Americans used a rubber band-track for their M3 halftrack, which went into production in 1940 and was still in use in the Arab-Israeli war of 1973. Halftracks were compromise designs without the clear advantage of either tracks or wheels. They were popular at a time when suitable designs for efficient tracked steering systems were not available, and they used existing truck production facilities. Now steering systems and other components for full tracked vehicles are available commercially in a wide range of sizes.

Soft ground ability has normally been quoted by reference to nominal ground pressure, the relationship of weight to area of track on the ground. Now, mean maximum pressure is used, applying to both wheeled and tracked vehicles, and taking account of actual pressure due to wheel size and spacing, and track link design. A tracked vehicle will have a better soft ground performance than a wheeled vehicle of similar size and payload. Tracks tend to be more expensive and noisier than wheels, and need more servicing. Wheeled AFVs—armoured cars—are chosen for cheapness and ability to patrol high mileages on roads. Tractors do not often justify tracks as they will usually pull on a draw-bar, transferring weight to the rear so the track does not develop full length ground contact. Agricultural tracked tractors have developed into slow unsprung vehicles of poor performance on long hauls, such as earthmoving, or carting between field and farm, so are only used in extreme cases. Tracks are needed for working on sharp stones or rock that would lacerate tyres, and on vehicles that cannot afford to fail due to soft ground, such as tanks and snow or marsh vehicles.

Electric vehicles Electric vehicles are often thought of as a new development, but as long ago as 1837 Robert Davidson of Aberdeen built an electric carriage powered by a crude iron-zinc battery and driven by a very simple electric motor, which contained all the basic elements of the modern electric vehicle.

The advent of the lead-acid battery allowed the first commercial battery-operated vehicle to be introduced in 1881 by the Paris Omnibus Company. London had its first electric bus in 1888, and also the world's first mechanically propelled taxicabs which were built by W C Bersey in 1897 for the London Electric Cab Co Ltd, and operated for two years.

Around the turn of the century the first land speed record was set in France by a battery driven car. In 1898 the flying kilometre was completed at an average speed of 63 km/h (39 mile/h), and in 1899 this was raised to 106 km/h (66 mile/h). by the Belgian, Camille Jenatzy, in his bullet-shaped electric car called *La Jamais Contente*. By 1902 technical improvements enabled Charles Baker, an American, to attain over 85 mile/h (137 km/h), but unfortunately on his third and

Above: a halftrack vehicle crosses an icefield. It has a conventional front axle with steered wheels.
Left: the grousers on the track sections of this dozer penetrate the ground, providing a high coefficient of traction. Optional shoes can be fitted for different conditions. Each track shoe is bolted to a pair of track links spaced by a pin and bushing which provide the necessary articulation to allow the chain to pass round the driving sprocket and front idler wheel.
Opposite page: the steel wheels run on a track in a fashion similar to a railway. A shock mounted front idler guides the track as it is delivered from the drive sprocket at the rear via the top idlers.

officially timed run the car suffered a mechanical failure and so no official world record was obtained. The most recent speed record for an electric car was attained by the 'Silver Eagle' developed by Eagle Picher Industries in the USA. In August 1971, at Bonneville, the car completed the flying kilometre at an average speed of 152.59 mile/h (about 245 km/h) and covered a mile from a standing start at an average of 146.437 mile/h (about 236 km/h).

By the early 1900s a high proportion of the cars on the roads of London, New York and Paris were battery electric, and electric cars became very fashionable. The rapid development of the internal combustion engine, however, accelerated still faster by World War I, meant that by the 1920s electric cars could no longer compete in terms of speed, acceleration and range. Between the wars many electric car companies came and went. One of the best known English makes was the Partridge Wilson 'Brougham' of 1936, powered by a 60 volt 324 ampere-hour battery which gave it a claimed maximum speed of 32 mile/h (51.5 km/h) and a range of 60 miles (97 km) per charge. This was typical for most electric cars of that time, and would be considered insufficient for most purposes by the average user today.

The factors which lead the private buyer to reject the electric car did not affect the commercial user in the same way. Electric vans and trucks were developed in parallel with the cars, and today in Britain for example there are some 55,000 electric vehicles in use performing a variety of tasks such as local authority duties, milk delivery and postal work, all of which involve a significant amount of stop-start driving. The electric milk truck is ideally suited for this arduous stop-start kind of work, and whereas the diesel-powered version lasts only 3 to 5 years the electric one has a useful life of over twenty years.

The electric vehicle is basically a very simple machine; it can be said to have only five basic components and eight moving parts, four of which are wheels. In essence it consists of the battery, a controller, the motor, the transmission, and the vehicle chassis and body, and it was for this reason that so much interest was shown in it in the early days of powered transport. Apart from the battery, controller and motor, the design of electric vehicles usually follows conventional lines.

Most batteries in use today are of the lead-acid type. Originally based on the flat-plate design, modern traction cells are now of tubular construction. This type of battery, provided it is regularly charged and the electrolyte level is maintained, has a guaranteed life of four years, which is regularly exceeded in practice. It is, however, rather heavy and the energy density (the amount of energy produced per unit weight) is only about 11 watt-hours per lb (approximately 5 Wh/kg), and this is the feature principally responsible for the poor road performance of most electric vehicles. Where vehicles are used comparatively infrequently, as in mining and tunnelling, nickel-iron batteries are often used as they can stand long periods of non-use without attention.

The motor in most electric road vehicles is a low-speed DC series-wound type driving the rear wheels through a conventional back axle arrangement. Speed control is carried out in two main ways. The simplest is to change the voltage applied to the motor by tapping the battery in a combination of series-parallel connections. For example, if a 48 volt battery is used, it is initially connected for 12 V to limit the high current needed for starting, and once the vehicle is on the move the controller switches the connections to give 24 V, then 36 V and ultimately the full 48 V. Most electric vehicles are controlled via a conventional type of accelerator pedal, the only other foot pedal being that for the brake. As

the only controls needed in electric vehicles are the accelerator, brakes, and steering, they are particularly easy to drive.

A more advanced form of control of the motor speed can be provided by the use of thyristors, which are solid state 'switches' that allow the current to be supplied in bursts. For starting and acceleration the thyristor conducts for short periods limiting the energy supplied. As the speed increases the length of the thyristors' conducting periods are increased so that maximum power can be developed.

Considerable interest has been shown recently in electric road vehicles, because the high levels of noise and pollution so common with petrol [gasoline] and diesel engines are absent. In terms of resources, the electric vehicle also competes on equal terms, being no less efficient and frequently more efficient in overall energy consumption than its internal combustion counterpart.

Extensive research programmes are under way in many parts of the world to produce designs which will provide road performances comparable to conventional vehicles. The major difficulty with the electric vehicle has been to find a battery which will give sufficient energy from a realistic weight. The standard lead-acid traction cell is really too heavy, although the most recent designs will now give 16 to 18 watt-hours per lb (7 to 8 Wh/kg) and working lives of up to 800 charging cycles are predicted. Using this type of battery, Chloride/Selnec in Britain and RWE/Varta/MAN in Germany have produced buses capable of carrying 50 passengers and operating conventional services, both types being capable of about 45 mile h (72 km/h) with a range of about 40 miles (64 km). To prevent the vehicles having to stand idle for lengthy periods while their batteries are recharged, they are designed so that the run-down batteries can be quickly exchanged for a fully charged set. The MAN bus carries its batteries in a trailer, and in the Chloride bus they are under the floor.

Many light electric vans are now being developed. In Germany a conversion of the standard Volkswagen van is so designed that the battery can be rapidly charged, and in Britain the Electricity Council and J Lucas Ltd have produced modified versions of British Leyland and Bedford vans. The former uses a mechanical gearbox and a high speed motor and the latter a variable speed motor controlled by thyristors. Both are capable of up to 50 mile/h (80 km/h). In the USA the Electric Vehicle Council have ordered a large number of vans

Above: the batteries for the MAN electric bus are carried in a trailer, so they do not take up any space in the bus itself. In this picture, the batteries are being changed for freshly charged ones.
Left: the batteries of the Ford 'Comuta' electric town car are carried on the central part of the chassis, and the two electric motors are at the rear.
Opposite page: Camille Jenatzy of Belgium with his record-breaking electric car La Jamais Contente, *in which he raised the world land speed record to 66mph (106kph) in 1899 at Acheres, France.*

from Battronic for use by electricity utilities throughout the country. Similar schemes are being undertaken in many other countries; Japan in particular is investing heavily in research.

The van is a particularly useful test vehicle in that the extra space available compared with a car allows new energy sources to be tested easily. In the USA for example General Motors have produced a fuel-cell-powered van with liquid hydrogen and oxygen being carried in cryogenic (low temperature) tanks. In Britain the Electricity Council has used one of its van conversions to demonstrate the first sodium-sulphur battery, which gave an energy density of 30 Wh/lb (66 Wh/kg). Later developments are expected to give as much as 70 Wh/lb (154 Wh/kg). Many other types of battery are also being tested such as the zinc-chlorine (Udylite, USA) and the zinc-air (Sony, Japan).

The electric vehicle development which creates most public interest is the battery car. To date no really successful product has been sold, and early visions of cheap electric town cars are receding. The majority of cars so far produced have been conversions of existing products, and although the major car manufacturers have produced or sponsored prototype designs they have as yet made no really significant contribution. In France, Electricité de France have produced conversions of the Renault R4 (reducing it to a two-seater in the process) and in Italy ENEL have converted a Fiat. In the USA there have been successful conversions of large-bodied cars which have speeds of up to 60 mile/h (97 km/h) and ranges of up to 100 miles (161 km), but no large-scale production has been started.

The electric car as a product presents several design problems. It is becoming more difficult to design small cars which will comply with new safety regulations, and the extra mass of batteries in the electric car present additional difficulties. The majority of purpose-built cars have so far been lightly constructed vehicles unlikely to pass stringent tests. The Enfield 8000, which has been ordered by the Electricity Council in England, has overcome these problems. This two-seater, which is designed to have a low aerodynamic drag and rolling resistance, is powered by electric batteries and has a top speed of over 40 mile/h (64 km/h), rapid acceleration and a range under normal conditions of 40 to 50 miles (64 to 80 km). This performance allows it to operate well in normal traffic, making it an attractive town car.

A recent development is the introduction of lightweight electric motorcycles, such as that produced by the Austrian firm of Steyr-Daimler-Puch, which may prove to be a useful form of personal transport in busy city centres.

The motorcycle

The first motorcycles, built at the beginning of this century, were essentially pedal cycles to which crude internal combustion engines had been added to assist the rider. Frequently, the rider had to assist the engine. (Nowadays, motorized pedal cycles, called mopeds, are used throughout the world; the engines are small but so efficient that the pedals are retained only for starting and to take advantage of licence tax concessions.)

Before World War I, motorcycles had achieved speed and stamina as a result of intensive development for racing. The frame became heavier, the seat was lowered, the front forks incorporated springing to protect the rider from road shocks, and the engine was mounted where the pedal cranks had been previously. Though in some cases the engine still turned the wheels by means of V-belts running in pulley wheels, up-to-date machines already had a gearbox providing two or three gear ratios between the engine and the rear wheel, a friction clutch, and power transmission by means of roller chains. The final refinement was a kick starter, a single pedal crank which enabled the rider to start the engine without having to push the machine.

Between the wars this basic design was improved in detail and performance, and Britain dominated the motorcycle industry with well-constructed machines of conservative design. The most popular type of machine had a one-cylinder engine; attempts to market more advanced designs with four-cylinder engines and shaft power transmission met with commercial failure.

Since World War II the motorcycle has become more popular than ever for transport and sport. After brief periods of influence by Britain, Italy and Germany, Japan emerged as the major producer of motorcycles, building highly developed designs in greater numbers than ever seen before, as greater prosperity in emergent nations made personal transport attainable. Japanese success stimulated design in Europe, with the result that the range of machines available is greater than ever.

Today motorcycles are built in sizes ranging from 50 cc (engine displacement expressed in cubic centimetres) models for beginners to 1000 cc and over for experts. Most motorcycles have two or three cylinders, except for small inexpensive models; a few have four cylinders and six-cylinder models have also been built. Rotary engined models are in the experimental stage.

Almost every machine has hydraulically damped springing for both wheels, which has greatly improved safe handling and road holding qualities; many other features have been taken from car design. Electric starters are provided on the more expensive models, together with disc brakes, in which friction pads are pressed by hydraulic pressure against a steel disc on the wheel, and flashing turn indicators. Instrumentation in general follows car practice, with matched speedometer and engine revolution counter (tachometer) and warning lights to monitor lubrication and electrical systems. Purchase and running costs frequently approach car levels; the more advanced machines cost more than a small car and compare unfavourably in fuel consumption. The attraction to the user is speed, 120 mph (193 km/h) being obtainable with stock machines and 150 mph (241 km/h) or more for racing models.

Transmission (gearbox) design is peculiar to motorcycles. Four gears are normal and the trend is to five or six. No synchromesh is supplied or demanded; a ratchet mechanism provides gear changes up or down one at a time by means of foot pressure on a gear pedal. It is much quicker and more positive than manual gearbox operation in a car.

Motorcycle engines employing the four-stroke principle with a power stroke on alternate revolutions use a circulatory lubrication system similar to that on a car. Two-stroke engines are lubricated by mixing a small amount of oil with the fuel, but these engines are available only on the cheaper utility models.

Development of engine design has resulted in machines of 250 cc capacity with more power than the largest machines of a few years ago, and machines up to this size are the most popular for road use. Four-cylinder models are built on racing engine lines, with the valves operated by overhead camshafts to enable them to run at speeds of as high as 10,000 rpm. In the case of two-stroke engines, rotary valves and vibrating reed valves are frequently employed to improve efficiency.

The popularity of cross-country motorcycle racing (called motocross) has resulted in trail machines, designed for off-road use, including increased ground clearance, lower gear ratios and deeply studded tyres.

Streamlined coverings are used on some road racing machines to reduce wind drag. Cigar shaped projectiles, sometimes fitted with two engines and enclosing the driver completely, have attained speeds of more than 250 mph (400 km/h), but their only resemblance to a motorcycle is in the disposition of the two wheels.

At the other end of the scale, variants with small wheels and footboards instead of pedal rests are called motor scooters. These often have shields to protect the driver from dirt from the road or from the machine.

Once popular on the road, the motorcycle sidecar, a passenger carrying attachment with its own load-bearing wheel, is rarely seen nowadays, but the wheel arrangement is used in specialized passenger machine racing.

Motorcycle controls have become standardized. Throttle control is effected by rotating the grip on the right handlebar. On the same handlebar is a lever for operating the front brake; on the other handlebar is a lever which operates the clutch. To satisfy international driving rules, the pedal operating the gear change is usually on the left side of the machine, and a pedal on the other side operates the rear brake. Switch controls for the electric lamps are fitted to the handlebars. Most machines have two stands, a prop stand for parking and a centre stand which lifts the wheels off the ground for maintenance work.

Above: brightly painted 'tricycles' (motorcycles with sidecars) are used as taxis in Manila, Philippines.
Opposite page: this early motorized cycle was built by Daimler in 1885. A belt from the motor drives a small gear inside a large annulus gear on the rear wheel.

dynamo
oil tank
rear shock absorber
chain drive
chain case
swinging arm frame
adjusters
footrest
kick start
stand

Motorcycles today, compared to those of 60 years ago, have lower seats, heavier frames and shock absorbers front and rear. These developments make them safer than they were before, by lowering the centre of gravity and providing better handling characteristics. Clutch and gearbox operation is quicker and more positive than that of a car.

FLYING MACHINES

In the legendary story of Icarus, wings made of feathers and wax were flown too near the Sun; the wax melted and the presumptuous man who would fly like a bird fell to his death. Leonardo da Vinci, a genius who might have been ahead of his time in any age, sketched flying machines; slowly, painstakingly, scientific data were collected and tested by men like Otto Lilienthal, photographed here a few weeks before his death in a glider accident in 1897. Finally, in 1903, the Wright brothers made the first sustained flight in a heavier-than-air craft. It is not often that such an achievement can be so precisely credited, but the two bicycle mechanics from Dayton, Ohio, had realized man's oldest dream.

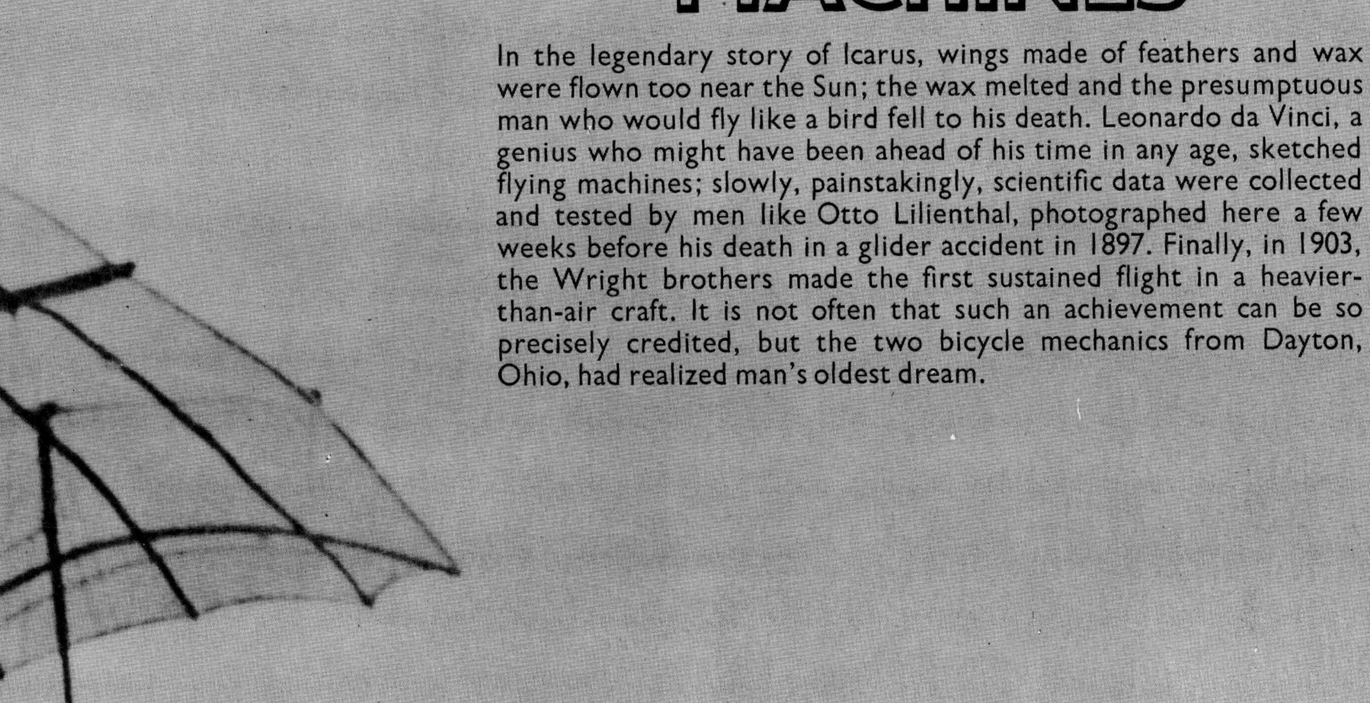

Balloons and airships

The first manned flight took place by means of a hot air balloon, on 21 November 1783. The balloon, made by the Montgolfier brothers Joseph and Etienne, paper manufacturers, was 75 ft (23 m) high by nearly 50 ft (15 m) in diameter—a frail affair of cloth backed with paper and heated by a furnace burning chopped straw. The pilots were Pilatre de Rozier and the Marquis d'Arlandes.

Only a few weeks later Jacques Charles, also French, made the first ascent in a hydrogen-filled balloon, and almost at once the gas balloon established itself as superior to the hot air version. Although it took longer to inflate it was quiet, easier to handle, lifted more, and above all could be used again. The hot air balloons had a tendency, not surprisingly, to set themselves alight or at least to finish their flights in a charred and brittle condition.

The use of hydrogen balloons grew unchecked until the development of airships. They proved useful in both research and war, in the latter particularly as military observation platforms, for which purpose they were used by both sides in the American Civil War and again in World War I. They were generally captive balloons, anchored to the ground by long cables. In sport, they reached their peak in the famous Gordon Bennett races which took place in Europe between 1906 and 1938, in which the greatest distance covered was 1368 miles (2191 km).

In 1955 the US Navy were operating blimps (small airships) for offshore patrols, and pilots were being trained on gas balloons. In an attempt to cut costs, the Navy sponsored a programme to develop hot air balloons and this culminated in a practical design being produced by a company in South Dakota, USA, in 1963. The Navy lost interest in the project, but the company decided to market the design as a sporting balloon, and the modern re-usable hot air balloon was born. Hot air balloons are now manufactured in varying sizes from 30,000 to 140,000 cu ft (850 to 3964 m³) to carry from one to six people.

A balloon is a lighter-than-air unpowered aircraft. Unlike heavier-than-air machines, which stay airborne by moving through the air to create dynamic lift, a balloon obtains its lift by displacement, which is a static force and does not require movement through the air to create it. An airship is fitted with one or more engines and also with controls (rudder and elevators), but a balloon has no engine and cannot be steered: it merely drifts with the wind.

A balloon is in equilibrium, that is, balanced in the air and not moving up or down, when its total weight is the same as the weight of the volume of air it is occupying, or displacing. Since the fabric, basket, crew, and equipment are all heavier than air, they must be balanced by filling the envelope with some gas lighter than air. When the difference between the weight of the gas in the envelope and the outside air is the same as the weight of all the components making up the balloon, the balloon will be in equilibrium.

Gases which are lighter than air, and therefore suitable for filling a balloon, include hydrogen, helium, and ordinary air which has been heated. Some gases are lighter than air because the weight of their molecules is less than the average weight of the molecules in the air, that is, they have a lower density than air. Hot air is lighter than cold air because any gas (and air is a gas) expands when heated. The molecules are driven further apart and therefore a given volume will contain a lower weight of molecules. At sea level 1000 cu ft (28 m³) of air at 212°F (100°C) will have a lifting capacity of 17.4 lb (7.9 kg) when the surrounding air is at 60°F (16°C). A similar amount of helium has a lifting capacity of 65 lb (29.5 kg), and for hydrogen the lift available is 70 lb (31.75 kg).

From the point of view of lift alone, it is obvious that the best thing to fill a balloon with is hydrogen and the worst is hot air. Ordinary coal gas can, and has been, used to fill balloons. Its disadvantage is that it varies considerably in purity, and the lift obtainable is therefore very unpredictable. The domestic gas of today weighs more than the gas available in the early days of ballooning. Gases are expensive, since they have to be extracted from the air and transported. It costs considerably more to inflate a hydrogen balloon than it does to inflate a hot air balloon. Hydrogen is highly inflammable, but helium, which is safer because it neither burns nor forms explosive mixtures with air, is much more expensive than hydrogen in most countries, except the USA.

These factors aside, there are other aspects of gas and hot air which have resulted in the development of two distinct types of balloon, different in design, performance, and to some extent in the manner in which they are controlled.

Gas balloons Hydrogen easily penetrates most materials and the fabric of a hydrogen balloon envelope is therefore quite heavy, about half the total weight of the balloon, made from fabric impregnated with rubber or neoprene. The envelope is spherical, the most efficient shape to contain a given volume, and is contained in a string net which distributes the load evenly over the fabric. Below the envelope, the net is drawn together at a load ring, from which a basket to carry crew and equipment is suspended.

The envelope is not sealed: at the bottom is a long narrow

open tube called the appendix. As the balloon rises, atmospheric pressure decreases and the gas in the balloon expands. The appendix allows gas to escape, thus preventing the balloon from bursting as a result of its internal pressure. When the balloon descends, the appendix closes (like a wet drinking straw if it is sucked too hard). This prevents air from getting into the balloon and forming an explosive mixture with the hydrogen.

To ascend in a hydrogen balloon it is necessary to reduce weight by discharging ballast in the form of sand, since lift cannot be increased. Similarly in order to descend it is necessary to reduce lift, since additional ballast cannot be obtained while airborne. This is done by opening a small valve in the top of the balloon; the valve is operated by a cord down to the basket and is held shut by springs or elastic. The height to which a balloon can rise is limited because the density of the atmosphere decreases (the air becomes thinner) as the height above sea level increases. Therefore, as the balloon rises, the air around it becomes less dense, until it reaches a height at which the atmospheric density is as low as the total density of the balloon. At this point, as the densities are equal, the balloon is in equilibrium with the air and will not rise any further.

The small valve and the ballast are the only forms of control in a gas balloon. Once the balloon is in equilibrium, they will not need to be used much unless the balloon is affected by outside factors, such as a general cooling or heating of the outside air, which will cause the lifting gas to contract or expand. Gas balloons, therefore, have very good endurance. On the other hand, every manoeuvre in a gas balloon is a 'wasting' process: even to descend it is necessary to get rid of gas.

Operationally, then, the main advantages of gas balloons are endurance and lifting power. The main disadvantages are the cost and inconvenience of inflating the balloon, a process which can take as long as two hours.

Hot air balloons Modern hot air balloons are almost the only really significant design development to have taken place since balloons were invented. Despite their relatively poor lifting power and endurance, they have been responsible for a tremendous upsurge in the sport because of their low running costs, their simplicity, and their safety.

Structurally a hot air balloon is quite different to a gas balloon. Hot air cannot penetrate fabrics in the way that hydrogen can: the envelope is therefore made of very light material, usually rip-proof nylon treated with polyurethane to reduce porosity.

The profile of a hot air balloon is termed 'natural shape': wide at the top tapering towards the bottom in the shape naturally created by the internal pressure. Loads are carried on tubular nylon tapes sewn into the integral with the envelope. From these tapes, steel wires lead down to the burner, which is in the same position as the load ring on a gas balloon.

The basket is suspended from the burner by steel wires or a rigid structure, depending on the manufacturer. At the base of the balloon there is a large opening to allow heat from the burner to enter. A modern hot air balloon burner uses propane, which is fed under its own bottle pressure to the burner jets. The heat generated may be anything from 3 to 5 million Btu (approximately 3 to 5 GJ) per hour, much more than many industrial space heaters.

Hot air balloons are fundamentally different from gas balloons in that it is possible to increase or decrease the lift simply by heating the air or allowing it to cool. Ballast and valve are therefore not strictly necessary. When flying a hot air balloon the pilot simply turns on the burner if he wants to ascend, and leaves it off and allows the air to cool if he wants to descend. The air in fact takes some time to cool, and the burner can be left off for quite long periods before the balloon starts to lose height. Inflation can be accomplished in a matter of minutes and this compensates for the relatively poor endurance of hot air balloons (up to about five hours depending on the load), since several flights can be made in a single day.

One control is common to both types of balloon. This is the ripping panel, a panel in or near the top of the balloon which can be opened quickly on landing to deflate the balloon rapidly. Unless this panel is opened the balloon acts like a sail and can drag the basket and its occupants a considerable distance. In the early days before the ripping panel was invented (in 1839), balloonists were on occasions dragged literally miles over the countryside before being stopped by a tree or similar obstacle. Hot air balloon ripping panels are held closed with a self fastening material such as Velcro, and can be secured ready for flight in minutes. Gas balloon panels have to be sewn and gummed to prevent leakage.

The airship There are, or have been, three categories of airship: rigid, semi-rigid and non-rigid. The rigid types consisted of a light metal framework containing several gasbags slung inside under nets, and with a separate outer cover. The German Zeppelins and most airships of the 1920s and 1930s were this type. The metals used were aluminium alloys, the outer skin was of cotton, and the gasbags were cotton lined with 'goldbeater's skin', a thin membrane taken from the intestines of cows.

The other types, semi-rigid and non-rigid, are known as pressure airships since their shape is maintained mostly by the internal pressure. The semi-rigid types had a metal keel along the length of the envelope. The Norge, an Italian airship which flew from Rome to Alaska over the North Pole in 1926, was a semi-rigid craft.

The only type still used today is the non-rigid or blimp, which has no internal framework. Modern airships are made in this way of a synthetic fibre, Dacron, coated with neoprene, a man-made rubber. Aluminium paint on the outside reflects the sun's light and heat, reducing the extent to which the interior is heated. Battens on the nose prevent the wind pressure from flattening it when the craft is moving.

Early airships used to control lift by releasing gas and replacing it with air, a wasteful method that caused a gradual reduction of lift as more and more gas was lost. This could be compensated for by carrying water ballast, which could be released to lighten the airship. But later airships replaced the system with ballonets, collapsible air bags inside the gasbag but connected to the outside air. By varying the amount of air in these with pumps, the volume of the gas in the rest of the bag can be changed. There are usually two ballonets, to the forward and rear of the gasbag, so that the balance of the ship can be adjusted.

The tailfins operate just like those on an aircraft, and are the control surfaces by which the ship is steered. Conven-

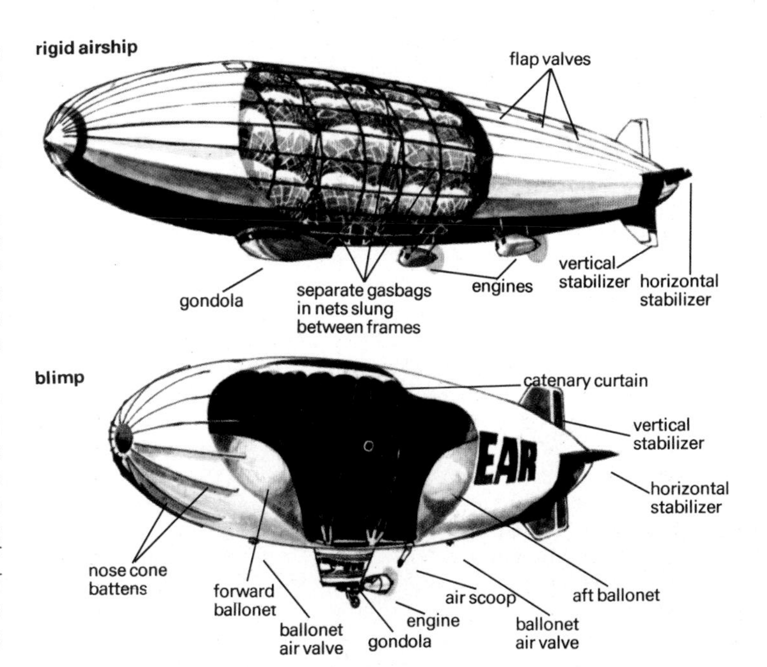

Above, top: diagrams of the rigid airship and the blimp. Inside its metal frame the rigid had several separate gasbags supported by nets. These were normally filled with hydrogen, which was vented through flap valves at the top to allow the ship to descend. The modern blimp is filled with helium, too expensive to vent, so it has internal ballonets into which air is pumped to compress the helium, making it heavier as well as bringing the blimp down by its own weight.
Above: a British airship frame of 1917 being weighed with spring balances (white discs).
Left: a Caquot balloon of the type used as observation and barrage balloons during World War II. Anchored to the ground with steel wires, they were a serious hazard to low-flying enemy aircraft.

tional elevators are used to change the altitude of the craft when it is moving; the change of atmospheric pressure with altitude is compensated for automatically by varying the amount of air in the ballonets.

The lightest gas is hydrogen, which is comparatively cheap to manufacture. But its extreme flammability has resulted in the much more expensive, slightly less effective, but completely safe helium being used in all modern airships. Helium is found in small amounts with natural gas in the United States, but is otherwise very expensive to produce in large quantities.

In the beginning it was France that led the way. After the invention of the balloon in 1783, ways were sought of making it independent of the direction of the wind. The problem was to produce a suitably light yet powerful means of propulsion, and it was Henri Giffard who first produced a 3 hp engine weighing 350 lb (160 kg). His 70,500 cu ft (2000 m³) hydrogen-filled craft ascended from the Hippodrome in Paris in 1852, and flew at 6 mile/h (9 km/h). There was an improvement on this in 1884 when the French built another airship, La France, which achieved a top speed of 15 mile/h (24 km/h) by means of a 9 hp electric motor.

Germany came into the picture in 1895 with the first rigid airship, built by David Schwarz. It was braced internally by a system of steel wires. Five years later, Count Zeppelin carried this idea further in his much bigger design, built at Friedrichshafen. This had an aluminium frame consisting of 16 hoops connected and kept rigid by wire stays longitudinally and diagonally. The design proved a success and although one was lost, more than 20 airships of the same type were built. On the power of two 15 hp Daimler engines, it made a speed of about 26 mile/h (42 km/h). In 1912, the latest of this class carried 23 passengers on a cruise of seven and a half hours. The Germans were thus well prepared to use airships for military purposes when war broke out in 1914.

The heyday of the giant rigid airships was in the late 20s and the 30s. The USA decided to use only helium in its airships, and banned its export. This meant that the large British and German craft had to rely upon hydrogen. The flammability of the gas and the lack of manoeuvrability of the ships often had appalling consequences. Many of the largest airships met with disaster, notably the British R101 in 1930, the American *Akron* and *Macon* in 1933 and 1935, and the Zeppelin *Hindenburg* in 1937.

The heavy loss of life in these crashes swung opinion against the use of airships, and they were no longer used for carrying passengers. But later, during the Second World War, the USA used large numbers of non-rigid airships without a single loss for sea patrolling. Their ability to operate for long periods of time at low speed and low altitude made them invaluable for detecting minefields and escorting convoys.

More recently, Goodyear have built four non-rigid airships, which are often used as vibration-free airborne platforms for television sports coverage.

Heavier-than-air craft

All heavier-than-air craft from a glider to a jet airliner rely on the application of mechanical energy to the air around to give an upward thrust, maintaining the craft in the air against gravitational forces. This idea is the same for autogiros, helicopters, vertical take off aircraft, and anything that might be described as an aircraft as opposed to an airship, which derives its lift by being lighter than the air it displaces.

In a glider, the energy is provided by a towing plane or a launching winch. The wings have a cross-sectional shape known as an aerofoil [airfoil] to derive lift from the forward motion, while a tailplane and fin give the machine added stability and let the pilot control the direction of flight. As soon as no further energy is supplied, the glider begins to sink, and must always come back to earth despite rising air currents—'thermals'—that might give temporary respite.

Above: In this United Press International picture, one of the most famous news photos ever taken, the giant German Zeppelin Hindenburg *burns at Lakehurst, New Jersey, on 6 May 1937.*
Right: the British airship R34 on a trial run in 1919. In July of that year the R34 crossed the Atlantic and returned, the first-ever double crossing by air. She carried a crew of thirty.

To maintain a heavier-than-air craft aloft requires a continuous input of energy—some means of maintaining the forward motion against wind resistance.

An aerofoil is a body shaped to produce 'lift' as it travels through the air. The most common example of an aerofoil is and aircraft wing, but the same principle is used to provide the driving force of the blades of fans, propellers and helicopter rotors. On some racing cars, aerofoils are installed upside down to press the car down, holding it firmly to the road at speed.

Seen in cross section, the upper side of an aerofoil is curved, and the lower side more or less flat. As it moves through the air, its leading (front) edge splits the air it encounters into two streams, one of which passes over the aerofoil, and the other under it. The streams rejoin each other behind the trailing (rear) edge of the aerofoil.

The curved upper surface of the aerofoil is longer from front to back than the straight lower surface. The air stream that takes the longer, upper route must therefore move faster relative to the aerofoil than the stream that goes underneath in order to reach the trailing edge at the same time.

The faster a fluid (such as a liquid or gas) moves, the lower its pressure—this is known as Bernoulli's principle after the 18th-century Swiss scientist Daniel Bernoulli, who discovered it. The fast air stream over the top of the aerofoil has a lower pressure than the slower one under it, and this pressure difference forces the aerofoil up from underneath.

Tilting an aerofoil so that its leading edge is higher than its trailing edge increases the distance travelled by the upper air stream, and so increases the 'lift'. The angle of tilt of an aircraft wing is called the angle of attack. The slower an aircraft flies, the greater the angle of attack its wings must have to create enough lift to keep it in the air. The increased 'nose-up' attitude of an airliner as it comes in slowly to land is quite noticeable.

The angle of attack cannot be increased indefinitely, however. This is due to the phenomenon of laminar flow. The friction between the wing and the air flowing over it causes the layer of air next to the wing (called the boundary layer) to move more slowly relative to the wing than the air further away. The same effect can be seen in rivers, where the flow near the banks is slower than in the middle.

As long as the flow over the wing remains smoothly laminar, it lifts well. But if the angle of attack is too great, the pressure above the trailing edge of the wing becomes so low that the boundary layer separates from it and the air flow becomes turbulent. As the angle increases, the point at which the boundary layer separates moves nearer the leading edge, and less and less of the wing produces lift. Finally, so little of the wing is functioning that the aircraft 'stalls' and goes into an uncontrollable dive until it regains normal flying speed.

Usually the wing will not form a continuous horizontal line, but will be divided in the middle with the tips raised by a small amount relative to the centre to give what is known as dihedral. Without this there would be nothing to keep the main axis of the wing horizontal during normal flight. As it is, dihedral results in greater lift from the lower wing when the aircraft tilts, thus producing a tendency to restore the wing to a horizontal mode.

The actual lift produced by a wing will vary with the speed of the plane. The faster it goes, the more lift will be produced; this is why aircraft have to attain a considerable speed on the ground before they acquire enough lift for take off.

At the same time higher speeds involve more wind resistance—more drag—so jets and other high-speed aircraft have thin wings to reduce drag. If a plane slows down to below what is known as stalling speed, it literally falls out of the sky, the lift being insufficient to keep it horizontal. With thin wings the stalling speed tends to be higher than with thick wings, so jet aircraft require higher take off and landing speeds.

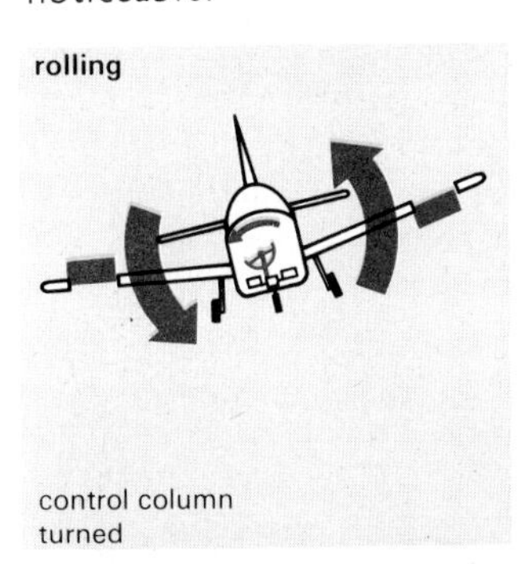

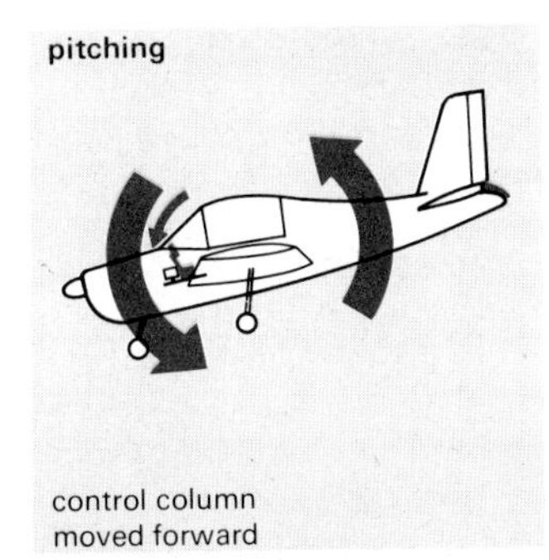

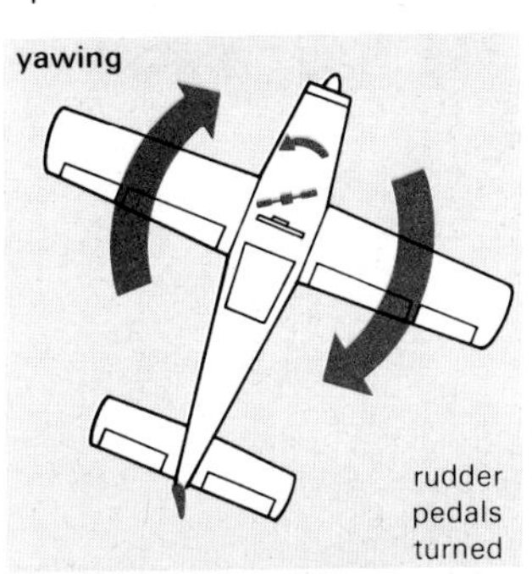

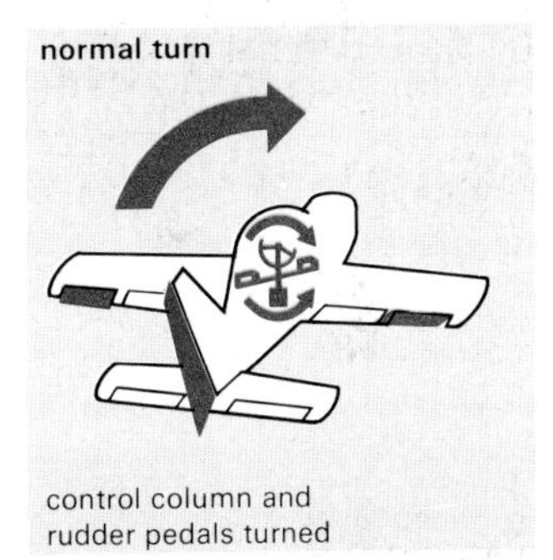

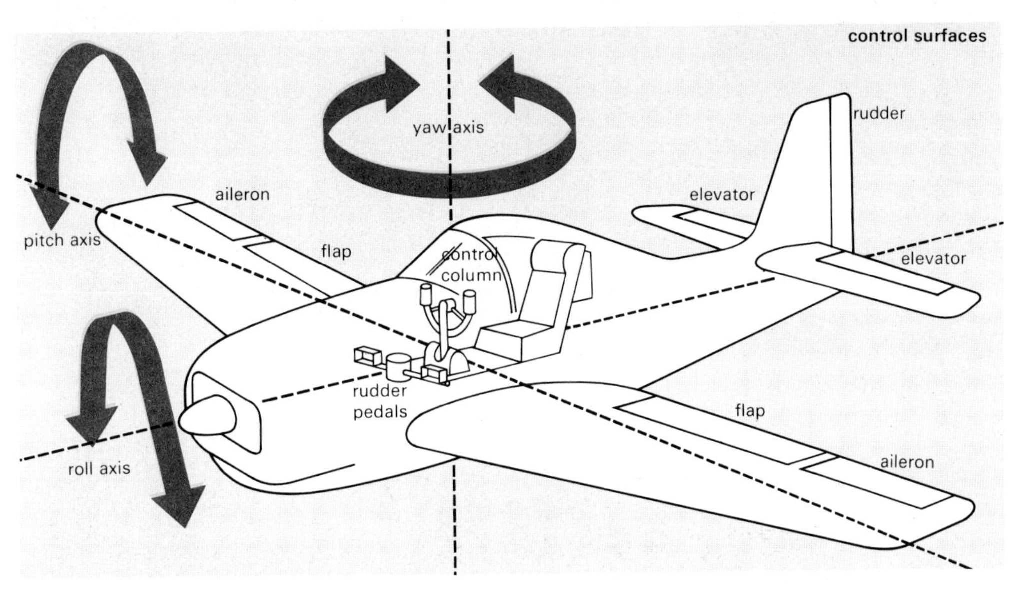

A conventional aircraft has three sets of control surfaces which tilt it about three axes. The ailerons, worked by turning the control column, cause it to roll; the elevators, worked by moving the column forward or backward, causes it to pitch (and thus to climb or dive); the rudder, worked by pedals, causes it to yaw or swivel. A normal banked turn is accomplished by using the rudder and ailerons at the same time; the inward tilt gives the aircraft stability as it turns.

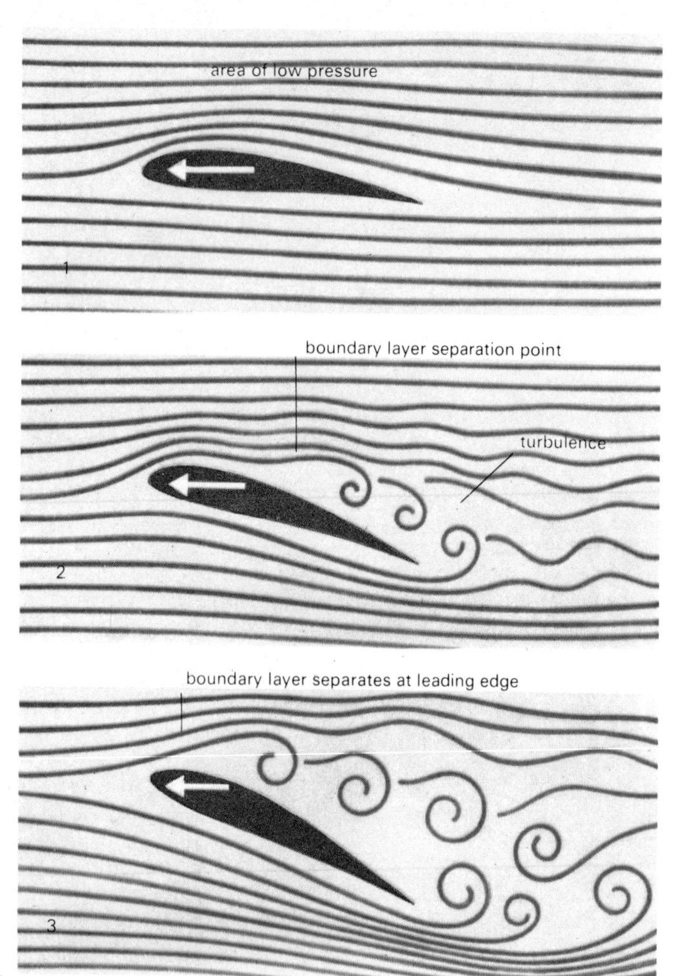

It would be difficult to control a plane if these factors could not be varied. Jet aircraft would need enormously long runways because of their high minimum speeds, while if these factors were taken care of through thicker wings, their maximum speeds would be severely cut.

Thus a device known as the flap has been developed to modify the wing section so that lift can be changed by the pilot. Part of the trailing edge of the wing, and sometimes the leading edge as well, is hinged downwards to exaggerate the aerofoil section and give more lift at lower speeds. The hinge is often arranged to open a slot between wing and flap through which air can flow to reduce turbulence. Fully extended flaps considerably increase drag, slowing the aircraft. This effect can be increased on some aircraft by opening out transverse flaps in the tops of the wings or elsewhere called air brakes.

Once an aircraft is in the air, it has to be capable of moving in three ways: in pitch—up and down; in yaw—side-to-side; and in roll.

Pitch is controlled by hinged surfaces on the trailing edge of the tailplane known as elevators. Moving these upwards curves the tailplane into an inverted aerofoil section, resulting in downward pressure on the tailplane and hence a tendency for the aircraft to adopt a nose-up or climbing attitude. Turning the elevators downwards has the opposite effect.

Yaw is controlled by a flap on the tail fin known as the rudder. If the rudder alone is used the aircraft slews sideways, but this way of turning is inexact and badly controlled. There is no counteracting horizontal force to prevent the aircraft continuing to turn regardless of the pilot's wishes. Additionally, the horizontal centrifugal forces would throw passengers and crew towards the outside of the turn.

By moving the ailerons, control surfaces at the wing tips, the aircraft can be made to bank or roll inwards at the same time as the rudder turns it, so that the aircraft tilts towards

Above: an aerofoil creates little turbulence when it is level, but as a plane slows its nose begins to rise and the boundary layer of air begins to separate from the top. Finally it separates the whole way along, no more lift is produced and the aircraft stalls.
Right: delta-winged aircraft have no separate tailplane, and so cannot have separate elevators and ailerons, but have elevons, single control surfaces to perform both functions. To roll the plane, one elevon is raised and the other lowered; to make it climb, both are raised. For a climbing turn, one is raised, the other is kept level, and the rudder is also used.
Far right: variable geometry or swing-wings are found on some supersonic planes. At low speeds the wings are extended for maximum lift; at high speeds, when too much lift would cause problems with the supersonic shock wave, they are folded back.

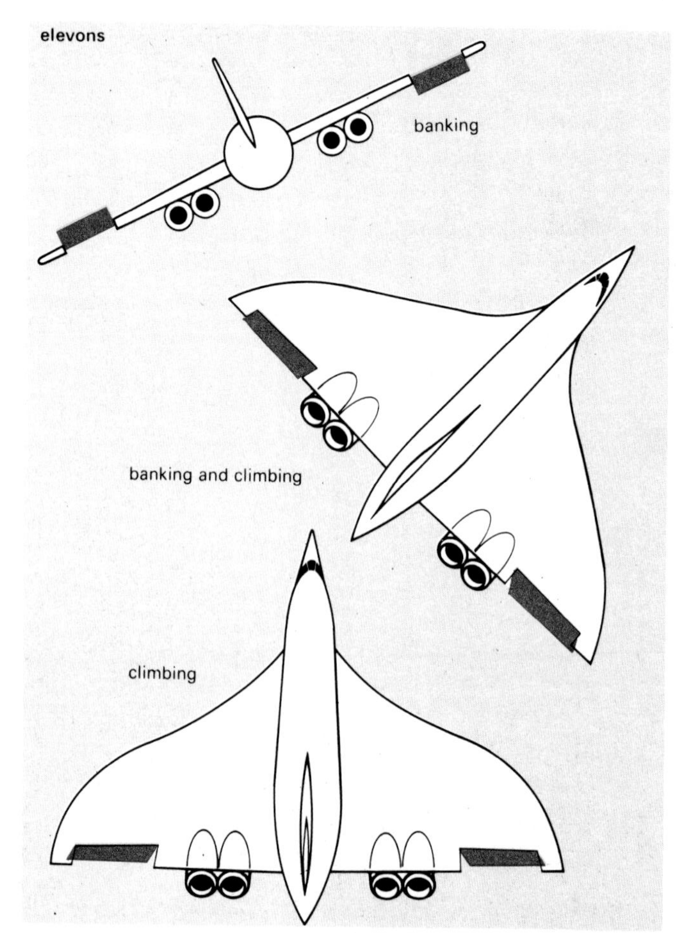

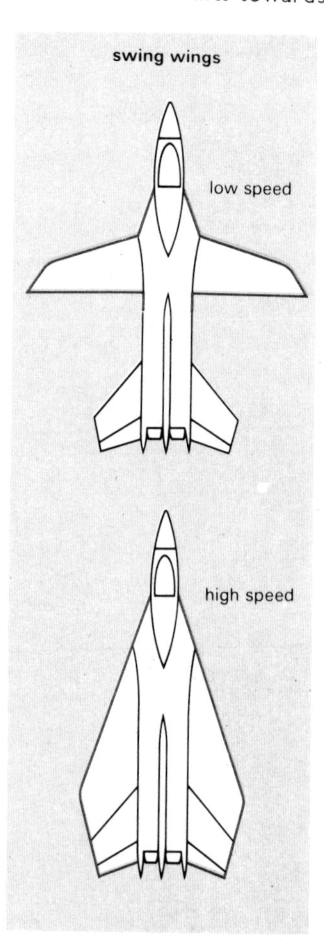

the centre of the turn like a bicycle. This is a more stable and comfortable way of turning.

In early aircraft the control surfaces—ailerons, elevators and rudder—were moved by the unaided exertion of the pilot through control wires. With today's high speed aircraft, the forces on the control surfaces are much too great for this, and so they are now generally moved by hydraulic cylinders, operated by the pilot through servo mechanisms. The arrangement works in a similar way to the power steering on a large car. This power assistance makes the controls of a modern aircraft very light, yet they are set to resist the pilot's action just enough to give him an indication that the surfaces are responding properly.

To move the elevators, he moves his control column backwards and forwards; to move the ailerons, he turns the control column. The rudder is activated by two pedals, leaving the hands free to operate the other control surfaces at the same time for banked turns.

Aircraft history The aircraft had a long period of gestation. Apart from its lack of a suitable engine, progress was hampered in the beginning by too much attention to bird flight. This led to a great deal of wasted effort on flapping-wing machines, known as ornithopters, although as early as 1804, Sir George Cayley had flown a model glider and, before he died in 1857, had flown at least two full-sized gliders with someone on board. Many of the early experimenters also made the mistake of concentrating on inherent stability in their aircraft instead of on controllability. Even Otto Lilienthal, who made gliding flights of 300 to 750 feet (90-230 m) in the 1890s, exercised control simply by moving his body, and only at the time of his death as the result of a flying accident in 1896, was engaged on the design of a body harness linked to a rear elevator on his latest glider.

During the second half of the 19th century a great effort went into achieving flight in England, France, the United States and Germany but it was diffused and fumbling. Little was known by one designer of what others were doing until Octave Chanute decided to collect and disseminate proved facts and other information to all who would listen to him. From 1896 onwards, he was engaged in building and flying his own gliders in the United States, and in spreading particulars about the principles involved in flight with fixed wings. In 1903 he visited Europe and lectured in Paris. Before that, he had given valuable help along the same lines to the Wright Brothers.

Some of these principles had been laid down by Cayley in 1809. He had outlined the forces of thrust, drag and lift, and had pointed out the value of the cambered, or arched, aerofoil wing shape in preference to the flat plate. As early as 1868, M P W Boulton in England had invented and patented the aileron. The Wrights' first glider of 1899 had wings that could be warped, or twisted, by cables for lateral control. Thus, by the time the Wrights made their first flight at the end of 1903, most of the devices for controlled flying were known and yet S P Langley in the United States and a string of pioneers in France and England were still meeting with little success. Even the Wrights in 1903 had not fully resolved the control situation; they found that wing-warping by itself was not enough. Their bright idea of linking the wing-warping with rudder movement was generously given away by Chanute to the Europeans, but too many of these still aimed at inherent stability through giving their wing tips a dihedral or upward-tilted angle.

Wilber Wright at the controls of the Model A in 1911. The pilot was now seated rather than prone, and the wing-warping and rudder controls separate, worked by the right hand; the other lever worked the front elevator. In France in 1908, the Model A made a flight lasting two hours and twenty minutes, which was astonishing. The nearest rival had achieved only 5 minutes.

There were many ambitious projects before real powered flight was accomplished by the Wrights. The biggest of them was the design by W S Henson in 1842 for an "Aerial Steam Carriage". It had a tail-piece to provide control and stability, box-kite wings, and a three-wheeled undercarriage for take off and landing. It was to have a wing-span of 150 feet (46 m) with wings properly constructed using spars and ribs, and it was to be propelled by two six-bladed airscrews. It was never built. Hiram Maxim staged an elaborate experiment to prove lift in 1894 with a device that weighed $3\frac{1}{2}$ tons and applied 360 hp through two steam engines. It was not intended to fly but it did lift off its rails, and there the project ended. In 1895–6 the Englishman Pilcher, a disciple of Otto Lilienthal, made several successful glider flights on the banks of the Clyde. In France, people like du Temple de la Croix, Pénaud and Ader worked hard with model aircraft driven by steam, twisted rubber and clockwork. In England, H F Phillips came back to Cayler's cambered wing and further showed the distribution of pressure and lift between the upper and lower surfaces of a wing.

This incoherent jumble of effort was given fresh interpretation by the Wright Brothers. Having learned all they could about the research and development up to their time, they proceeded to put the most promising ideas to the test. At the same time, they worked out their own calculations concerning not only the relation of thrust to lift and of the efficiency of propellers but also stresses and methods of construction. By their third glider they had developed a satisfactory airframe and began looking to the newly-arrived automobile engine for their power. Failing in their attempt to get from the new industry an engine of a suitable power to weight ratio, they set to work to build their own engine, and on 17 December 1903 the first flight was made. Two months earlier S P Langley's latest aircraft had fouled the launching mechanism and plunged into the Potomac. The triumph of Wilbur and Orville Wright was complete and exclusive. Orville, who made that first flight, was in the air only 12 seconds and travelled a distance of 120 feet (37 m). Three more flights were made that day and the last covered 852 feet (260 m). After that they went home to Dayton, Ohio from their flying ground at Kill Devil Hills in North Carolina and in 1904 built their second aircraft, again a biplane, with reduced wing camber and a more powerful engine.

Little attention was paid to them and their achievements by the rest of the world but they continued their experiments in control and in 1905 they produced their third

Below: a poster for the first ever aviation fair, in 1909. Bottom of page: the first transatlantic crossing was made by A C Read in 1919 in this flying boat, the NC-4, with frequent stops for fuel.

aircraft. This was extremely successful and before the year ended, a flight of 24 miles (38 km) had been made at an average speed of 38 mph (61 km/h). For the next two and a half years, they did no flying. Futile negotiations with the US and British Governments and anxiety about the risk of having their secrets pried into were at the bottom of this inactivity. In 1908, Wilbur visited Europe in search of business while Orville stayed at home, preparing for the military trials to which the American authorities had at last consented.

Meanwhile, Europe had been moved to fresh effort by gliding pioneer Chanute's encouragement. In France this was led by Esnault-Pelterie, later to be a prominent figure, and yet the first copies of the Wright glider were failures. In England, S F Cody had worked forward from man-lifting kites to a relatively inefficient glider. Slightly earlier, in France, Leon Levavasseur had built a monoplane with bird-like wings. It was a failure, but this pioneer, together with Louis Bleriot, were to give monoplanes a place in competition with the currently favoured biplanes.

The first aircraft factory was set up in 1905 at Billancourt, France, by the Voisin brothers, who had already built two float gliders towed by motor boats. They built for themselves and other designers, but the fashion in Europe was still to aim at stability rather than control. Soon the Brazilian pioneer aviator Santos-Dumont had turned away from airships to experiment with monoplanes and biplanes and also with tractor (front mounted) airscrews. In 1906, he flew 720 feet (220 m) in a tail first pusher biplane. In 1909, A V Roe in England produced a tractor biplane and J W Dunne built the world's first swept-wing aircraft, again a biplane, and again aimed at inherent stability. A year later, F W Lanchester did for aerodynamics what Newton and Bernoulli had done for hydrodynamics, when he put forward his theory (never disputed) of the circulation of air over the wing surfaces.

By 1909, the Wrights had made flights in public on both sides of the Atlantic, and the cause was given a healthy impetus. In the United States, Glenn Curtiss came to the fore as the designer of both aircraft and engine in the *June Bug*, which had wingtip ailerons for lateral control. European designers, with the exception originally of Henri Farman, followed the Wrights in aiming at good control either by wing-warping or by the fitting of ailerons. A string of new types now appeared and at the Rheims aviation week in August 1909, more than 30 aircraft were on show, six of them built to the Wright specification. A Curtiss successor to the *June Bug* won the speed contest at 47.85 mile/h (77 km/h) an Antoinette won the height award at 508 feet (155 m) and a Farman the distance at 112 miles (180 km). The same year Bleriot had staggered across the English Channel in his underpowered monoplane and the aircraft had ceased to be regarded as an erratic and essentially dangerous toy. Its progress was helped by the development of more powerful engines.

Above, top: the Supermarine S.6B seaplane which won the Schneider Trophy in 1931. It later set a speed record of 407mph (655kph).
Above: the Supermarine Spitfire was first designed in 1936, and owed much to the S.6B. For almost the entire war the Spitfire was the fastest plane in service, thanks to design modifications. The Spitfire XVI in the picture is a late model, with a four-bladed prop, square cut wingtips to reduce drag and a 'bubble' canopy.
Left: the Douglas DC-3 is one of the most successful aircraft ever designed, dating from the early 1930s. During the war it was called the Dakota, the C-46 and 'the work-horse of World War II'. Many are still in use today.

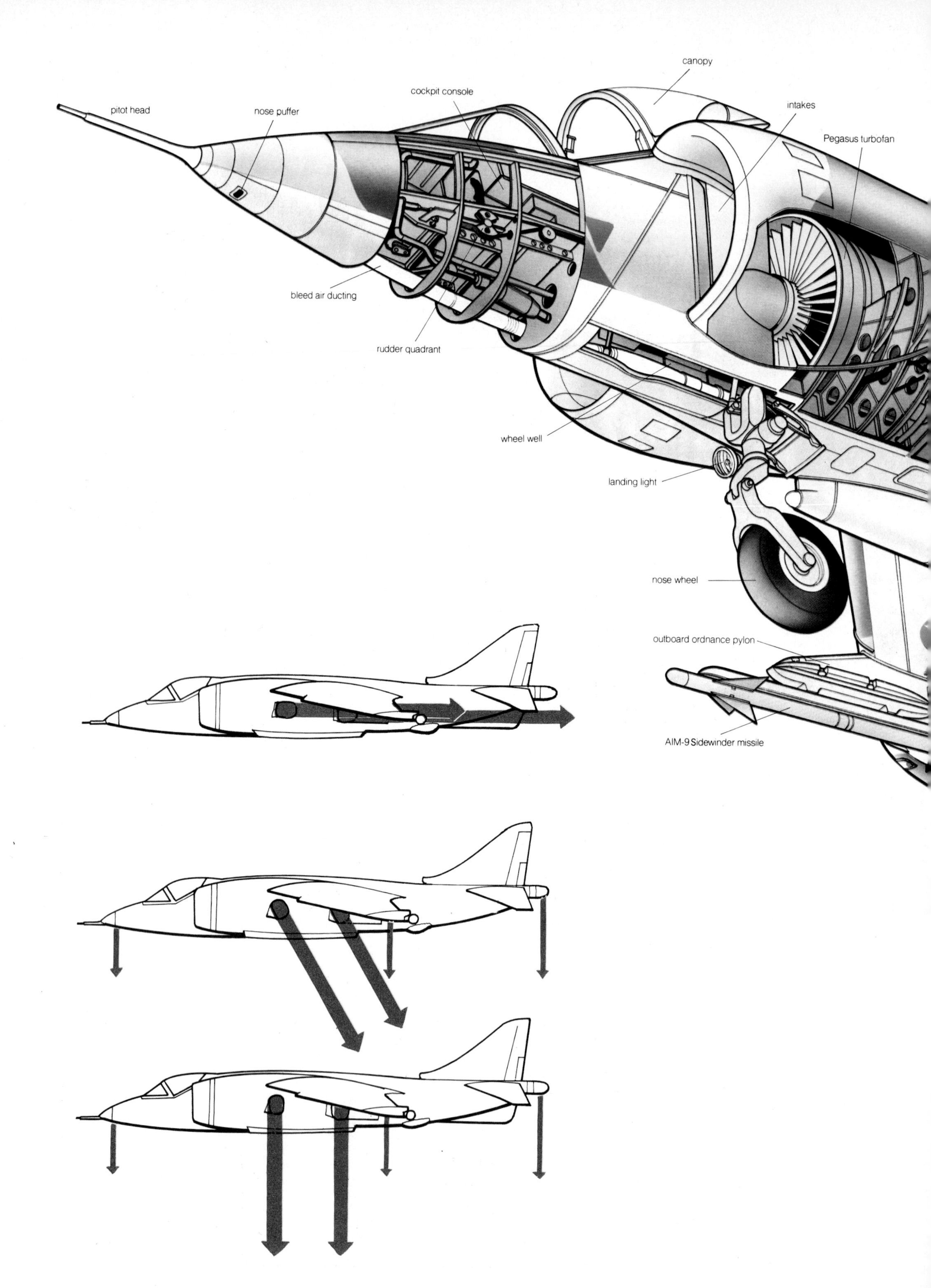
canopy
cockpit console
pitot head
nose puffer
intakes
Pegasus turbofan
bleed air ducting
rudder quadrant
wheel well
landing light
nose wheel
outboard ordnance pylon
AIM-9 Sidewinder missile

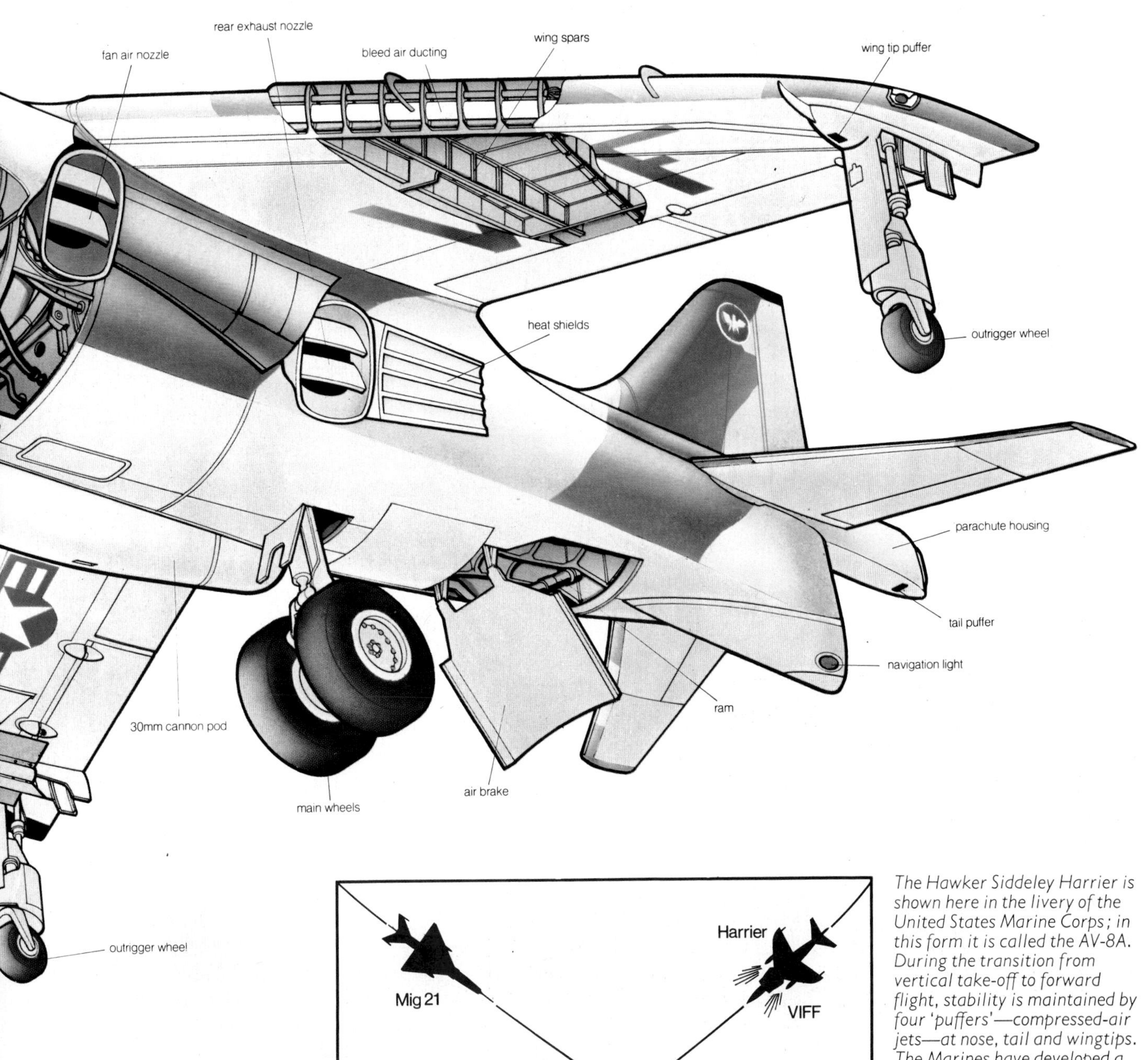

The Hawker Siddeley Harrier is shown here in the livery of the United States Marine Corps; in this form it is called the AV-8A. During the transition from vertical take-off to forward flight, stability is maintained by four 'puffers'—compressed-air jets—at nose, tail and wingtips. The Marines have developed a technique of vectoring in forward flight (VIFF) to make the aircraft more manoeuvrable than enemy aircraft of similar size and power; by swivelling the nozzles in flight the plane can be made to decelerate suddenly, another enormous advantage on top of its ability to land and take off without conventional runways.

The Wright aircraft types remained popular, but by 1910 the influence of the Wrights on development had virtually ceased and a vigorous independent line was being pursued by designers in various countries. The best performers continued to be biplanes though a good deal of work was done on monoplanes, and the germ of the cantilever wing, supported only by the fuselage instead of a system of wires, was contained in a patent registered by Junkers in Germany as early as 1910. A year later, another improvement appeared in a German device for raising the undercarriage legs on hinges to lie flush with the fuselage in flight. At the same time an oleo undercarriage leg (a telescopic leg incorporating an oil-filled shock absorber) was designed at the Royal Aircraft Factory at Farnborough, England. All these advances marked the movement from the wood-and-wire structure to the use of metal in aircraft. They also accompanied the increasing popularity of the tractor airscrew and the universal adoption of a tube-shaped fuselage to connect the wings, tail and landing gear, and to provide a less exposed position for pilot, passengers and power plants.

The possible use of the aircraft for military purposes had now become clear. The service view was that it might be made to serve reconnaissance purposes; but passengers had already been carried and other loads were obviously possible. Also the ability of the aircraft to perform aerobatics had been demonstrated well before World War I broke out. Throughout that war, the aircraft made great progress in Germany, France and England and the flimsy, underpowered box-kite aircraft were superseded by sturdier designs. Fokker came to the aid of Germany with a series of monoplanes which owed something to the Morane Parasol, a high wing design with the fuselage suspended beath it, and something to the Junkers idea for a cantilever wing. England clung obstinately to biplanes and, after 1917, produced some remarkable efficient specimens not only in the single-seater field but also in the big bomber class. This distrust of the monoplane by the British persisted into the early thirties while Germany added to the cantilever wing the equally revolutionary method of stressed skin construction, which was also early adopted in the USA. When England did at last come to monoplanes, she had her own structural contribution to make, in the form of geodetic construction. This was a sort of open basketwork construction of aluminium strip so designed as to direct stresses to the surface and there distribute them over the whole surface. This was proved to be extremely resistant to war damage by the World War II Wellington bomber. Its disadvantage was that it could not compete in weight with the stressed skin method unless it used a fabric covering, and this led to its disappearance as the stressed skin technique proved itself.

The designs of the middle 1920s were not very different from those at the end of World War I. Wire-braced biplanes and monoplanes with external strut or wire bracing, both types powered usually by water cooled engines, continued to be popular. A prize of £25,000 had been offered in 1919 by an American, Raymond Orteig, for the first nonstop flight from Paris to New York. Charles Lindbergh, an American airmail pilot, finally won the prize in 1927. His Ryan monoplane, the *Spirit of St Louis*, flew 3600 miles nonstop in 33 hours and 39 minutes. From this point onwards air cooled engines were used extensively.

The last few years of the 1920s provided the impetus to explore more fully both the military and commercial aspects of flight. All the contests for endurance and speed records had brought modifications to aircraft design, but on a haphazard basis. New and reliable aircraft engines were being developed, which in turn led to improvements in other features in aircraft.

During this period, much work was done on other types of machine, such as the helicopter, pioneered in the USA by the Russian-born Sikorsky and in Germany by Focke and Achgelis. At the same time, the Spaniard Juan de la Cierva invented the autogiro, which has an unpowered rotor turned by the backwash from an ordinary airscrew.

Retractable landing gear was developed for use on amphibious aircraft by Loening and independently by Sikorsky but it was considered too costly and heavy for light aircraft. All this changed in the early 1930s when most

aircraft were built with fold-up landing gear in the interests of more efficient air performance.

Apart from the acceptance of the necessity for retractable landing gear, wind tunnel studies produced new cross-sectional shapes for wings and other surfaces. These included the Handley Page automatic slot, an air-pressure operated device mounted on the leading edge of the wing to warn of an approaching stall, and also delay the actual stall by inducing an extra flow of air through the slot to smooth the air flow. At the trailing edge, movable flaps were attached to increase lift at low forward speeds by increasing the wing area and accentuating the camber of the wing.

New materials were tested and by the 1940s the wood, fabric and wire construction had almost disappeared. Most of the larger aircraft in the earlier 1930s were twin-engine types and it was not until the end of the decade that four-engine designs became usual. Aircraft such as the Boeing 247 of 1933 and the Douglas DC series of 1934 began to change the face of air transportation because they were fairly large and sound-proof and they carried enough electronic navigational and landing aids to make for safety in most weather conditions. For sea crossings large flying boats were developed by Latécoère of France. There were also Sikorsky's S-42 flying boat and later, flying clippers built by Glenn Martin and Boeing.

Designers in this period also began experimenting with the jet engine. The first flight of a jet aircraft was in August, 1939 by the German Heinkel He 178.

World War II led to rapid advances in design. Although nearly all aircraft were still driven by piston engines, work on rocket and jet propulsion intensified. Almost two years after the first jet flight, in May 1941, the British Gloster aircraft powered by a Whittle jet engine took to the air. The USA followed in October 1942 with the Bell XP 59A twin-jet aircraft. During World War II the Germans Busemann and Betz and R T Jones in England worked on designs for supersonic flight, though this was finally achieved in the USA by the Bell X-1 in 1947. By the end of World War II around 1300 Messerschmitt Me 262 twin jet fighters had been built and Britain was producing the Gloster F-9-40 Meteor.

Air transportation relied on propeller-driven aircraft until the emergence of the high-speed jet-powered designs, which first appeared in the form of the de Havilland Comet in 1949. The hybrid turboprop engine, which uses jet power to drive a propeller, was invented later than the pure jet, but the first airliner to use it was built in 1948, just a year before the Comet. This was the British Vickers Viscount.

Above: the nosewheel gear of the BAC TSR2 supersonic strike and reconnaissance aircraft. The scissor-type mechanism was designed to extend the strut at take-off to tilt up the nose.
Left: the propeller hub of a Turbomeca Astazou XIV-C turboprop engine, showing the pitch change linkage, the bronze blade counterweights, and the black de-icing elements which are supplied with current through straps of rubber.
Opposite page: the BAC-Sepecat Jaguar is the type of modern jet with wings so thin that the landing gear must fold into the fuselage.

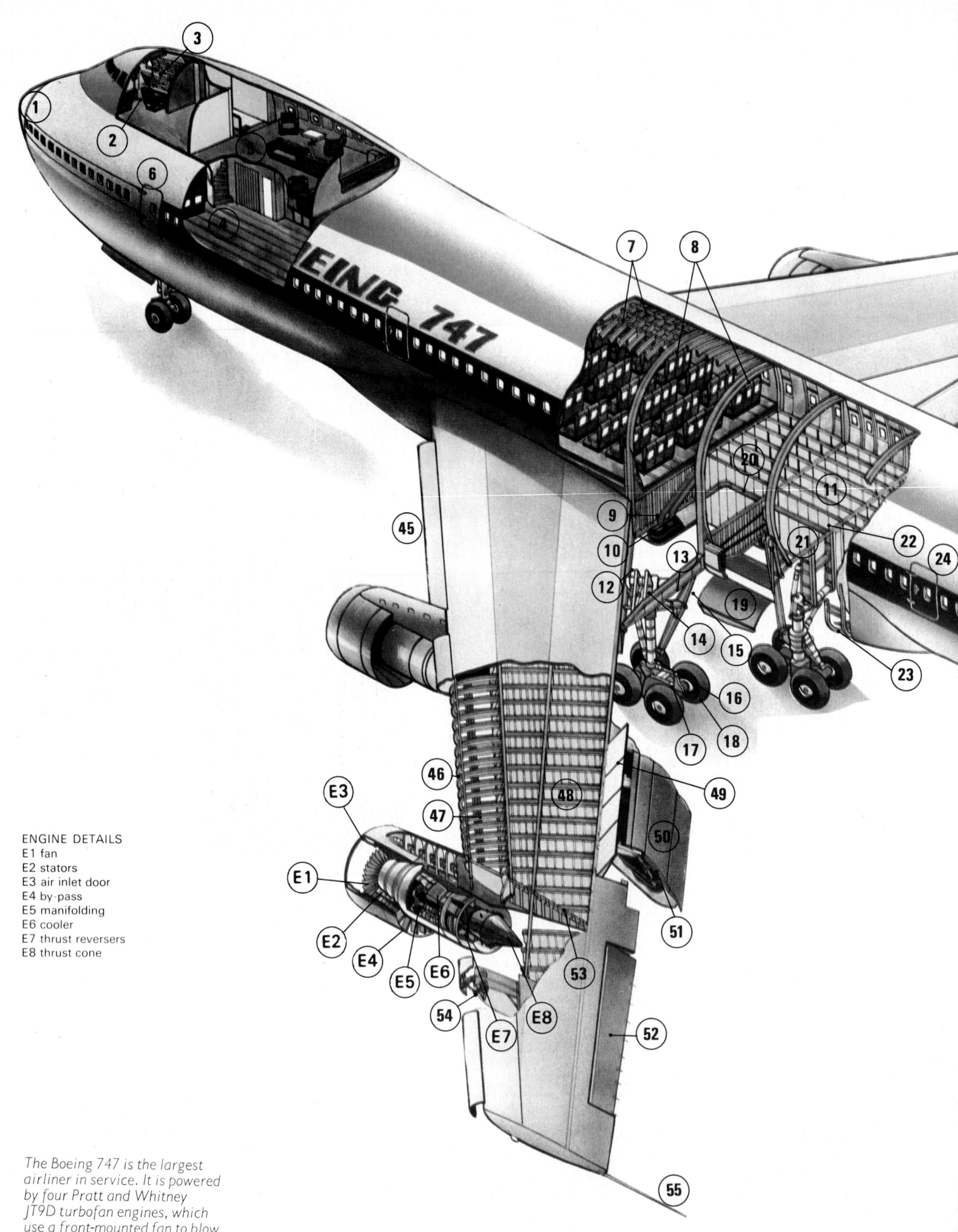

The Boeing 747 is the largest airliner in service. It is powered by four Pratt and Whitney JT9D turbofan engines, which use a front-mounted fan to blow air through ducts around the jet exhaust, improving both quietness and fuel economy.

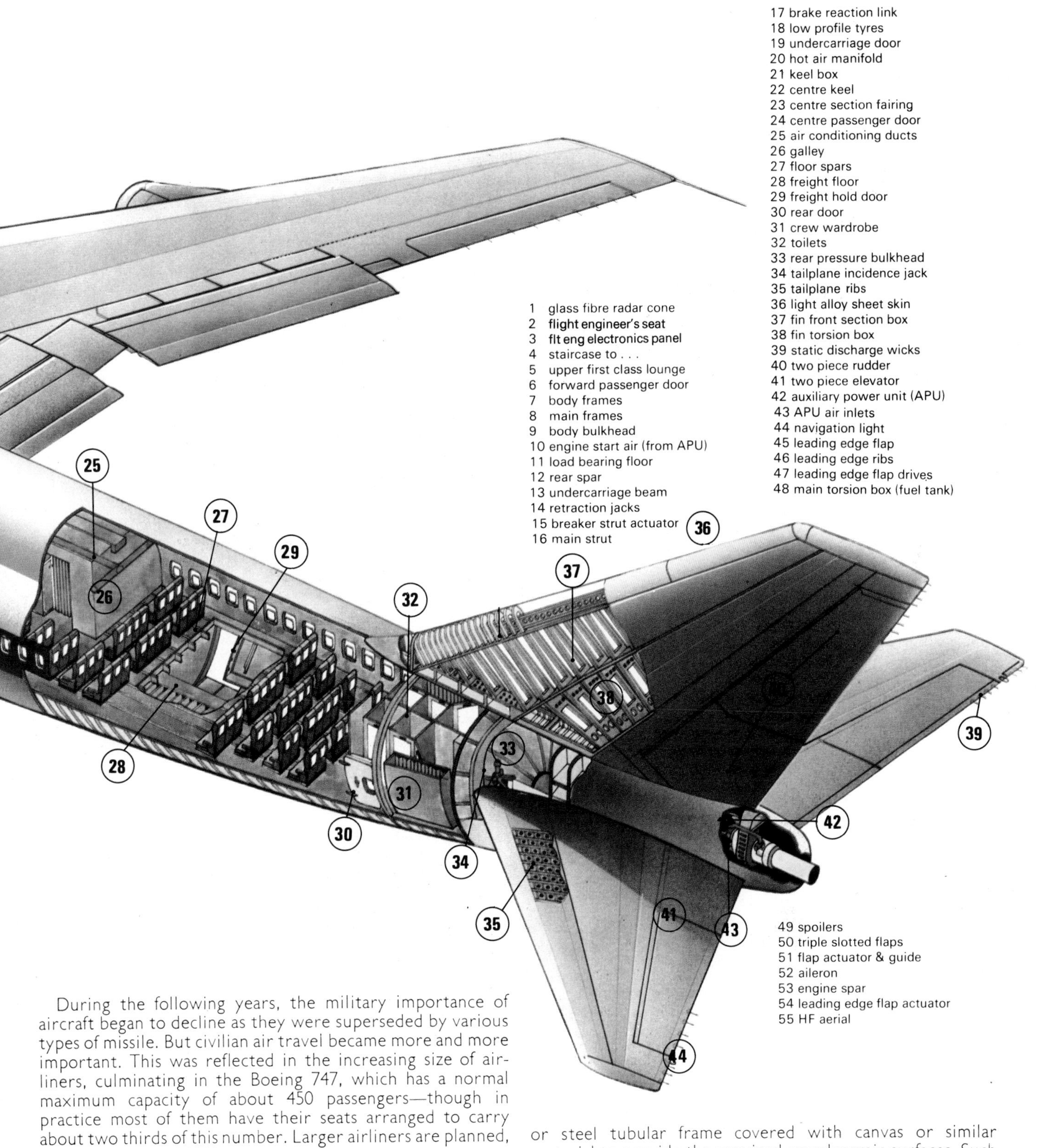

During the following years, the military importance of aircraft began to decline as they were superseded by various types of missile. But civilian air travel became more and more important. This was reflected in the increasing size of airliners, culminating in the Boeing 747, which has a normal maximum capacity of about 450 passengers—though in practice most of them have their seats arranged to carry about two thirds of this number. Larger airliners are planned, with a capacity of 1000.

Supersonic airliners are also being built by the British, French and Russians. It will not be possible for some years to say whether they will be a commercial success. The design problems were solved some years before, but the main restraints on aircraft design are now economic rather than technical.

The construction of all aircraft from simple glider to swing-wing supersonic craft has to be carried out with one principal aim in view: to reduce weight as much as possible. Primitive types at the turn of the century adopted a wooden or steel tubular frame covered with canvas or similar material to provide the required aerodynamic surfaces. Such construction methods are still used for some simple light aircraft, though the canvas may also be replaced by glass fibre materials or aluminium sheet.

But for faster high-powered craft, where dynamic forces are involved, such constructional practice would be unsuitable. Most larger aircraft use a reinforced monocoque construction, in which the outer shell takes a lot of the stress but is backed up by a suitable frame of light alloy. The shell is sometimes made of solid light alloy but nowadays tends to be a sandwich of two thin layers glued to a metal 'honeycomb' mesh to give high stiffness with low weight. The aim here is to produce a material akin to corrugated cardboard in cross-section.

For supersonic aircraft, where the stress problem becomes even more acute, wings and other components have to be made by machining outer skin and frame together from solid pieces of alloy. Though giving the strength and heat resistance that are necessary, this method is very expensive and has contributed considerably to the high cost of, for instance, the Concorde project. It seems likely that it will be reserved strictly for military types in future.

Supersonic flight Common to all forms of transport is the emphasis on speed. The longer the distance to be covered, the greater the speed necessary to accomplish the journey in a reasonable time for commercial economy, personal comfort and military safety. Since the birth of powered aviation in 1903, speeds have increased from 40 mph (64 km/h) to well over 4500 mph (7242 km/h) and the technology is now available to support any projects envisaged. Costs rise steeply with increases in speed, however, and there exists, particularly for commercial aviation, an economic limit associated with the velocity of sound waves in the air.

Sound is a wave disturbance in the atmosphere, rather like the ripples which spread out when a stone is dropped into still water. The velocity of sound waves in air is proportional to the square root of the absolute temperature. At sea level, in temperate zones, where the average temperature is about 59°F (15°C), sound waves travel at 761 mph (1225 km/h). The velocity falls with increasing height (because the temperature is dropping) until at just over 36,000 ft (10.97 km) it is only 691.3 mph (1112.5 km/h). Further increases in altitude,

Above: the Bell X-1 rocket-powered aircraft was the first to fly faster than the speed of sound. Launched from a B-29, it reached Mach 1.06 on 14 October, 1947. The rocket motor produced 6000 pounds (2722kg) of thrust.
Right: this RAF Lightning was flying at Mach 0.98 when the picture was taken, but the airflow over the canopy and wings reached supersonic speed locally, causing moisture in these regions to condense.
Opposite page, top: the North American X-15 rocket-powered research craft being launched from a B-52. A plane of this type reached a speed of Mach 6.72 (4534mph; 7297kph) on 3 October 1967. At this speed parts of the airframe became red-hot from friction.

at least up to about 60,000 ft (18.29 km), have no effect since the temperature of the atmosphere remains constant.

During the last century an Austrian scientist, Professor Ernst Mach, studied the propagation of sound waves, and in recognition of his work the speeds of aircraft flying close to the velocity of sound are described by their Mach numbers. At sea level an aircraft travelling at 761.5 mph (1225.5 km/h) is said to have a speed of Mach 1, and if it is making 1320 mph (2124 km/h) at 40,000 ft (12.19 km), its speed is Mach 2. Flight up to Mach 1 is described as subsonic; that above Mach 1 is supersonic.

During World War II, piston-engined fighters were sometimes flown at speeds considerably above normal during test or in combat, and their pilots would report severe buffeting or even loss of control. The greater performance of the early jet fighters brought with it increasing experience of these compressibility effects, as they were called, though their maximum speed still fell short of Mach 1. The struggle to fly ever faster, and the difficulties involved, gave rise to the popular though erroneous idea of a 'sound barrier'.

At low speeds air behaves as if it were incompressible; the passage of a slow aircraft through it does not result in an

Opposite page: an assembly line in the British Aircraft Corporation's factory at Weybridge, Surrey, England, producing fuselage sections for the BAC/Aerospatiale Concorde supersonic airliner. Final assembly is carried out in England or France, from components supplied by more than 800 sub-contractors. Below, right: the Concorde has a total of ten wheels on its undercarriage; the brakes are made of a new carbon composite material.

increase in the pressure of air ahead of it because the molecules of air have plenty of time to move out of the way. But as the speed of the aircraft increases the air molecules have progressively less warning of its approach, and therefore less time to move. If the aircraft is travelling at the speed of sound, its speed is the same as that of the motion of the molecules which warn of its approach. Under these conditions the arrival of the aircraft compresses the air and the disturbance is propagated as a shock wave and the resistance of the air to the motion of the aircraft increases very rapidly. If the aircraft is to travel even faster, more power has to be applied to overcome this drag, or resistance to motion. Associated with the formation of shock-waves is a breakdown in the airflow behind them, which may destroy the effectiveness of the aircraft's controls.

The first aircraft to fly faster than sound was America's little rocket-powered Bell X-1 which, carried by a converted B-29 Superfortress bomber into the stratosphere in order to conserve fuel, attained Mach 1.06 on 14 October 1947. A few years later, employing the German idea of wing sweepback to reduce drag, jet fighters were able to exceed Mach 1 in a dive, though their engines were still insufficiently powerful to take them to the speed of sound in level flight. America's F-86 Sabre and Britain's Hunter were examples of these transonic aircraft.

Supersonic aircraft The first truly supersonic aircraft was the North American F-100 Super Sabre of 1953, which could reach Mach 1.25 in level flight. It had an even more streamlined shape than its predecessors with, in particular, an extremely thin wing. Such was the pace of development that only five years later the Lockheed F-104 Starfighter showed sustained speeds of Mach 2 to be not only technically possible but militarily realistic. It had an exaggeratedly small wing, only a few inches thick, and with such sharp edges that they had to be covered with felt when the aircraft was on the ground to prevent people from injuring themselves.

Nowadays there is no particular difficulty in designing a fighter to fly at up to Mach 2.5, though speeds above this call for the use of special metals to resist the high temperatures caused by the friction of air molecules passing over the structure. At the beginning of 1975 the fastest aircraft in the world were the Lockheed SR-71 reconnaissance vehicles and Russia's MiG-25 intercepter, code-named in the West 'Foxbat'; both cruise at heights of over 80,000 ft (24.38 km) at Mach 3, approximately 2000 mph (3218 km/h) at that altitude. Fastest of all, however, was the North American X-15 research aircraft which, on 3 October 1967, flew at 4534 mph (7297 km/h), equivalent to Mach 6.72. The friction heating was so severe that parts of the airframe became red-hot, and so tough, very expensive Inconel (nickel-based) alloys were necessary to withstand the heat loads. Speeds above Mach 5 are described as hypersonic, to indicate that the nature of the airflow over the plane has changed again.

Wings designed to fly efficiently at supersonic speeds are invariably inefficient at low speeds, generating high drag and low lift. Landing speeds are much higher than for subsonic layouts, calling for longer runways. In an effort to couple low drag at high speed with high lift at low speed, a number of the newer combat aircraft have variable-geometry wings, the sweepback of which can be varied continuously to provide the best efficiency at any particular speed. But delta-shaped wings are still the cheapest ones to build, and they are also lighter. They were chosen for the Anglo-French Concorde supersonic transport (and the Russian Tu-144), where the weight penalty and the technical risks involved in variable-geometry designs were considered unacceptable at the time of the original design.

The autogyro

An autogyro is a heavier-than-air flying machine which derives its lift from a rotor system mounted above the machine, with blades rotating more or less horizontally.

Autogyros differ from helicopters in that their rotor blades are driven by the air flowing upwards past them (the principle being known as autorotation), whereas the helicopter has mechanically driven blades set at a greater pitch angle, so as to 'screw upwards' through the air.

To maintain height, or climb, an autogyro needs a propulsive system such as an engine and a propeller to drive it forwards. Then, by tilting the lift rotor system slightly backwards, the rotor blades will lift the aircraft, even though the air flows up and through the rotor.

A simple autogyro needs some forward speed to maintain height, so unlike the helicopter, it cannot hover or take off vertically.

The motion of the rotor and the resulting upward thrust, or lift, depends entirely upon autorotation, resulting from the air flowing up and through the slightly tilted rotor blades as the machine moves forward.

Nature has applied the principle of autorotation for millions of years, seen in the whirling flight of the sycamore seed as it falls to the ground. Autorotation slows its descent and the wind has greater opportunity to disperse the seeds over a wider area.

The windmill was probably the first human invention which used autorotation, by harnessing the wind to produce rotary motion. The idea of a flying windmill, where rotating sails produced a wind to lift the machine, had a certain fascination with inventors, and among Leonardo da Vinci's thousands of drawings is an idea for flight along these lines. The real possibility for achieving such a machine was, however, delayed until the development of the aerofoil and the aircraft which embodied this device.

A windmill is basically an airscrew or propeller working in reverse, such that the air flowing over the sails is deflected by them, and exerts a force on the sails pushing them round. The sails effectively 'give way' to the wind and are pushed round by it.

As early as the Middle Ages, however, it was realized that if the sails were set at a very flat angle to the wind they would be made to rotate against the airflow and thus be 'pulled' round into the wind. The principle here is the same as with a sailing ship which can 'tack' close to the wind, meaning it can move forward against the wind, at a shallow angle to it, if the sails are properly set. In much the same way a glider moves forward as it descends through the air.

The rotor blades of an autogyro are shaped to achieve the same effect, and set at a shallow angle of about 3° to the horizontal plane in which they rotate. The shape is that of an aerofoil which enables the blades to turn into the airflow rather than be pushed round by it.

When turning fast these rotor blades offer considerable resistance to the upward airflow, and it is this resistance that

Concorde
195
Concorde

can be used to provide lift. The amount of lift created depends upon a compromise between the airspeed of the rotors, and the resistance the rotating blades offer to the airflow past them. In practice, the desired lifting force is only produced when the blade speed greatly exceeds the forward speed of the machine.

To take off the rotor must produce adequate lift and it is necessary, therefore, to bring the rotor up to the required speed. This can be done in two ways.

The first and simplest way is to propel the machine forwards and, by tilting back the rotor system, use the airflow through the blades to build up the rotor speed. This, however, requires a suitably long runway. The second method involves more complex machinery but makes possible very short take-off distances. Here the rotor is brought up to speed by a linkage to the engine used to provide the forward motion. When the rotor has the correct speed, the linkage is disengaged. The machine is then allowed to move forward, and take-off is achieved by tilting back the rotor system. Some autogyros can 'jump-start', by over-speeding the rotor using the engine. The drive is then disengaged, and the rotor pitch increased. The aircraft jumps, using the stored energy, and continues then in autorotation.

When the engine and propeller speed are reduced the forward speed will decrease and the autogyro goes into a steady descent path. The autorotation principle still applies, as the air flowing up and through the rotor maintains the rotor speed. A lifting force is therefore produced which, although insufficient to maintain the machine altitude, prevents it from falling like a stone. Even when the propeller is stopped, the autogyro will descend safely.

In this respect the autogyro is at some advantage over the helicopter, since, in the case of helicopter engine failure the 'climbing pitch' angle of the rotors (about 11°) would quickly stop them, with disastrous results. To keep his rotors turning the pilot will have to quickly reduce the pitch angle of his blades to that which provides 'autorotation' for a safe forced landing, but some valuable height may be lost in the process.

The first successful autogyro was designed by Juan de la Cierva and was flown on 9 January, 1923 at Getafe Airdrome near Madrid. This was his fourth design, the other three suffering from an alarming tendency to roll over.

The instability was due to the use of rigid rotor blades. With the machine moving forward and the rotor turning, the blade turning into the airstream experiences a greater lifting force than the opposite blade moving downstream. With rigid rotor blades, this imbalance is transmitted to the whole machine producing a rolling motion. To overcome this instability, Cierva designed a rotor system with blades suitably hinged at the root, so that rather than transmit the imbalance to the whole machine, it was taken up by the individual blades which could move accordingly.

At the root of each blade he inserted two hinges. One allowed the blade to flap up and down and was called the flapping hinge and the other permitted sideways movement and was called the drag hinge.

The autogyro (or helicopter) rotor blade is not, by itself, stiff enough to carry the weight of the machine. It is the enormous centrifugal force of rotation that keeps the rotors moving in an almost flat path, and even though they have 'flapping hinges' at their roots, the weight of the machine is carried here.

The helicopter

Helicopters and autogyros are superficially similar to one another in that both are wholly sustained in flight by the lift generated as a result of the rotation of long thin wings, or rotor blades, in a horizontal plane.

The blades of an autogyro, however, are caused to rotate by the action of air blowing through them, while those of a helicopter are driven by an engine. Autogyros cannot therefore land or take off vertically in calm air. Helicopters, on the other hand, can take off or land vertically, hover, fly forwards, backwards or sideways irrespective of the wind.

The principles of helicopter flight have been known for centuries. Many helicopter models were made by early flight pioneers such as Sir George Cayley (in 1792). The first helicopter capable of carrying a man was built by Paul Cornu in France in 1907, powered by a 24 hp engine, but insurmountable stability and engineering problems held back

Opposite page: This Kellett K-3 autogyro was taken to the Antarctic by Admiral Byrd on his second expedition of 1933. The pilot who accompanied it, William S McCormick, is shown at the controls.
Below: the cockpit of the Bell 206A helicopter, showing the cyclic control sticks, the rudder pedals, and the collective pitch lever, which is at the right side of the seat.

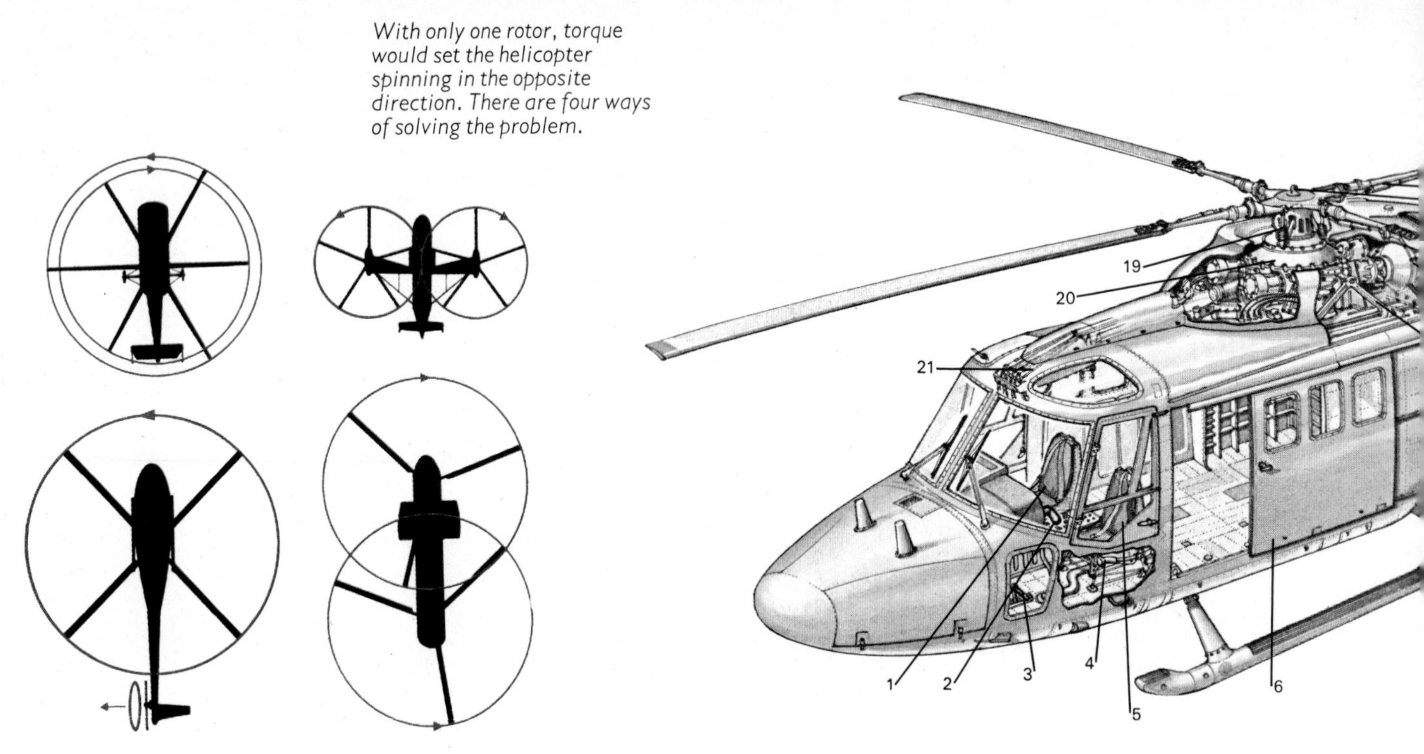

With only one rotor, torque would set the helicopter spinning in the opposite direction. There are four ways of solving the problem.

helicopter development compared with that of conventional aircraft.

It was not until January 1942 that the world's first practical helicopter, the VS-316A, was built by the Russian-born American engineer Igor Sikorsky. This machine, designated the R-4 by the armed services of the USA and Britain, had the simplest possible configuration for a helicopter and one that is still the most widely used today.

The main structural element of a helicopter is the fuselage, housing the crew, payload, fuel and powerplant, which until the mid-1950s was a piston engine but is now usually a gas turbine. The output shaft from the engine, turning at several thousand revolutions per minute, is connected to a main gearbox, which steps down the speed to between 300 and 400 revs/min to drive the rotor (the assembly carrying the hub and attached blades).

The reaction of the rotor spinning in one direction would cause the rest of the helicopter to rotate uncontrollably in the opposite direction. In order to prevent this a secondary rotor, of smaller diameter, is mounted on the rear end of the fuselage and driven by a second shaft from the gearbox at such a speed that it exactly neutralizes the turning action of the main rotor.

Each of the blades of the main rotor (modern helicopters have any number from two to seven) is inclined (with its leading edge upwards) so that it meets the air at a small angle to the horizontal. This is the pitch angle, analogous to the pitch of a propeller or a screw thread.

When in hovering flight the combination of rotor speed and the pitch of the blades provides a lift force which exactly balances the weight of the helicopter. In order to climb, the rotor has to generate more lift, and this is achieved not by increasing the speed of the rotor (the rotor speed of a particular helicopter at all times remains virtually constant, irrespective of what the aircraft is doing) but by increasing the pitch of the blades. More lift, however, also means more drag (wind resistance) and so extra power is needed from the engine.

The pitch of the blades is controlled from the cockpit by means of the collective pitch lever, so called because it changes the pitch of all the main rotor blades by the same amount. This lever is mounted on the floor, and is one of the very few differences between the cockpit of a helicopter and that of a conventional fixed-wing aircraft. Operated by hand, it is moved up to gain height and down to descend.

Since most manoeuvres, including climbing and descending, necessitate changes of power, the collective pitch lever has a twist-grip throttle control at the top so that engine power and blade pitch can be controlled and co-ordinated with one hand.

Helicopters have no wings or tailplane and so the main and tail rotors are required to generate between them not only the forces needed to provide lift but also those needed to control it. In order to make the helicopter travel forward,

rotation
flapping hinge
lag hinge
hub
rotor blade
feathering hinge
control link
rotor shaft
flap
rotating plate
non-rotating plate
connection to pilot's controls

Changes in blade pitch are transmitted by the swash plate, which moves up or down for collective pitch change, and tilts for cyclic change. If it is tilted, the blade pitch is increased as it passes the high side.

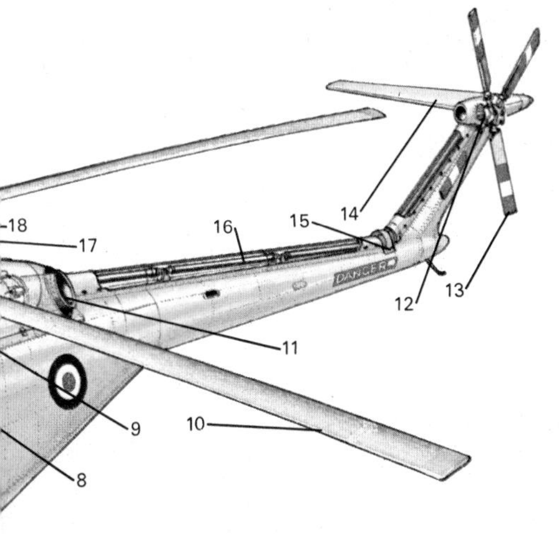

WESTLAND/ AEROSPATIALE LYNX

1. pilot's seat
2. cyclic pitch stick
3. tail rotor control pedals
4. co-pilot's collective-pitch lever
5. co-pilot's seat
6. main door
7. wheel attachments
8. forged frame
9. Rolls-Royce BS.360 gas turbine engines
10. rotor blade
11. exhaust outlet
12. tail rotor power unit
13. tail rotor
14. tailplane
15. intermediate gearbox
16. tail rotor drive
17. rotor hub
18. drag hinge damper
19. pitch control rod
20. main gearbox
21. collective pitch linkages

The rotor blades flex upwards when moving toward the front of the machine, and down when moving toward the rear, in order to make the lift equal on each side. They are at their highest when pointing straight forward, and lowest when pointing toward the tail.

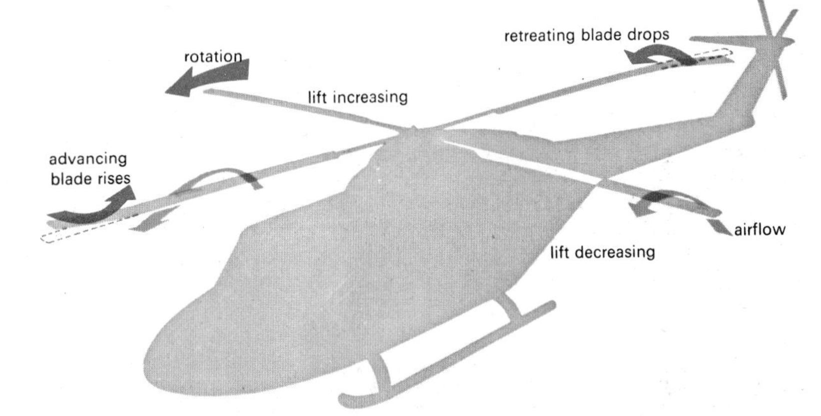

or in any other horizontal direction, the rotor has to be tilted in that direction. Its reaction, or total lifting force, is then inclined away from the vertical and can be considered as being made up of two components: one acting vertically to balance the weight, and the other, much smaller force, acting along the direction in which the pilot wishes to travel.

To make the helicopter travel faster, the rotor blade has to be tilted further so that more of the reaction of the rotor acts in the desired direction. At the same time the vertical component of lift must be maintained, so more engine power has to be applied.

The rotor is made to tilt by arranging that one half of the disc traced out by the rotating blades generates more lift than the other half, and this is achieved by increasing the blade angle on one side and decreasing it on the other. Thus the pitch of a blade goes through a complete cycle, from maximum to minimum and back again, during one revolution.

The cyclic pitch control, which determines where on the rotor disc the variations of lift shall occur to perform the desired manoeuvre, or change the speed, is commanded by a conventional control column in the cockpit. Changes in heading (the direction in which the helicopter is pointing relative to north) are made by collectively altering the pitch of the tail rotor blades by means of conventional rudder pedals.

As the forward speed of the helicopter increases, the velocity of the forward-moving blade is increased by an amount equal to the speed of the aircraft, while that of the rearward-moving blade is decreased by the same amount. Eventually a situation is reached when the forward-moving blade is approaching the speed of sound over a considerable portion of its travel. Undesirable aerodynamic effects then cause the drag of the blade to increase rapidly and its lift to decrease. At the same time the relative velocity of the retreating blade, travelling momentarily in the opposite direction to that of the helicopter itself, is too low to provide useful lift and it may become stalled. These effects limit the speed of a conventional helicopter to about 250 mile/h (400 km/h).

If engine power is suddenly removed, the rotor slows very rapidly, lift is lost, and the helicopter begins to drop. To prevent loss of rotor speed the collective pitch lever has to be rapidly lowered, so as to set the blades at a negative pitch angle. This means that the leading edges of the blades are inclined slightly downwards from the horizontal, but as the air is moving upwards through the rotor, the blades still meet the airflow at a small upward angle.

As the helicopter begins to descend, usually in a forward glide, the air blowing up through the rotor disc generates forces on the blades which keep them spinning. This is called autorotation. As the helicopter nears the ground the collective pitch lever is raised slightly so as to reduce the rate of descent, using the kinetic energy stored in the rotor to provide, for a short time, the extra lift necessary to decelerate the machine for touchdown.

The use of two smaller main rotors in place of a single large one can be an advantage, particularly with bigger helicopters. The rotors are arranged to spin in opposite directions so that the reaction torque of one cancels that of the other. There is thus no tendency for the fuselage to rotate, so the tail rotor can be dispensed with. This reduces the weight somewhat and enables all the power developed by the engines to be put to useful work in lifting and moving the aircraft.

The rotors may be mounted transversely across the fuselage as on the Mil-12, the world's largest helicopter, or they can be located at either end of the fuselage as on the Boeing Vertol CH-47 Chinook. They may be set one above the other and driven by concentric shafts, as on the Kamov Ka-26, or they may be carried on separate shafts mounted at a slight angle to one another; in this case (as with the Mil and the Boeing) the rotors intermesh with each other and the drives to them have to be synchronized so the blades do not collide with one another.

The general adoption of gas turbine propulsion for helicopters in place of piston engines in the mid-1950s resulted in a tremendous improvement in performance. Nowadays piston engines are only to be found on the older or very smallest helicopters. Instead of generating a high-velocity stream of hot gases to thrust the aircraft through the air, as in the jet engine, the power of the engine is extracted mechanically by fitting it with extra turbines and connecting these to the shaft which drives the rotor. This type of powerplant is known as a turboshaft engine, and only a small proportion of the gas energy emerges as thrust.

The seaplane

The practice of flying aircraft off water is almost as old as aviation itself and results from the number of good natural runways available from the world's rivers, lakes and harbours. It releases the aircraft from its greatest limitation, the need for specially prepared strips of ground for take-off and landing. If, however, the seaplane is fitted with retractable wheels to exploit landing strips as well, it becomes the truly liberated amphibian. Even today, when airfields proliferate, there are more seaplanes and amphibians in service than ever before.

The true water plane is the flying boat, with the fuselage itself designed to operate on water with most of the characteristics of a boat. Most small landplanes can be fitted with mini-hulls or floats instead of wheels and, as such, earn the separate designation of floatplane.

The first recorded successful flight from water was made in March 1910 by Henri Fabre of France, just over six years after the Wright brothers achieved the first sustained powered flight of a heavier-than-air aircraft. The next step, in 1911, introduced one of the great names of seaplane development when Glenn Curtiss of the USA flew a novel craft with a single float—to which wheels were soon added to produce the first amphibian. His first real flying boat came out the following year, and suddenly seaplanes gained acceptance. In 1914 the world's first scheduled airline began in Florida, operating between St Petersburg and Tampa.

1913 to 1931 were the years of the famous Schneider Trophy races for seaplanes, whose role in stimulating high performance technology is reflected in the progress of winning average speeds from 47.75 mph (76.8 km/h) for the first meeting to 340 mph (547 km/h) for the last. The final winner, Britain's Supermarine S.6B, later set a world record of 407 mph (655 km/h), and evolved directly into the Spitfire of World War II fame.

Britain's efforts, in fact, led seaplane development of all kinds through the 1920s and early 1930s, culminating in the huge Short Sarafand of 1935 which, with its 150 ft (65.7 m) wingspan, was the largest biplane built. But, by then, such devotion to biplanes had already cost Britain the lead, as faster monoplane flying boats were developed in Europe and America. Germany's 12-engined Dornier Do.X, although never entering service, introduced the age of the giants, dominated by the great Sikorsky boats and Martin Clippers of America. Finally came the Boeing 314 Clipper which, in 1939, established scheduled transatlantic passenger services. Largest of all was Howard Hughes' 460,000 lb (208,650 kg) Hercules, with a 320 ft (97.54 m) wingspan, which made its first and only flight in 1947, and could have carried 700 passengers. It is the largest aircraft ever to fly, with an overall length of 219 ft (66.75 m). Among the most famous aircraft of World War II are America's Consolidated Catalina and Britain's Short Sunderland flying boats, used unceasingly for maritime patrol in both the Atlantic and the Pacific.

Small seaplanes still play an important role, particularly in Canada and Alaska and other countries with many lakes and harbours and limited surface access such as Norway, Finland and the USSR. Seaplanes are the only practical means of transport around the Canadian north where they serve in hundreds as taxis, freighters, ambulances and fire-fighting water-bombers.

Water is as demanding an element as the air. Seaplane pilots must also become sailors. Likewise seaplane designers must understand hydrodynamics as well as aerodynamics. Both must appreciate the complex forces of water and wind

A short Sunderland maritime rescue and anti-submarine flying boat used during World War II. It has a top speed of 210mph (338kph) and a range of 3000 miles (4828km). The flying boat has obvious advantages when it comes to rescuing downed flyers; anti-submarine aircraft carry sonar equipment for the search.

on water. The variations and conflicts in design, both past and present, are as numerous as those of the boat industry.

The aerodynamic factors differ little except for a preference for keeping wings, tailplane and engines high and as far as possible from spray. The seaplane's greater bulk forward usually calls for a larger vertical tail area to control it. But the 'landing gear' requires unique considerations such as good flotation and stability, ruggedness and lightness plus hydrodynamic lift with minimal spray.

Like an ordinary boat, a flying boat must have its centre of buoyancy beneath its centre of gravity (cg), enough displacement for its gross weight, enough freeboard to prevent swamping and a high bow for low-speed taxying. It differs only in the greater stability offered by wing-tip floats or other outrigger-type stabilizers and which compensate for a high cg and a minimal keel. Retractable water rudders are usually fitted to assist low-speed taxying.

For take-off, the hull must rise quickly out of the water and start planing like a speedboat if flying speed is to be attained, and so the hull bottom is designed to push the water downwards. A shallow 'V'-shaped bottom is now almost standard, often slightly concave to flatten out the spray and improve lift. Fluted bottoms with an intermediate chine or ridge running between keel and side improve these characteristics and are now popular on floats and small flying boats.

Unlike a speedboat, where the cg is near the stern or transom, an aircraft also needs hull support well behind the cg, to cope with displacement when at rest and to give lift during early acceleration. So slightly behind the cg, the flying boat hull has a sharp up-break called the 'step', corresponding to the transom of a boat. The step reduces skin friction on the hull afterbody while planing and allows the aircraft to tilt up at lift-off to achieve a suitable flying angle for the wings. When planing, the aircraft is described as being 'on the step' and can also taxi at speed like this with low throttle setting while manoeuvring tightly yet safely.

The floats fitted to landplanes to convert them into floatplanes are little more than small, sealed hulls. Modern floats also have some aerodynamic shape to gift lift and reduce the weight penalty. Twin floats are now standard, although a single float was popular before the war and a tail float was carried on some early seaplanes. The high stance of floatplanes, however, gives them a high cg and sensitivity to crosswinds.

Helicopters, not needing hydrodynamic gear, are often fitted with light, inflated-rubber pontoons or have these strapped to landing skids ready for inflation by compressed gas. Some larger modern helicopters even have full boat hulls or amphibious operation. A new concept now under test exploits vertical sponsons to provide a smoother ride in heavy seas for air-sea rescue craft. Fixed-wing aircraft could swivel such floats to the horizontal for take-off and landing. Possible use of lightweight water skis and hydrofoils has also been receiving attention in recent years.

Above: an amphibious version of the Canadian de Havilland DHC-2 Beaver light utility transport, typical of the hundreds of small seaplanes which serve isolated communities in Canada and Alaska.
Left: a Convair R3Y Tradewind naval patrol turboprop flying boat. Variants of this craft were built for troop transport and for use as in-flight refuelling tankers.

Aircraft Instruments

The instrumentation system of an aircraft is a means of providing the pilot and crew with the information necessary to control the aircraft. This information basically concerns navigation, engine performance and airframe controls.

The physical quantities being measured, such as temperature, speed and pressure, are known as parameters. Most modern systems are electrically based, and therefore the parameters are sensed by transducers which produce signals (data) analogous to the parameter values.

Many of the electrical indicator movements are of the moving coil (D'Arsonval) type and another frequently used system is the ratiometer type of instrument. This instrument has two coils wound on a common former, and its deflection is proportional to the ratio between the currents flowing in the coils, each coil being connected to a separate signal. As the instrument is responding to current ratios, it is relatively independent of supply voltage fluctuations.

The cockpit of an airliner has many of the instruments duplicated, with separate displays for the pilot and the co-pilot.

The ratiometer type of instrument is used in conjunction with variable transducers (typically used for oil and air temperature measurements), or the many pressure and displacement transducers which have a potentiometer output. Many circuit variations are possible, for example engine pressure ratio (EPR) measurement where a pair of transducers measure the inlet and outlet air pressures of a jet engine. Although the data is in fact the difference between the two pressures, the instrument is calibrated in terms of ratio.

The moving coil principle may be reversed by using a moving magnet, mounted between fixed coils, in place of the moving coil. An extension of this principle is the desyn indicator, which responds to the ratio between the currents in a three wire DC circuit. This type of instrument has the ability to indicate through a full 360°, making it particularly useful for navigational bearing indications. A more modern version is the synchro indicator which works from a three wire AC system where the data is in the form of varying phase relationships between the AC waveforms in the circuit. Both of these types are relatively insensitive to supply voltage variations.

Servo mechanisms are used in many advanced instrument designs, particularly those used with air data computers. This method can give greater accuracy, which is particularly useful when the readout is directly in numeric form. This type of instrument usually works from the aircraft 400 Hz electrical supply, and the data is in the form of phase variation rather than voltage or current. The data is sensed by synchros or selsyns, devices which produce signals proportional to the angular position of a shaft, and which signal the servos within the instruments to move the indicator mechanisms to the correct readings. Some instruments, such as encoding altimeters and airspeed indicators, use servos in conjunction with shaft encoder digitizers which generate data in digital form.

Fuel management Small aircraft generally require only a simple instrument set for their fuel systems, comprising a float type fuel contents gauge and perhaps a fuel pressure indicator. In a large aircraft, however, the fuel load is much greater and it is stored in a number of irregularly shaped tanks throughout the structure. It is important to control the fuel level in each tank to maintain the aircraft's correct centre of gravity. The fuel may surge about during manoeuvres, and to maintain a high degree of accuracy it is usual to measure the contents by means of a number of probes in each tank.

The important parameter is fuel mass and therefore the change in specific gravity with temperature has to be taken into account. Fuel flow is generally sensed by a small impeller within the fuel feed system which transmits electrical pulses at a frequency proportional to the fuel velocity. These pulses are electronically processed and presented in terms of pounds per hour (on a meter) and pounds used (on a numeric counter).

Engine instruments The basic engine instrumentation of a light piston-engined aircraft comprises gauges showing the cylinder head temperature, oil pressure and temperature, and engine speeds in revs/min. The jet engine, however, needs additional instrumentation covering jet pipe temperature (JPT) and engine pressure ratio (EPR). The JPT has to be critically balanced between the engine running efficiently hot or completely burning out. The outlet gas temperature, usually between 500 and 850°C (932 and 1562°F), is measured by thermocouples, their signals being used both for indication and for automation engine control.

Engine power is a vital parameter during take-off, and with propeller aircraft this is frequently indicated by a hydraulic pressure gauge operated by a transducer in the drive mechanism, or by engine inlet manifold pressure gauges. Jet engine power measurements are derived from EPR, JPT and engine speed measurements.

Hazard warnings Hazard warnings usually take the form of lamp or audible warnings rather than instruments. Stall warning, for example, is frequently presented by an obvious warning light plus a mechanism which shakes the pilot's control column. Undercarriage faults and fire warnings are usually indicated by lights and an audible warning.

Artificial horizon The artificial horizon provides indication of aircraft attitude in both pitch and roll and is particularly valuable for 'blind' flying conditions. The heart of the instrument is a vertical free gyroscope acting as a fixed reference about which the aircraft rotates.

In the basic instruments the gyro is inbuilt, and presentation is usually in the form of a moving bar attached to the gimbal assembly, which is read with reference to a fixed 'gull's wing' type of symbol that relates to the banking position of the aircraft's wings. Additional information, such as roll angle in degrees, is often presented.

In aircraft with more advanced equipment, where the gyroscope is remote from the instrument, further informa-

Below: the cockpit of a DC9, with the following indicators: 1) hydraulic brake pressure; 2) airspeed; 3) flight director; 4) altimeter; 5) clock; 6) gyrocompass; 7) course; 8) vertical speed; 9) turn and slip; 10) total fuel quantity; 11) left and right main fuel quantity; 12) centre fuel quantity; 13) ram air temp.; 14) engine pressure ratio; 15) engine speed; 16) exhaust gas temp.; 17) fuel flow/fuel used; 18) oil pressure; 19) oil temp.; 20) hydraulic control system pressure; 21) hydraulic quantity

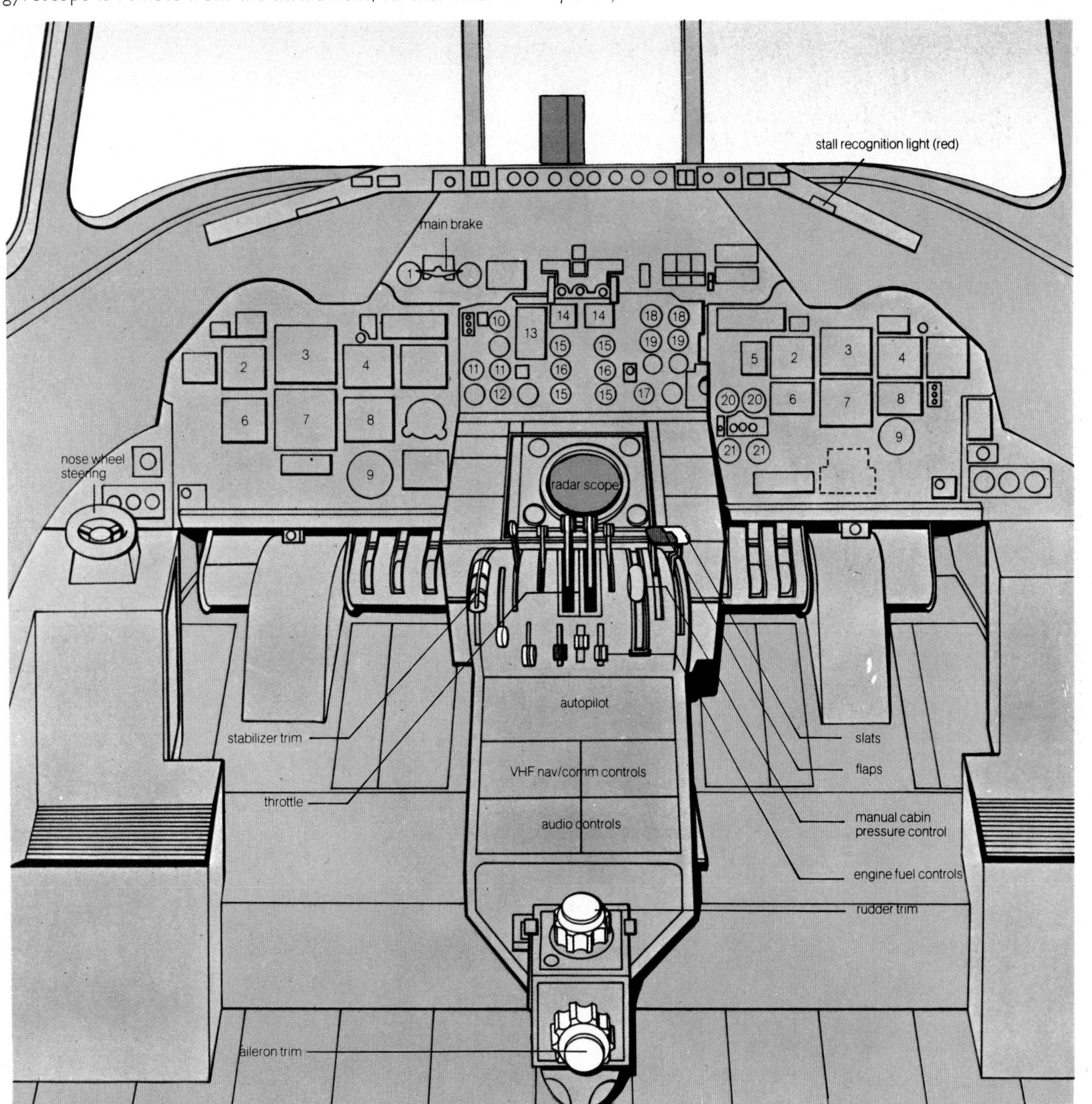

tion may be combined, often with the horizon as a background shown in the form of a moving sphere coloured appropriately to represent earth and sky.

Turn and slip indicators These are two-part instruments which enable the pilot to first set a required rate of turn (for example a 'rate 1' turn is 180° in one minute), and then to relate this to the correct bank angle. When turning, there are two components of acceleration, one due to the Earth's gravity (vertical) and the other due to the turning motion (lateral). The correct bank angle of an aircraft is when the resultant of these accelerations is acting perpendicular to the aircraft floor. If the aircraft is wrongly banked, it is said to be 'slipping' or 'skidding', and apart from this being an inefficient flight path, an uncomfortable sideways force is exerted on the occupants.

Although the more complex instruments derive their signals remotely, the basic instrument relies on an inbuilt gyroscope coupled to a pointer to indicate the rate of turn, and an inclinometer consisting of a ball in a fluid-filled curved tube to indicate the axis of acceleration. The inclinometer is similar in principle to a spirit level except that a mass is used rather than an air bubble, so that it is sensitive to acceleration forces.

Flight director systems This is a generalized title for a family of instruments where the purpose is to combine many of the navigational and flight control readings into a common display. Superimposed upon this display is a number of director bars and pointers which automatically move into alignment when the aircraft is brought on to a predetermined flight path.

This type of equipment is also known as an Integrated Instrument System, Flight Control System, or Pictorial Navigation System. These systems consist of two basic instruments, the Horizon Flight Director, which represents a rear view of the aircraft, and the Course Flight Director which represents the aircraft viewed from above.

Above: these are basic aircraft instruments. Left to right: a fuel meter, an engine speed indicator, and a hydraulic pressure gauge.
Opposite page: the cockpit of the Hawker Siddeley vertical take-off aircraft called the Harrier. It uses the miniaturized inertial guidance platform pictured on page 217 as the basis of its navigation and attack system, part of which is the map display in the centre of the instrument panel.

The data driving the instruments is obtained electrically from other systems including the aircraft compass, radio navigational aid and altimeters, and remote gyroscopes similar to the artificial horizon or rate of turn indicator. Height error sensing devices may also be incorporated, or other aircraft systems which may be of simple nature or may include a complex air data computer and inertial navigation system.

The Horizon Flight Director is generally based upon an artificial horizon display represented by a moving sphere, and to this may be added a rate of turn indicator, radio altimeter, an inclinometer and command bars indicating speed and altitude.

The Course Flight Director is principally a compass indicator, with the addition of course director bars and pointers which correlate magnetic compass headings with the relationship to radio beacons. The system may also include distance measuring equipment (DME) and flight path indication.

Speed indication The speed of the aircraft may be expressed in terms of distance travelled per hour or, in the case of high speed jet aircraft, by the Mach number (the ratio between the aircraft speed and the speed of sound). These instruments compute the aircraft speed from the difference between static air pressure and the pressure of the airstream, using a pitot head transducer to determine the airstream pressure.

The vertical speed indicator (or rate of climb indicator) shows the rate of change of altitude in thousands of feet per minute, and is a pressure-operated device which responds to the changes in atmospheric pressure as the aircraft's altitude changes. The simplest form of altimeter, which measures the aircraft's height, is also pressure-operated and virtually identical with an aneroid barometer, but this has now been largely superseded by the radar altimeter except on light aircraft.

Avionics

The word avionics comes from AVIation electrONICS, the technology of electronics used in aircraft communications, navigation and flight management. In military aircraft it also covers electronically controlled weapons, reconnaissance and detection systems. In its broadest sense, avionics includes the ground equipment used with aircraft, such as radar, test and training equipment.

Avionics equipment has undergone a revolution in design since the invention of the transistor and, later, the integrated circuit. Space, weight and power consumption are a fraction of that required for earlier equipment using valves [vacuum tubes]. As a result, today's equipment is far more complex and reliable, and more can be carried in the aircraft.

Communications Short distance air to ground communication is usually on VHF (very high frequency) channels in the aviation frequency band 118 to 135.975 MHz. The power transmitted by airborne equipment is usually up to 25 watts. The signals travel only in the line of sight, so the range depends on the aircraft's altitude. With aircraft flying at 30,000 feet (9000 m), ranges of 250 miles (400 km) are normal. Military aircraft also use UHF (ultra high frequency) channels in the range 225 to 399.95 MHz giving a similar air to ground range and up to 600 miles (965 km) air to air range.

World wide communication is allocated to aircraft in certain channels in the band 2 to 30 MHz. Powers of up to 400W are used with a choice of telemetry methods, such as voice and telegraph. Low frequency (60 to 160 kHz) radio teleprinter equipment is sometimes fitted on large commercial aircraft.

So that the flight crew does not have to listen constantly to the radio system for incoming calls, a selective calling system is used. In the SELCAL mode, the calling station sends out a two-tone signal, coded for the particular aircraft being called. The airborne receiver is left tuned to the calling frequency, and can be heard all the time. When the aircraft code is received and decoded, the flight crew are alerted by a visual or audible signal, and only then need give their attention to the radio.

A crew intercommunication system is generally linked to a passenger address system on passenger aircraft so that the flight crew can communicate with passengers.

Automatic pilot The automatic pilot was demonstrated as early as 1914 when Lawrence Sperry, son of Dr Elmer Sperry the gyroscope pioneer, won a substantial prize offered for the first 'hands-off' flight. The autopilot senses any deviation from an aircraft's flight pattern and automatically adjusts the ailerons, elevator, rudder and trim tabs (these are small extra surfaces mounted on the other control surfaces) to compensate for the deviation. The basis of the system is a gyrocompass which controls the aircraft's direction and a vertical gyro which controls pitch and roll. Early autopilots had air-driven gyroscopes with the aircraft controls activated pneumatically or hydraulically. Today, nearly all autopilots have electrically driven gyros. Even the simplest autopilot

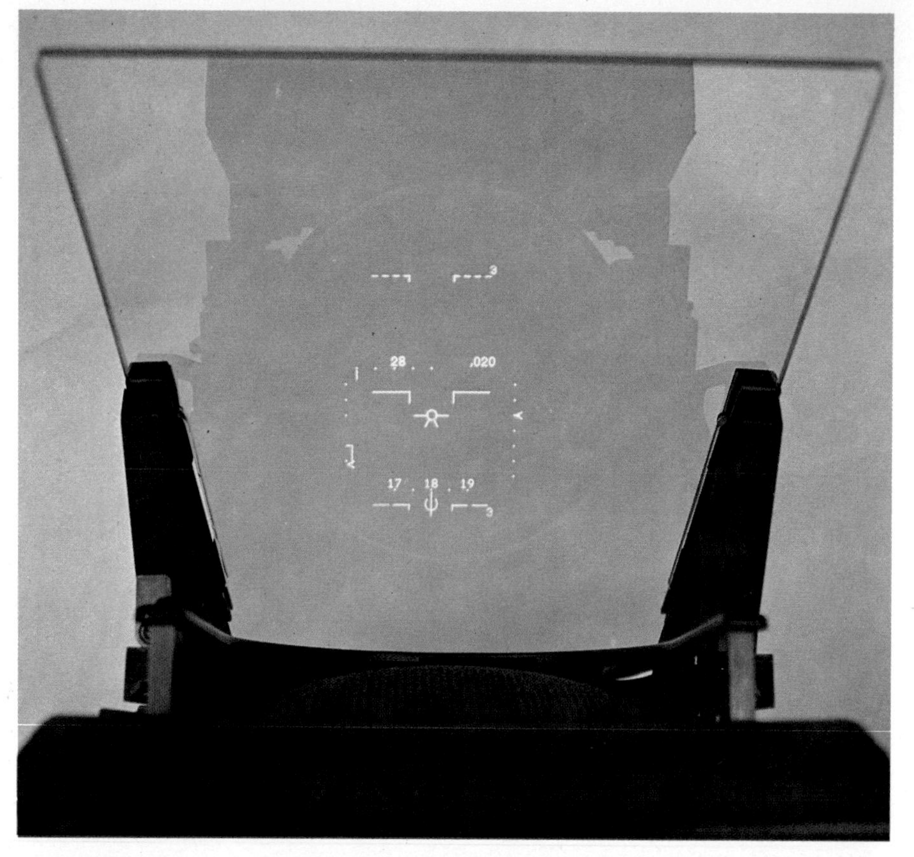

A head up display (HUD) projects essential flight information optically on to the aircraft's windscreen in the pilot's line of sight, so that he can read it without taking his eyes away from the outside world. The optics are arranged to make the image appear to be at infinity, so that the pilot has no need to change the focus of his eyes. This is important for safety when flying the high-speed aircraft of today.

will keep the aircraft on a selected heading in level flight far more accurately than a human pilot and this is still the autopilot's main function, but through the years many refinements have been added. Turns may be selected by the pilot through his autopilot: if a new altitude is chosen, the aircraft will climb or descend automatically until it has been reached; and the autopilot may be coupled to radio navigation systems (see below) so that the aircraft will automatically 'home' on to a radio beacon or lock on to an instrument landing system.

A modern addition to the autopilot is the autothrottle which provides automatic speed control throughout the cruise, descent and final approach to landing. Not only does the autothrottle reduce the pilot's workload during the critical moments before landing, but it also improves flight path holding accuracy, especially with swept wing jet aircraft where the lack of stability at low speeds can be very demanding on the pilot.

Navigation devices The navigation equipment on an aircraft may be entirely self-contained, operating entirely without external aid, or it may work in conjunction with ground-based aids such as radio beacons and area radio navigation systems. Most navigation equipment can be coupled to the automatic pilot to fly the plane, the flight crew watching the overall system and taking manual control only during take-off and landing, or in an emergency.

When an aircraft is flying between airports, it may use a system such as Loran, Decca Navigator or Omega. Each of these employs its own network of fixed ground stations which radiate a pattern of radio signals. From this, positions can be determined by measuring time or phase differences between the radio stations.

Loran (Long Range Air Navigation) is widely used by aircraft on long-distance routes. The most refined version can give positions accurate to within a few hundred yards or metres even in mid-Atlantic. The Decca Navigator can give a higher accuracy but has a shorter range. Planes on short haul flights and helicopters tend to use Decca Navigator rather than Loran. Both systems use frequencies of around 100 kHz.

Omega uses a much lower frequency, 10 to 14 kHz, and is being brought into use mainly because its signals will penetrate underwater, so it can be used by submarines. A few aircraft are equipped to make use of the system.

Doppler navigators have been developed for use in addition to Loran-type systems. The Doppler effect is the change in pitch of waves—such as sound, light or radio waves—as their source approaches or recedes. It is most often noticed when a car goes past at speed, its pitch being higher as it approaches and lower as it goes away.

In the case of a Doppler navigator, a microwave radio signal of known frequency is sent downwards to the earth's surface. It is reflected back at a slightly different frequency, since the ground is moving relative to the aircraft, and this difference is measured to give the speed. Typical systems have four fan-shaped radio beams which are transmitted in sequence forward, aft and on each side of the aircraft to provide measurement of ground speed, drift angle and total miles travelled. The transmitted frequency is centred around 8.8 GHz (8,800,000,000 cycles per second) and the power transmitted is typically one watt.

Doppler systems are entirely self-contained and can be programmed to provide information on the distance and time, to go to preselected points on the route to show latitude and longitude, to operate the pilot's steering indicator, or to drive moving map displays.

Inertial guidance Inertial guidance is a navigational technique employing inertial sensing devices whereby a vehicle is guided, without external aids or influence, from one place to another. The term 'Inertial Navigation' is also frequently used to describe this process.

Inertial sensing devices measure the vehicle's acceleration, direction and attitude by making use of Newton's laws of motion. Additionally, simple computations derive speed and position from the acceleration measurements and thus complete the information required for navigation and guidance. Accelerometers, as the name implies, are used to measure the vehicle's accelerations and very sensitive gyroscopes indicate changes in vehicle and direction.

Two designs of inertial guidance unit platforms. They carry gyroscopes and accelerometers mounted in a gimbal system. The unit on the left has associated electronics on printed circuit boards plugged into the back; the one below is a miniature design weighing 25 pounds (11.3kg).

The foundation for inertial guidance was laid in 1852 by Leon Foucault who showed that a gyroscope, because it tends to remain fixed in attitude with respect to the stars, can measure the rotation of the earth. This discovery lay dormant for fifty years awaiting development of the gyroscope to a precision suitable for navigation. In 1908 a German inventor named Anschütz-Kaempfe produced the first sea-going gyrocompass, but difficulties were experienced with the device during ship manoeuvres. Gyrocompasses indicate true north but act like pendulums when upset. Max Schuler, a German professor, recognized that the gyrocompass would not be affected by the ship's motion if its swing, or period, were deliberately adjusted to conform with the earth's radius and gravitational attraction. He proposed the combined use of a gyroscope and velocity-measuring device for this purpose.

Other gyrocompasses were produced in the United States and England, but the first true inertial guidance system was invented for the German V2 in World War 2. This was crude but effective. It consisted of attitude-controlling gyroscopes and a single accelerometer which was constructed to cut-off the rocket engine at a preset speed. By 1948 the United States had taken the lead by developing gyro-accelerometer principles into full inertial navigators. Ten years later, one of these complex systems accurately navigated the USS Nautilus under the Arctic ice-cap.

It is sufficient to travel in any form of transportation to experience the phenomenon of inertia. As the vehicle accelerates from rest, brakes or corners, the passenger experiences a force accelerating or decelerating him in sympathy. If the force is absent, or too small, the passenger may be thrown about within the vehicle.

As physical objects, by virtue of their mass, require a force to accelerate or decelerate them, the property of inertia, or sluggishness, is attributed to mass because it resists changes of motion. Using this principle, a simple accelerometer may be constructed of a metal weight held centrally in a case by two springs, one in front and one behind. As the case (and vehicle) accelerates, the inertia of the weight compresses the trailing spring. This compression is proportional to the acceleration and is easily detected electrically. Provided that the accelerometer is pointing in a known direction, this electrical signal can be used for computing speed and distance travelled in that direction.

Acceleration accompanies a change in the vehicle's speed. In other words, if the speeds at two instants are different an acceleration has occurred. This may be calculated from the speeds by subtraction. Conversely, and this is the order adopted in inertial guidance, speed may be calculated from acceleration by a process of addition—the opposite of subtraction. The method of continuous addition called integration is used.

Similarly, speed is accompanied by changes in the vehicle's position. So, just as acceleration is integrated to give speed, the speed is itself integrated to give position.

A single accelerometer mounted lengthwise in a vehicle can only sense motion in one direction at a time—north-south or east-west, for example. Furthermore, it must be kept level, or otherwise compensated, to remove the effect of gravity. Useful information is available from such an arrangement but complete inertial guidance systems need two more accelerometers in order to detect motion in all directions. One of these points to the side and the second points downwards, at right angles to the others like the corner of a cube. The three mutually perpendicular accelerometer directions are called axes.

A support for the accelerometers, called a stable platform (or inertial platform) ensures that the three axes point constantly in set directions, isolated from the changes in

Below: the Davall 'red egg' flight recorder. The resin-bonded glass-fibre case is designed to withstand extremes of pressure and temperature. Information from the flight deck is converted to digital form and recorded on stainless steel wire. There are two sets of reels and heads so that while one set is recording the other is rewinding; when one cycle is completed, the platform is swung over and the process reversed.

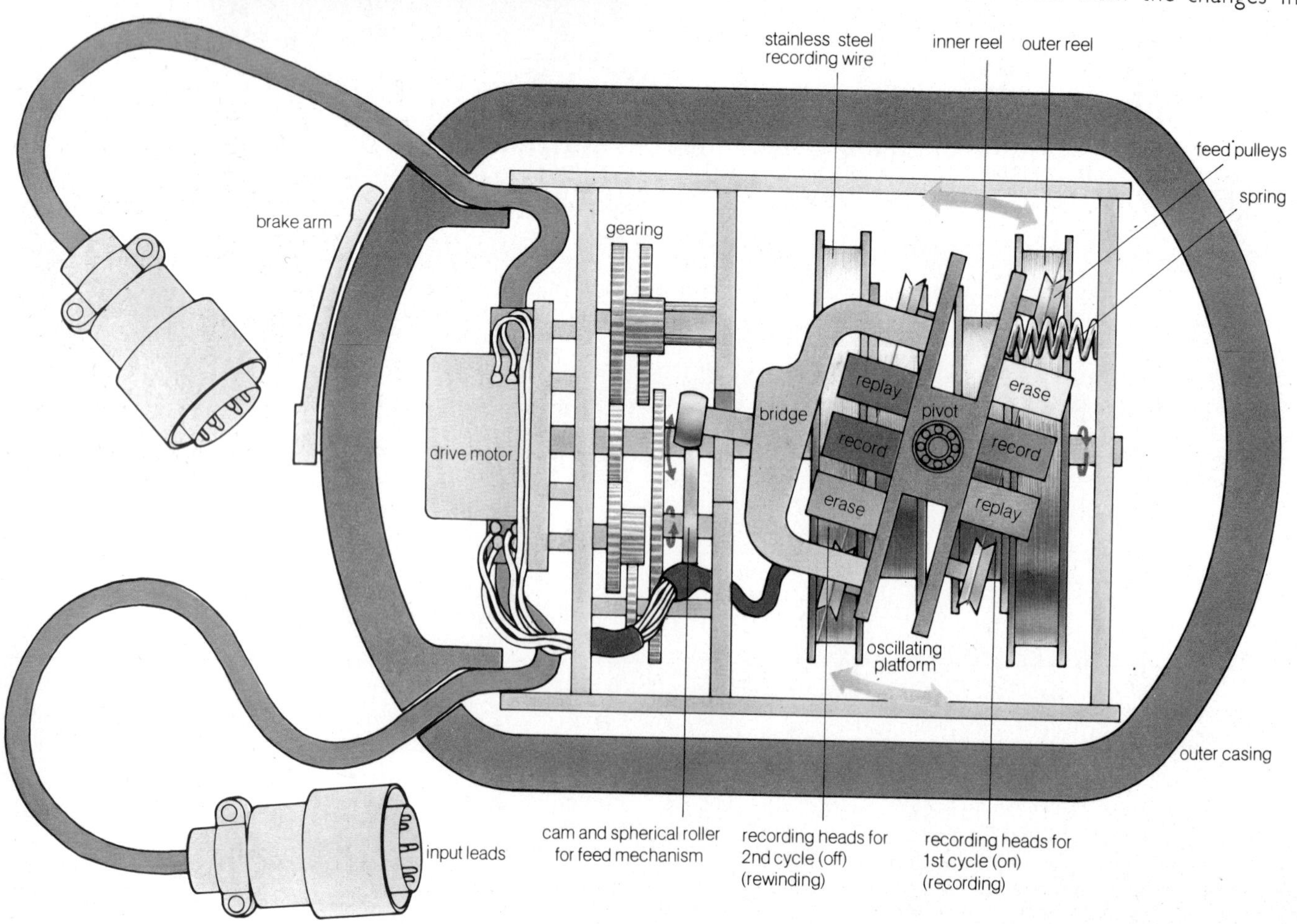

attitude of the carrying vehicle. For this purpose, the platform is mounted in gimbals. These are bearings which allow almost perfect rotational freedom about the platform.

Three gyros are fixed to the platform in order to maintain its alignment with the Earth's north and east directions, or with the Sun and stars in the case of a spacecraft. If the platform tilts out of alignment for any reason, the gyros sense the movement and activate motors which return the platform to its original position.

The stabilized accelerometers, then, provide the speeds and distances along known fixed axes, while electrical instruments mounted on the platform gimbals measure the vehicle's pitching, rolling and turning. These quantities are used to guide the vehicle to its correct flight path and destination without aid or interference from any external source.

Sadly, no measuring instrument is perfect; all to some extent suffer from errors. The accuracy of an inertial guidance system is almost totally dictated by the sensitivity of the three gyros. Numerous small imperfections, which cannot be removed from the gyros during manufacture, cause the gyros to wander, or drift, thereby tilting the platform out of the desired orientation.

The resulting inertial guidance errors are very complicated but they can be reduced by ingenious alignment techniques and good systems navigate accurately to within a few miles for many hours. Additionally, some applications of inertial guidance employ position and speed references from external navigation aids to correct the guidance errors.

Direction finding Most aircraft still carry Automatic Direction Finding (ADF) equipment for finding their position in areas where there is poor coverage by other systems. They use a simple loop aerial which picks up transmissions from radio beacons. The loop picks up the strongest signals from stations which are at right angles to it, so it can be used to find their direction.

An improvement on this very simple system is the VOR or VHF Omnidirectional Range. VOR transmitters radiate a VHF radio beam which rotates in the same way as the light beam from a lighthouse. It does this 30 times a second, while transmitting a signal which varies at 30 Hz, so that each rotation of the beam corresponds to a single cycle of the transmission. Another transmission, which does not rotate, is sent out also at 30 Hz arranged so that the two are in phase when the rotating beam points to magnetic north, on which all navigation systems are based. Equipment on the aircraft is tuned to the VHF transmission, and receives the two signals which will be out of phase to an extent which depends on their direction. By coupling this direction information to the autopilot, the plane will fly automatically to the beacon. VOR beacons are located at suitable intervals along established air corridors.

Distance Measuring Equipment (DME) and Tactical Air Navigation (TACAN) are also available on much flown-over routes, and use airborne interrogator-receiver equipment. Simple DME systems transmit a signal to a ground radio beacon, automatically triggering a reply signal. The time interval between the original transmitted pulse and the reply is directly related to distance and is read off by the receiver in nautical miles. TACAN is a military form of DME and normally gives the beacon's direction as well as its range.

TACAN is also used air-to-air to give range and direction between suitably equipped co-operating aircraft. Ranges of DME and TACAN are typically 200 miles (320 km).

Landing Near an airport are the ILS (Instrument Landing System) beams used in air traffic control. Equipment on the plane uses the ground-based localizer, marker and glide path transmitters to guide the pilot to the airport and help in the landing procedure. Outer, middle and inner marker beacons give the distance from the runway threshold, and glide-path signals provide guidance on the correct angle of descent. The airport localizer beacon is picked up some 20 to 30 miles (30 to 50 km) away and the aircraft approaches to start the landing run at an altitude of some 2000 ft (600 m). When the glidepath signal is picked up the pilot follows this down, his progress being indicated by successive marker beacons. Visual and audible indications are given of the sequence of events and of any corrective action to be taken.

Weather radar Most large commercial aircraft are fitted with weather radar which provides the flight crew with a picture of cloud formations and other atmospheric disturbances ahead. The equipment consists of a forward looking radar mounted in the nose of the aircraft and a display unit in the flight deck. The radar scanner can be tilted downwards to provide ground mapping as an extra navigation aid when crossing coast lines, estuaries or other prominent geographical features. For weather radar, ranges of up to 300 miles (480 km) are possible. Good interpretation of the radar picture depends to some extent on the skill and experience of the crew.

The aerial of an air traffic control system radar. Such a device is part of an extensive network of electronics which enables the controller to keep track of dozens of aircraft which may be in the air at one time, entering and leaving the area or 'stacked' above the airport.

Airports

At one time, airports were located in convenient fields as close as possible to the city they were to serve, without much regard to environmental problems. It was difficult in those days to envisage the growth potential of civil aviation and the complexity of air services and airports to handle them. Today, the spread of cities and the greater importance of air travel mean that airports have to be planned to meet a careful balance of aviation and environmental needs.

A major airport requires a good highway and rail links with the city centre. Passengers should be able to park their cars within a short walk of the berth in which their airliner is docked. They do not want to have to park a long, confusing walk away, struggling with baggage through rain, traffic and disembarking passengers to the terminal building.

In between the car and the airliner, the airport, airlines and control authorities like immigration and customs must provide the embarking passenger with, in order: ticket and check-in counters, passport checkpoints for international flights, concourses with lounge and general consumer services such as duty-free shops and a pier connecting the terminal with the door of the aircraft. The disembarking passenger wants to get out of the airliner as quickly as possible and either leave the airport or get to another berth to catch a connecting airliner. If he is catching a connecting flight he wants to get to the appropriate berth—though it might be literally miles away—without a long walk and without bothering about his baggage. All this has to be accomplished without mixing inward and outward-bound passengers.

To these passengers' requirements the airport designer must now add those of the airlines. An airport like London's Heathrow employs about 50,000 people whose jobs include dealing with passengers and their baggage, servicing and refuelling aircraft, air traffic control and so on. Nearly as many workers may go into and out of an airport each day as passengers. They want car parks and offices, and separate

access for service vehicles to the airliners.

The aircraft themselves make great demands on space. Runways 12,000 feet (3.7 km) long and 150 feet (45 m) wide are required for modern commercial jets. Ideally there should be at least two runways, each aligned with the prevailing wind and with turn-offs and taxiways for the least taxying time to and from the terminal.

A modern airliner berth requires a terminal frontage of at least 300 feet (90 m). During rush hours at big airports there may be as many as 50 airliners in dock. This represents a frontage of some three miles. If each airliner has 350 seats it may generate a hundred cars; half a dozen airliners a day may dock in that berth and so parking for 600 cars has to be found near each berth.

The aircraft also need parking and maintenance space. Planes these days are designed for 'on-condition' or 'on-wing' maintenance, in which replacements for faulty units are simply fitted while the plane is parked.

Airliners have to be 'turned round' as quickly as possible between flights to reduce costly time spent on the ground, when they are not actively earning money. The time between flights may be as little as twenty minutes, during which passengers have to be disembarked, the aircraft refuelled, maintained, cleaned and re-provisioned and the next lot of passengers embarked. These activities have to be carefully scheduled to avoid clashes, or the fire risk of passengers embarking during refuelling.

Finally, the design has to allow for the rapid growth of air transport, permitting terminals to be added or extended without the 'alterations as usual during business' that characterize so many airports.

Air traffic control Air traffic control, ATC, is one of the most vital factors in air safety. It is a system for preventing collisions between aircraft in congested areas, particularly in the neighbourhood of airports, where the air is full of aircraft of different sizes travelling in various directions at various speeds and heights. ATC also keeps air traffic flowing smoothly.

Above: a night view of Heathrow airport, showing the row of aircraft berths connected by passenger jetties.
Far left: an aerial view of New York's John F Kennedy airport. Nearly all the central space is devoted to car parking. This area is surrounded by a hollow ring of airport buildings, then by parking and taxying space for aircraft, then by the runways. The famous curving shells of Eero Saarinen's TWA terminal can be seen at the right (eastern) corner of the car park.
Left: Heathrow's passenger reception in 1946.

On those parts of long-distance routes which are uncongested, the pilot uses the built-in navigational aids of his aircraft, and sometimes electronic aids on the ground, to make his own way and avoid collisions without help. But as soon as he approaches a much flown-over area, or nears an airport, he enters a control zone, where he is obliged to follow a course at a given speed and height, all prescribed to him by the air traffic controller.

The air traffic controller is the decision maker. It is he alone who has complete information on all aircraft movements within his control zone. He must exercise powers of discretion on the minimum safe spacing between aircraft, both vertically and horizontally, and determine priority in take-off and landing within the framework of flight schedules.

The minimum information required by a controller is the current height and position of all aircraft under his control, the intentions of all aircraft under, or soon to be under, control, and the identity of each aircraft. He gets this information from a flight progress board which tells him intention, identification, vertical position and timing, and from a plan position radar which gives the exact position and distance of all aircraft within his control zone.

The controller's work involves continuous updating of information as new situations develop and earlier ones pass from his control. He receives advance information of traffic about to enter his control zone from adjacent zones and informs adjacent zones of traffic leaving his zone. He also monitors and controls all traffic within his zone.

The controller communicates with the aircraft normally by VHF (very high frequency) radiotelephone with a range of up to 200 miles (300 km) when the aircraft is at high altitude although the range decreases as the aircraft descends. A radio direction-finder (RDF) system is frequently employed with the VHF radio link to supply compass bearing of any call received.

Wind speed and direction, visibility, cloud base, air temperature and barometric pressure data is fed to the controller from local sources and from meteorological centres. Runway visibility can now be accurately measured by electronic means rather than by someone's personal estimate, as is still usual today.

Flight plans With certain exceptions, each flight requires a flight plan which includes aircraft identification, airport of departure and destination, route plan, desired cruising level, departure time and estimated time of arrival. The data is transmitted, generally by land line rather than radio, to the ATC control centre from airports within the controller's zone or from adjacent zones. The information is always to a standard format.

Radar control systems The basic radar system gives a continuous plan, as seen from above, of all aircraft within radar range. The plan position indicator (PPI) radar display shows an aircraft 'target' as a bright spot with the range (distance) of the aircraft indicated by its distance from the centre of the screen and its bearing by the angle to the centre. An electronic means known as video mapping makes it possible to permanently superimpose fixed features such as defined airways on the screen. It is also possible to eliminate all unwanted permanent radar echoes from stationary objects and display only those which are actually moving (moving target indicators).

In yet another refinement the radar echo from a particular aircraft can be 'tagged' with its identity or other information as a code of letters and numbers, the identity tag slowly moving across the screen in synchronization with the movement of the aircraft.

PPI type radars are in three broad categories, long range surveillance, airfield control and airfield surface movement. Long range surveillance radars have typically up to 300 nautical miles (550 km) range from power of the order of two megawatts peak power. Airfield control radars operating at less power have typically 50 to 150 nautical miles (95 to 280 km) range. Airfield surface-movement radars are designed for very high definition and range is normally confined to runways, taxi-ways and aprons of the immediate airfield. Modern surface movement radars have sufficient picture resolution to identify individual aircraft types by their shape and size.

These radars are all of the 'primary' type, which obtain information from a reflection of the radar beam from the aircraft or other 'targets', and require no co-operation from the aircraft. Another important type of radar system is known as secondary surveillance radar (SSR) in which equipment carried on the aircraft receives the transmitted ground signal and transmits a reply. The reply is entirely automatic and generally includes a coded message giving identity of the aircraft and present altitude, both of which can be integrated into the main PPI display and 'tagged' to appropriate aircraft on the display. The airborne equipment of the SSR is called a transponder, because it responds to a received signal by transmitting another signal.

Instrument landing system The controller normally controls aircraft up to the final approach to the airfield when the pilots can lock on to the instrument landing system (ILS). This system provides a fixed radio beam so that an aircraft can align itself with the runway and adopt the correct descent path. The equipment comprises two ground transmitters, one emitting a beam to guide the aircraft in azimuth or

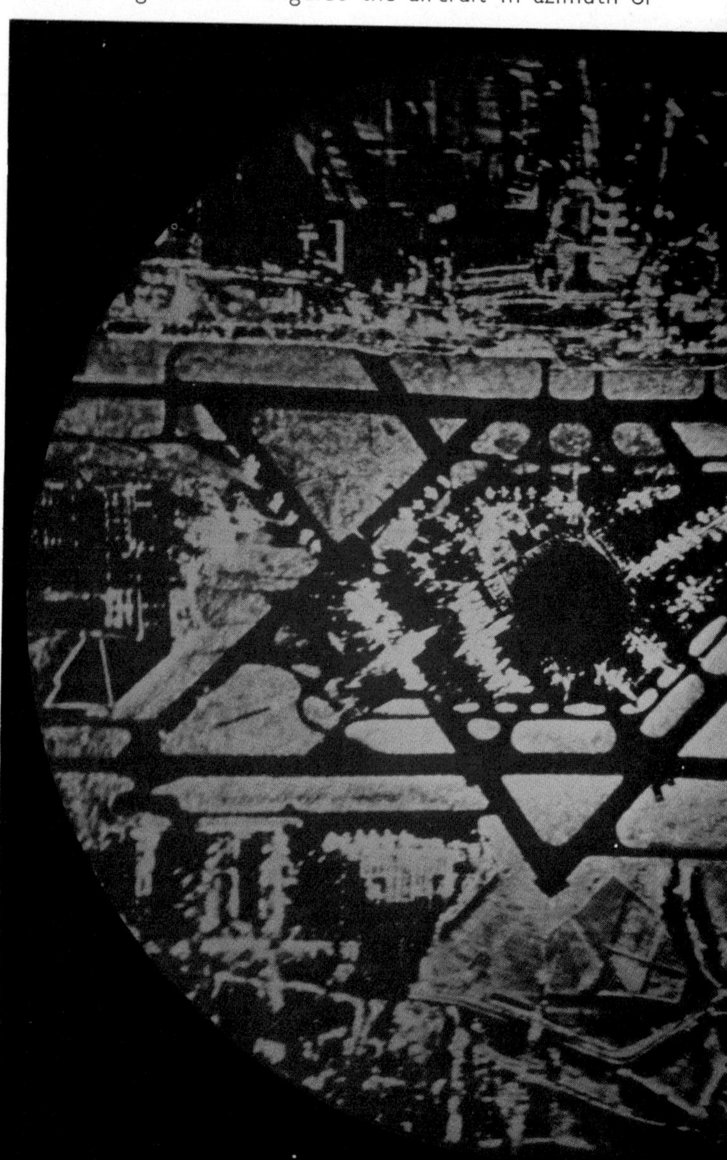

compass bearing, the other a beam to guide the aircraft in altitude. The beams are known respectively as the localizer and glidepath. Both beams are modulated with tones at audio frequency (at a pitch which enables them to be heard) which are used to activate instruments in the aircraft flight deck (or indicate audibly to the pilot) whether he is deviating to the left or right of the centre line and above or below the glidepath.

Along the approach centre line are three vertically transmitted fan shaped beams known as the outer, middle, and inner marker beacons. Once brought to the position for final approach, the marker beacons indicate the distance to go, and the ILS system proper shows any deviation from the centre line and glidepath. If he keeps to the centre line and glidepath he will be brought accurately to the threshold of the runway at about 200 ft (60 m) altitude and can then complete the landing visually. The controller directs the aircraft to the appropriate runway exit for parking and discharge of passengers.

Blind landing Suitably equipped aircraft can use ground based ILS (provided it is of exceptional accuracy) combined with the autopilot and additional airborne electronic aids (principally highly accurate radar altimeters) to land in near zero visibility. The main complexity in such a system is the duplication and even triplication of airborne equipment to secure acceptable reliability.

Blind landing has been achieved thousands of times in commercial practice but always in acceptable visibility as part of the proving trials. The pilot has the system engaged but monitors the landing throughout and is ready to take manual control at any instant. Completely blind landing will become a reality when the equipment is fully proved.

Less busy airports use VOR (VHF Omni Range) beacons, which are also used as en route radio beacons. As an approach aid VOR is less satisfactory than ILS because it gives heading guidance only.

Visual landing aids are still important. The visual approach slope indicator, or VASI, which operates day and night, is not a substitute for ILS, being an 'airfield-in-sight' aid. Bars of red and white lights on either side of the runway are angled to show the pilot all red lights when he is below the glidepath, red and white lights when he is on the correct glidepath, and all white when he is too high. These lights are on either side

Above: the inside of the control tower at Chicago's O'Hare airport, one of the largest and busiest in the world. The sheer number of aircraft coming and going combine to make the air traffic controller's job uniquely demanding. Although he is backed up by radar and computer technology, in the final analysis it is his judgement that ensures the safety of every aircraft.

Left: a picture of London's Heathrow airport shown by the airfield surface movement radar. Individual aircraft are clearly visible, their long 'shadows' radiating outwards from the central aerial on the control tower.

of the touchdown point of the runway.

The approach to the runway is indicated by Calvert approach lighting with a white centre line crossed by five white bars, getting narrower as the runway approaches.

The runway itself has white centreline lights and bars to mark the touchdown area. At the end of the runway, the centreline becomes all red. The edges of the runway are marked with white lights. The latest design taxiways have green centreline lighting, replacing the blue edge lights that were formerly used.

The lights set into the runway are carefully designed to withstand the 300 tons or so exerted by a landing aircraft, and yet present no obstacle. The light from a 200 W tungsten-halogen bulb shines through an aperture no more than ½ inch (13 mm) high.

Stacking At times of great congestion it is necessary for aircraft to queue, awaiting instructions to land. The controller directs aircraft to a holding or stacking area where aircraft fly round and round one above the other but separated by a safe vertical distance of about 1000 ft (300 m). The lowest aircraft is called off first after which the remaining aircraft descend one stage lower according to the safe stacking separation height. Aircraft arriving at a stacking area take the uppermost position.

Electronics The electronic computer now plays a central role in ATC in information processing and storage and supplying data to individual controllers in a large complex. Its main function is to reduce the workload on controllers so that they can concentrate on supervision of aircraft movements and decision making.

The Eurocontrol Maastricht Automatic Data Processing system (MADAP), which will be fully operational by 1975, is a good example of a modern ATC system as it is multinational in the equipment used, the ATC officers who work in it, the countries that support it and the aircraft which fly through its area of operation.

MADAP will control the upper air space of a region covering Belgium, Luxemburg, the Netherlands and the northern part of the Federal Republic of Germany. Its design incorporates all the features outlined above with the

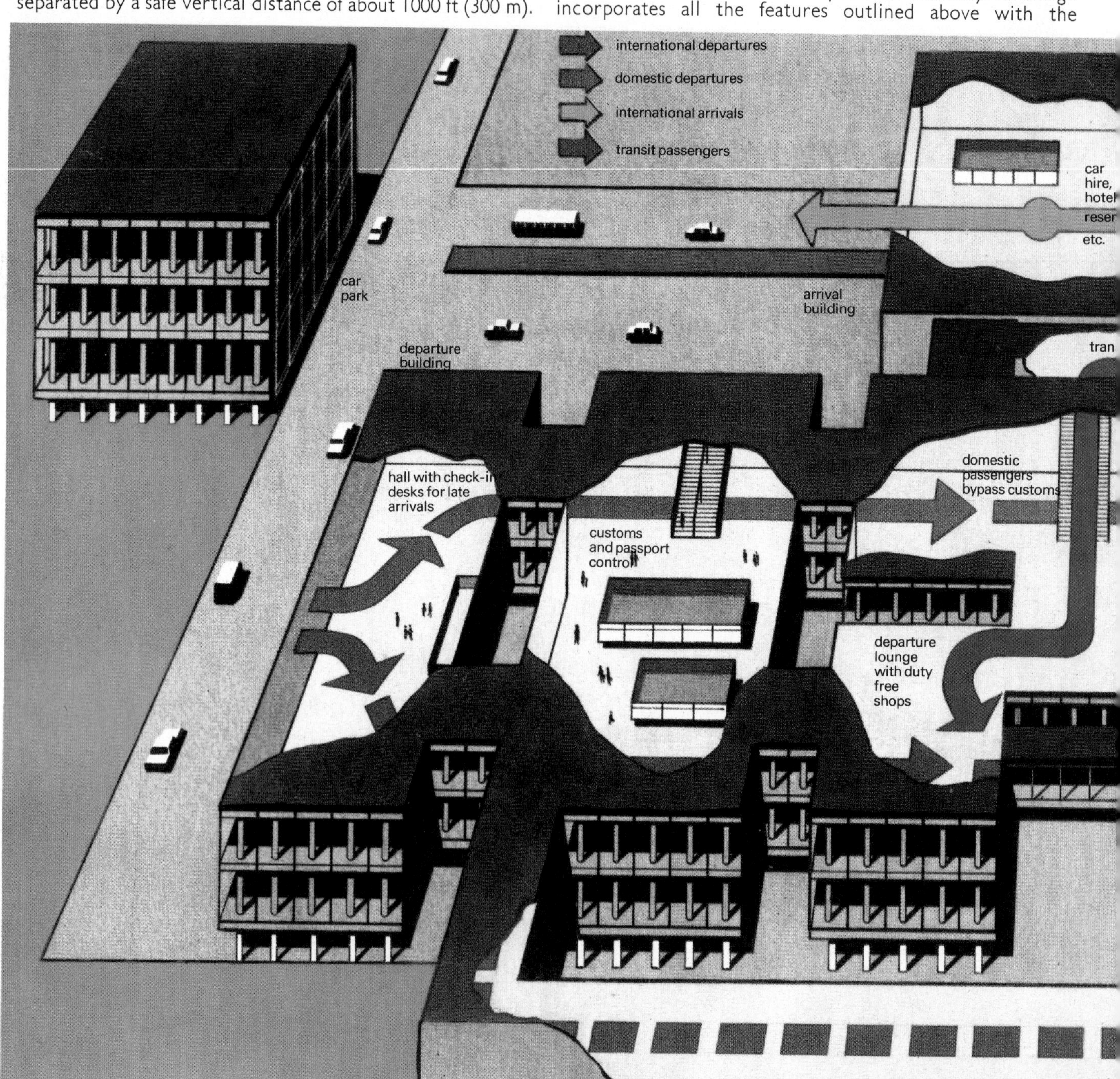

exception of ILS and other ground aids which do not apply to upper air space.

The ATC centre will receive data from four radar centres and a continuous stream of flight data from airports and adjoining areas. All inputs, including radar data, are to be processed through high power computers to provide controllers with the information they want as soon as it is needed. For example, incoming flight plans will be held in the computer store until the aircraft concerned are entering the area. The computer complex also performs such tasks as printing out events such as an aircraft passing over a reporting point as they occur, and predicts conflict conditions. But automation is kept firmly in place. The equipment is only there to aid the controller, who is still the final decision maker. The MADAP installation has eight computers, more than 80 operating and training positions for controllers and some 140 radar and data display units; it is designed to deal simultaneously with 200 flight plans and 250 aircraft tracks. For the foreseeable future, however, no amount of hardware will replace human judgement.

This schematic diagram of a modern airport shows how the buildings are carefully organized so that streams of passengers going in different directions do not mix, which would allow people to dodge ticket, baggage and passport checks. Each passenger must go through the stages in the proper order, but the layout must be flexible enough to be rearranged if necessary, for example during rebuilding. The main requirement is space, which creates its own problems by making necessary great distances from one part of a building to another.

INTO SPACE

The history of transport has required the solution of one technical problem after another. Now the most difficult problem of all—that of overcoming the gravitational pull of the Earth itself—has been defeated. In the summer of 1969, men set foot on the moon, fulfilling an age-old dream. Now that space travel has been proved possible, the next question is whether the enormous distances between planets and between galaxies can be conquered. It is difficult to believe that the answer will not ultimately be 'Yes!'.

The orbit

The path of a body under the gravitational influence of another is known as an orbit. It is the balance between the inertia of the moving body and the strength of the gravitational field, and can be illustrated by a bullet being fired horizontally: a bullet from a normal gun will eventually fall to the ground, but if it could be given a high enough initial velocity, the downward curvature of its path would become equal to the downward curvature of the Earth's surface. It would stay at the same height, although falling continuously.

Strictly speaking, the gravitation of the bullet must affect the Earth as well, and the orbit of the bullet is not about the centre of the Earth, but about the centre of gravity of the two bodies (the barycentre). Although this correction is negligible for something as light as a bullet it is important when considering the orbit of the Moon, whose mass is 1/81 of the Earth's. Both the Earth and the Moon orbit about their barycentre, which is 3000 miles (5000 km) from the centre of the Earth, only 1000 miles (1600 km) below its surface.

Shapes of orbits Sir Isaac Newton showed in 1687 that as a consequence of his inverse-square law of gravitation all orbits must be one of a group of curves known as the conic sections, and in fact some 80 years earlier Johannes Kepler had found that the planets move in the elongated circles called ellipses (which are conic sections), the Sun being at one of the two focuses. Kepler also discovered that a planet moves fastest when at its closest to the Sun (perihelion) and slowest when furthest away (aphelion). The average speed of the planets decreases with increasing distance from the Sun: Mercury, the innermost planet, moves at 108,000 mile/hour (47.9 km/sec), the Earth at 67,000 mile/hour (29.8 km/sec) and the outermost planet, Pluto at only 10,800 mile/hour (4.8 km/sec).

The other conic sections are the circle, the parabola and the hyperbola. Circular and elliptical orbits are called closed because the orbiting body returns to its starting place; parabolic and hyperbolic orbits extend to infinity and are open, that is, the orbiting body never returns. Some comets' orbits may be hyperbolic, but it is very difficult to distinguish between hyperbolic, parabolic and extremely elongated elliptical orbits from the few observations available when the comets are close to the Earth.

Spacecraft The principles governing the motion of the planets about the Sun also apply to the orbits of artificial satellites around the Earth. The lowest orbits possible are about 125 miles (200 km) above the Earth's surface, because at lower altitudes the drag of the atmosphere slows down the satellite, and it spirals downwards until it is destroyed by friction with the dense lower atmosphere. At a height of 125 miles the orbital period is about 90 minutes and the speed 18,000 mile/hour (29,000 km/hour). Spacecraft are usually launched eastwards, because then they start with the velocity of the Earth at the latitude of the launch site (about 1600 km/sec at Cape Canaveral) and so they need less expenditure of fuel to reach the orbital velocity. The rotation of the Earth has no other influence on orbital velocities—even if it were to spin ten times faster, the orbital details would be the same.

A particularly important Earth orbit in the present case, however, is that whose height is 22,300 miles (35,700 km), where the period of the orbit is exactly 23 hours 56 minutes, so that a spacecraft appears to remain overhead a particular spot on the Earth's surface. This orbit is used by communication satellites (such as Intelsat) so that the ground station can be continuously in contact with them.

A spacecraft travelling at 25,000 miles/hour (40,000 km/hour) will escape from the Earth's gravitational influence and move into an orbit around the Sun. Planetary probes are put into orbits of this kind: for example, a spacecraft to Mars would be put in an elliptical orbit whose perihelion is at the Earth's orbit, and its aphelion at the distance of Mars' orbit.

Right: during the Apollo missions, the craft were able to orbit as low as 10 miles (16km) from the surface, because the Moon has no air to cause drag. Opposite page: Newton's work on gravitation was inspired, he maintained, by seeing the fall of an apple from a tree. From this he reasoned that the force which attracted the apple to the Earth's centre was the same force which held the Moon in its orbit around the Earth. This giant leap in thought culminated in the Apollo Moon missions, 300 years later. This photograph of the Earth was taken on the Moon.

This Hohmann transfer ellipse is achieved by accelerating the spacecraft along the direction in which the Earth is moving in its orbit. As a result the probe starts with the orbital velocity of the Earth, and requires only a relatively small increase. The alternative approach, namely to accelerate outwards from the Earth's orbit directly towards Mars, would not use the Earth's velocity as a contribution to the spacecraft velocity, and hence would require a much greater expenditure of fuel. The disadvantage of using the Hohmann ellipse is that the journey takes a long time, 260 days in the case of an Earth to Mars flight, and in practice some extra fuel is used to speed up the flight. The journey of Mariner 9 to Mars in 1971, for example, was reduced to only 192 days.

The most economical method of travelling to the planets between the Earth and Sun, Mercury and Venus, is to use rockets to decrease the spacecraft to a speed less than the Earth's: it will then travel in an ellipse with the distance of Earth as aphelion. In any of these transfer ellipses the target planet must be at the right part of its orbit when the probe arrives there; and hence the launch date of the spacecraft is restricted to a few days each year (the launch window) when the Earth and target planet are suitably placed relative to each other.

The satellite

Since the Earth's first artificial satellite, *Sputnik* I, was launched by the USSR in 1957, several hundreds of satellites of increasing size and complexity have been put into orbit, mostly around the Earth. Craft have been orbited around the Sun, Moon and Mars, but these are generally known as space probes.

The use of Earth satellites has many practical benefits to mankind. The world's meteorological services depend increasingly on satellite photographs of cloud cover and on measurements of atmospheric properties made from space. Communications satellites are of great importance, not just for relaying live pictures of sporting events but mainly for providing telephone and data links for governments and industries. The data gathered by scientific satellites has greatly increased man's knowledge of the world and its surrounding environment. Major powers use satellites to gather military intelligence.

Because satellites cannot yet be serviced in space they have to be built to be very reliable. With the advent of the space shuttle, a reusable manned space vehicle, they will be launched more cheaply directly from a space platform, and it may even be possible to service or recover a failed unit.

Many different types of orbit can be achieved, depending on the mass of the satellite, the launch rocket capability and the angle of the orbit to the equator. The satellite must reach a velocity of 18,000 mile/hour (30,000 km/hour) relative to the Earth's surface to achieve a low orbit, and 25,000 mile/hour (41,000 km/hour) to escape from the Earth's pull and go into deep space. So that the satellite can be as large as possible, the launch vehicle is staged: as each section of the rocket uses up its fuel it is jettisoned.

Working environment Satellites have to operate in a very harsh environment which cannot totally be simulated in ground tests. Equipment must work in zero gravity, under high-vacuum conditions and with wide temperature variations. During launch, the vibrations transmitted from the launch vehicle and shock loadings as the upper stages take over are severe. To add to the problems, some materials can emit gas or give off particles of their substance under a vacuum, and have to be avoided, as the particles could confuse star sensors or contaminate solar arrays. Two similar metal surfaces may weld together under pressure in a vacuum, and conventional lubrication systems would not work as the oil would evaporate. Surfaces which have to touch each other are made of dissimilar materials, and solid film lubrication systems using lead or PTFE ('Teflon') are needed.

Construction The design of a satellite involves a number of sub-systems: structure, thermal control, attitude control, power, electrical distribution, telemetry and command, and an operational payload. The payload may be a number of scientific instruments, or cameras to photograph the terrain below.

So that the satellite can carry the maximum payload, its construction must be as light as possible, yet it must maintain its integrity under the strains of launch. Aluminium alloy and conventional aerospace building techniques are generally employed. Floor panels, side walls and solar array frameworks are made from aluminium or glass-fibre laminate-faced honeycomb panels. Threaded inserts fitted to the honeycomb are used to mount the equipment. Machined parts are made from aluminium alloy, titanium or beryllium—materials chosen for their non-magnetic properties as well as for their strength and thermal stability. The satellite has to stay correctly balanced under all configurations and conditions, since it is usually designed to spin about one particular axis. It should not be easier for a satellite to spin about any other axis.

Electronic equipment is sensitive and has to be kept within the temperature range −40°C to +50°C (−40°F to 122°F). The satellite's environment, however, can vary from full sunlight to full eclipse, and it may or may not be illuminated by light reflected from the Earth. The apparatus on board can vary in its heat output as well, and yet the interior of the satellite must be kept in thermal balance. If the experiments are not too demanding a *passive* control system may be used:

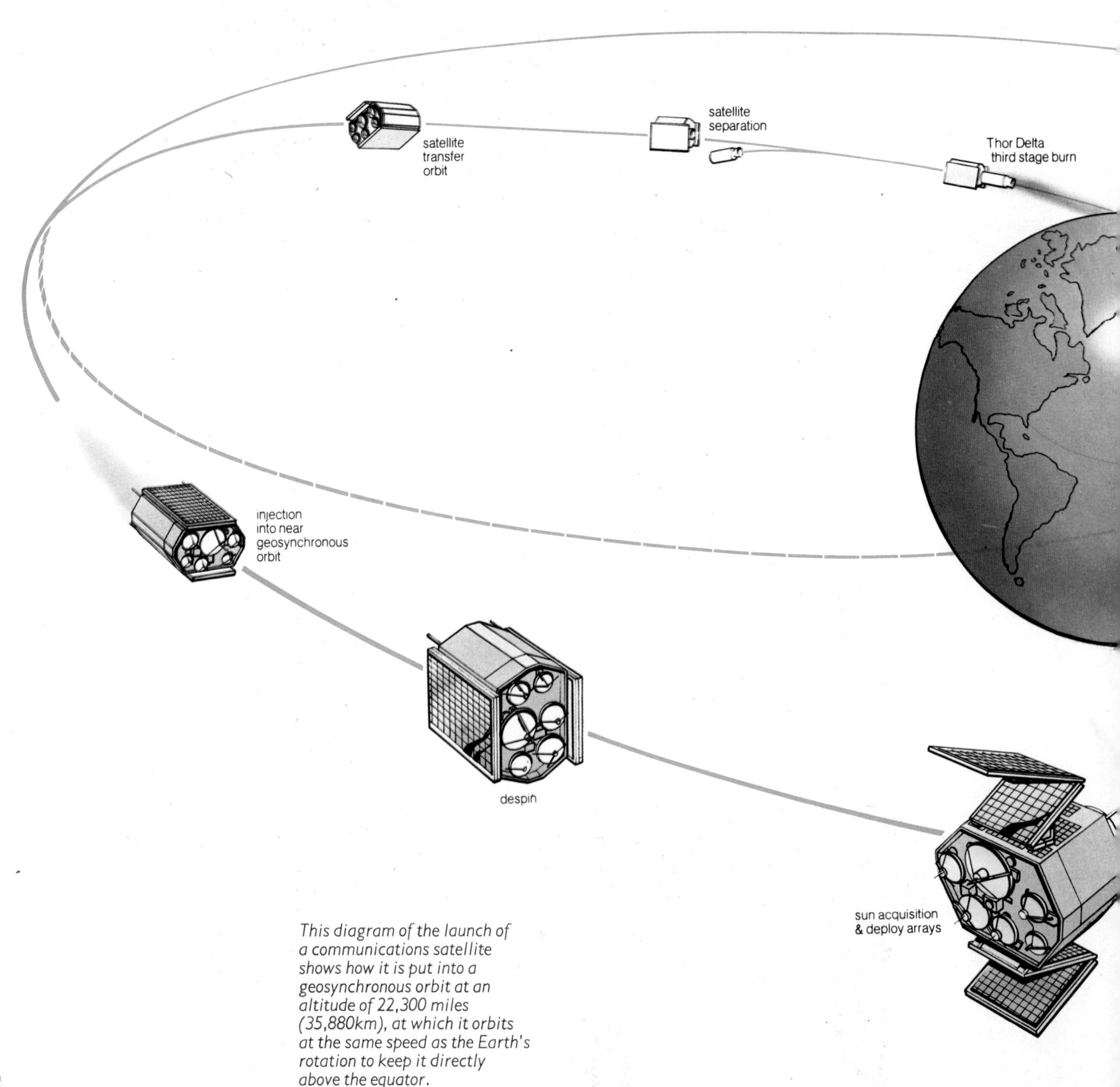

This diagram of the launch of a communications satellite shows how it is put into a geosynchronous orbit at an altitude of 22,300 miles (35,880km), at which it orbits at the same speed as the Earth's rotation to keep it directly above the equator.

the exterior finish of the satellite is chosen to absorb or emit radiation and no further temperature control is needed. For example, a light or shiny finish will neither absorb nor emit very much radiation, whereas a black finish will. Thermal blankets of crinkled multi-layer aluminized plastic film are used to insulate some areas.

For complex satellites, an *active* thermal-control system is used. A louvre system on the spacecraft wall can control the inside temperature, or alternatively heat pipes may carry excess heat away from localized hot spots to the wall of the satellite.

Attitude control Satellites usually have to be pointed in space in some way so that the solar arrays are always pointing at the Sun. Simple satellites are spin-stabilized like a gyroscope about one axis, while others which need to be pointed accurately are oriented by a system of control jets fixed to the body—three-axis stabilization. With this system, the satellite's orientation has to be measured, so that corrections can be made, and this is carried out by sensors which are designed to observe the position of astronomical bodies such as the Sun, Earth and certain stars. The attitude control system receives this information or an overriding ground command and demands pulses from the control jets to re-align it as required. These control jets may use as fuel hydrazine—a single propellant which does not have to be mixed with anything else, but which decomposes on contact with a catalyst; or, less commonly, propane. Future satellites may use ion propulsion for the same purpose. Using these systems, a satellite can be pointed with high accuracy—an orbiting craft can be pointed at a selected spot on the ground below to an accuracy of 10 metres, for example.

Electrical power Satellites in Earth orbit have almost unlimited supplies of power available from the Sun. Solar cells are either mounted around the body or are on panels which can be extended from the sides of the satellite on light spars. Since there is no air in space, and since the satellite is

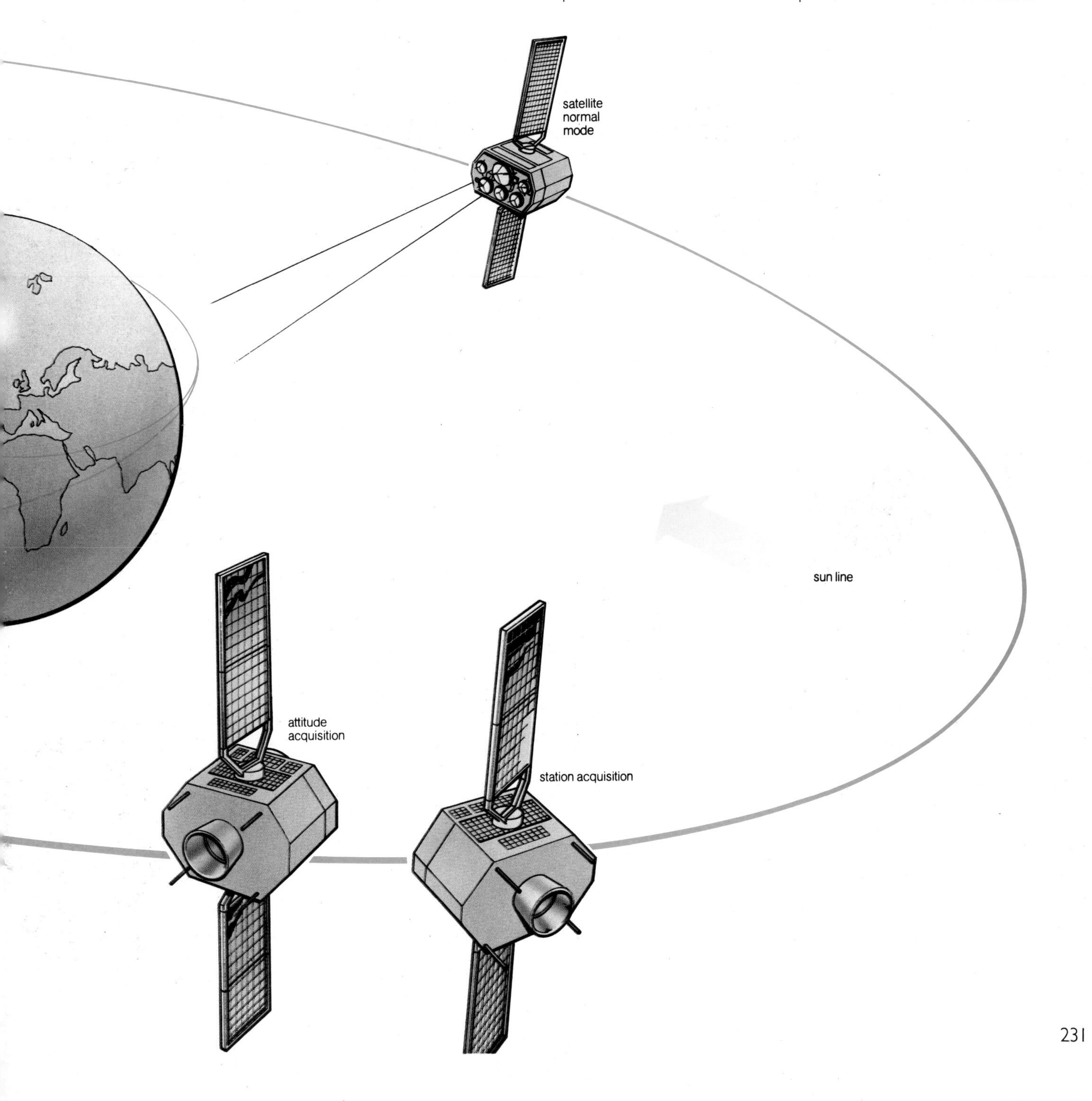

Right: Skylab, seen from the command module during the final inspection after a mission, was highly successful despite incidents such as the loss of a solar panel which led to overheating. The Mylar parasol rigged up by the astronauts to reduce the Sun's heating is visible in this picture.
Opposite page: the Intelsat satellite, which has a life of about seven years, after which its systems are likely to fail. This design will be capable of handling 13,000 telephone channels. It is spin-stabilized, with most of the electronics and antennas on top, rotating opposite to the body, which carries the power systems and attitude control rockets.

in free fall with hardly any strain imposed on it, these panels can be very lightly constructed. A panel of two square metres, the size of a tabletop, can generate 250 watts of power all the time it is in sunlight. A battery provides power for emergencies and for when the craft is in the Earth's shadow.

Satellites have a limited lifetime for a number of reasons. The output of the solar cells decreases with time, a fact which can be allowed for by producing excess power to start with which is re-radiated as heat into space from special panels. The attitude control systems have only a certain amount of fuel; and the satellite itself may encounter sufficient air resistance as to bring the craft out of orbit to burn up in the atmosphere in a comparatively short time. Lifetimes of two to ten years are common.

Telemetry Commands from the ground and details of how the satellite is functioning are transmitted to and from the satellite by telemetry links, generally operating at high frequencies. Many scientific satellites collect data continuously but are only in view of the ground stations for a short time. In this case, the data will be recorded and played back at higher speed when in contact with the ground station.

Applications Besides the well–known communications satellites, weather satellites are in constant use providing pictures of cloud cover over the Earth. By observing infra-red wavelengths, they can operate at night, since there is a temperature difference between cloudy and clear areas. One system transmits pictures continuously so that anyone with a suitable receiver on the ground can pick up signals and convert them on a facsimile machine to a photograph of the region as seen from space. Other satellites have been proposed which will carry out tasks more economically than ground systems—for example, to help oceanographers construct accurate profiles of currents, temperatures and densities of the oceans, a task which would be almost impossible by conventional means. Such information, together with satellite data on the polar caps, is of great value to the meteorologist, hydrologist and deep–sea fisherman.

Air–traffic control can be made safer by a system of satellites monitoring aircraft positions, so that more planes can be flown along crowded routes in safety.

The Earth's resources can be studied from space far more effectively than from ground level. By using cameras which take images in several different colour bands, potential diseased crop areas can be spotted as a result of the effect on the reflectivity of the plants. Mineral resources, cattle grazing densities, forestry and water supplies can all be surveyed rapidly from space. An experimental satellite, the Earth Resources Technology Satellite, has proved remarkably successful in this respect: its images show ground features as small as 300 ft (90 m) from an altitude of 570 miles (914 km), Such a satellite can also detect pollution sources: oil tankers are now under surveillance and can no longer risk prosecution for illegal tank washing and dumping oil at sea.

Military operations demand the very finest detail possible from satellite pictures, and standard scanning or television systems are not usually of high enough quality for this. Consequently the satellites use fine grain photographic film which then has to be recovered. Either a capsule containing the film is ejected from the spacecraft or the craft itself comes out of orbit when all the film has been used up. With suitable design a small package will not burn up as it re-enters the Earth's atmosphere and can be caught by nets towed behind aircraft as it falls by parachute, or recovered from the land or ocean. There is almost no theoretical limit to the detail that can be seen on such photographs, which are equivalent to the quality seen from a low flying plane. Such devices have done away with the need for 'spy planes', which caused much political trouble during the 1960s.

Large space stations, with a permanent human staff, will eventually be placed in orbit. The present range of satellites in orbit exploits only a fraction of the potential uses which will eventually be discovered.

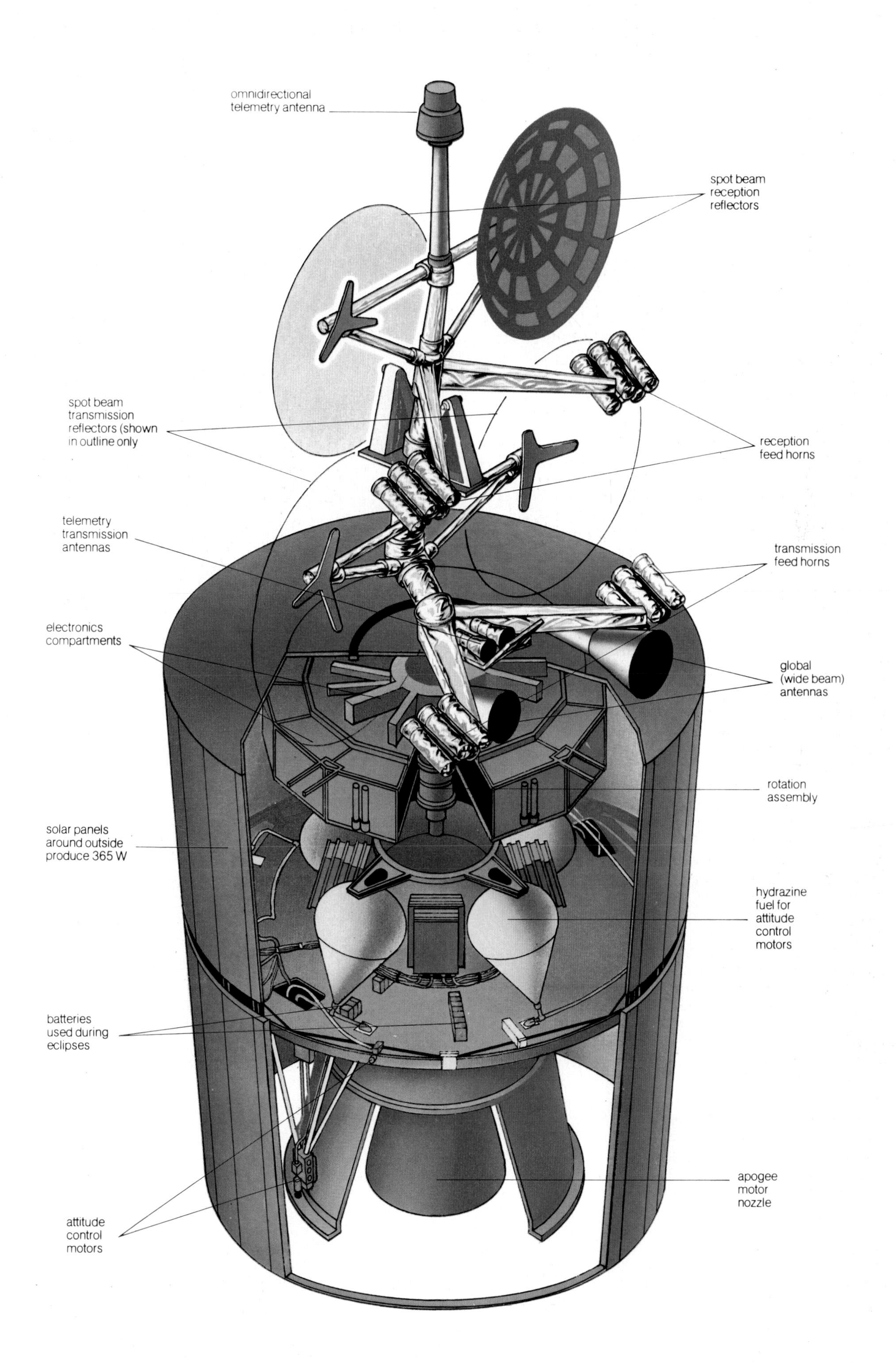
omnidirectional telemetry antenna
spot beam reception reflectors
spot beam transmission reflectors (shown in outline only
reception feed horns
telemetry transmission antennas
transmission feed horns
electronics compartments
global (wide beam) antennas
rotation assembly
solar panels around outside produce 365 W
hydrazine fuel for attitude control motors
batteries used during eclipses
apogee motor nozzle
attitude control motors

The orbital laboratory

Man's further progress in space depends on establishing laboratories in Earth orbit in which scientists can experiment under weightless conditions with access to a 'hard' vacuum.

Research already carried out in America's Skylab space station—in which three-man teams spent periods of 28, 59 and 84 days after docking from Apollo spacecraft—suggest that we shall see 'space factories' in which ultrapure metals can be produced in electric furnaces free of contact with containers. In a gravity-free environment it should be possible to manufacture extremely lightweight foamed steel with many of the properties of solid steel; also to combine dissimilar materials like metal and glass and metal and ceramics. Crystals of high purity may be produced for the electronics industry. The weightless environment of the space laboratory should also be ideal for extremely delicate methods of isolating specific biological materials, as in the

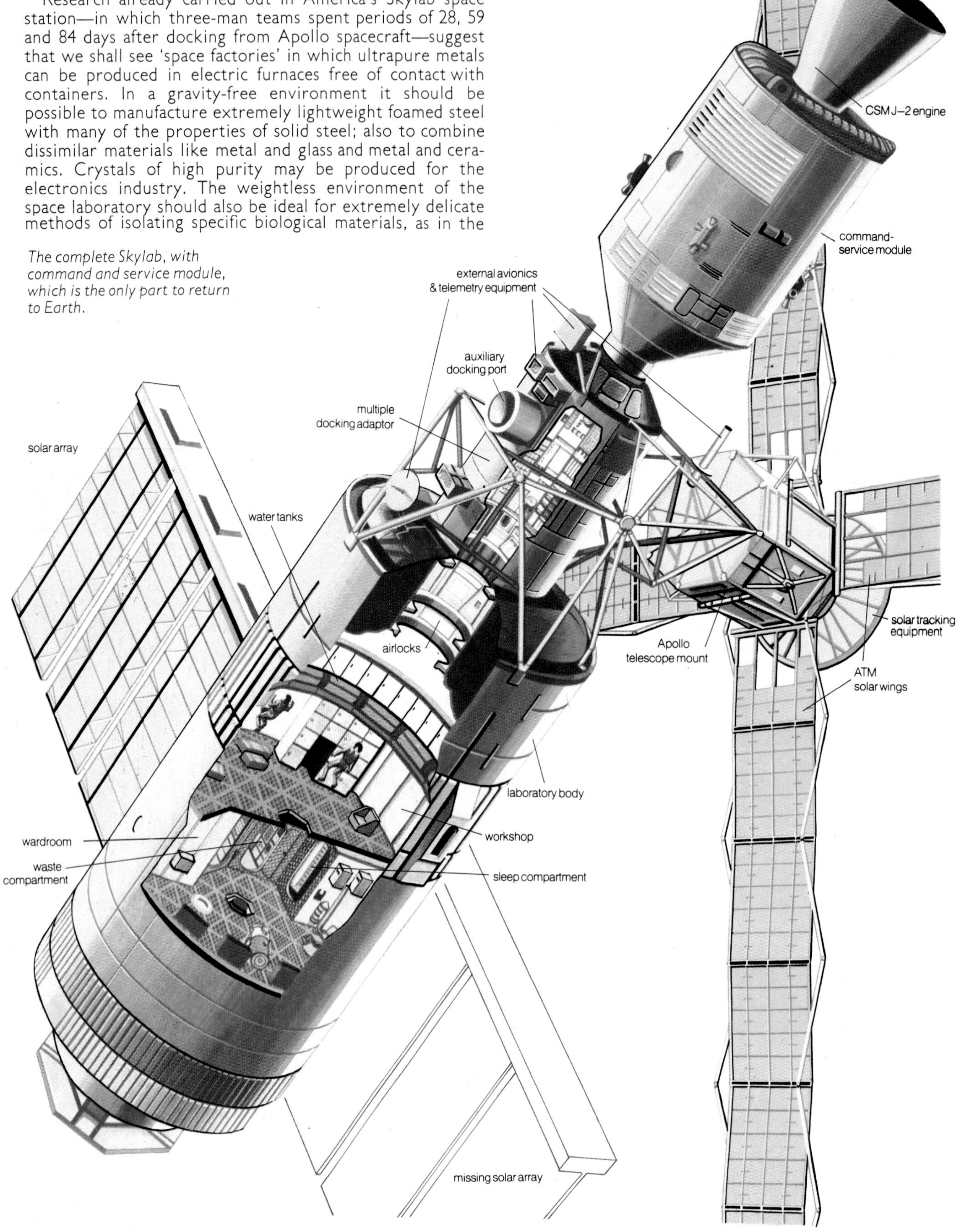

The complete Skylab, with command and service module, which is the only part to return to Earth.

preparation of concentrated antibodies for the treatment of certain diseases, and in the production of high-purity vaccines.

Orbital laboratories can also be used as stable platforms to examine the surrounding Universe across the entire spectrum of electromagnetic radiation reaching the Earth, much of which is cut off from ground observatories by the atmosphere. Detailed examination of the Sun—the nearest star—was made by the Skylab astronauts using the station's Apollo Telescope Mount (ATM).

Inspection of large-scale features of the Earth by multi-spectral (false-colour) photography has already led to the discovery of unknown mineral deposits, and forests and agricultural areas have been surveyed for general condition, soil quality, moisture content and disease damage. Healthy vegetation shows red or pink on the photos; diseased or impoverished vegetation blue-black.

It has been possible to register thermal patterns in the sea and locate areas of plankton where fish feed. Air and sea pollution have been tracked on a wide scale.

The further development of orbital laboratories depends on the ability to transport people and material more cheaply between the ground and Earth orbit in space vehicles which can be re-used many times. This is the task of America's space shuttle being developed for use in the 1980s which will take off vertically like a rocket and land like an aircraft. The winged component, called the orbiter, will have a crew of four astronauts. As big as a medium range airliner, it will fly a heavy cargo—up to 65,000 lb (29,484 kg)—into a low orbit, and is capable of being used as an orbiting laboratory in its own right.

Inside the orbiter's 60 ft (18.3 m) long by 15 ft (4.6 m) wide cargo bay will fit the European Spacelab in which four scientists can perform experiments as they orbit the Earth for up to 30 days. This will have two main sections: a pressurized laboratory module in which the research team will work in a 'shirtsleeve environment' (that is, without the need for spacesuits), and an external platform, or pallet, for mounting large man-directed instruments such as telescopes or antennae requiring direct exposure to space. The Lab itself will have standard laboratory instruments, equipment racks, work benches, data recording and processing equipment, and extendable booms for remote positioning of certain experiments in the space environment.

According to the National Aeronautics and Space Administration, the US space agency, the researchers who fly in Spacelab will be healthy men and women, qualified in their field, and requiring only a few weeks of specialized training.

At the end of the mission, Spacelab will be flown back to base with the research team to land on a conventional runway. On the ground—while the orbiter is being prepared for another flight—the Lab can be lifted out and exchanged for another containing different experiments.

Beyond Spacelab is the opportunity to assemble large space stations in orbit from prefabricated modules. The modules would be ferried up piece by piece by the shuttle and docked together.

Space probes

Space probes are some of the most complicated robots devised, since they have to replace the abilities of man on long journeys through space to the Moon and planets and, on arriving at their destination, have to act as eyes and senses to detect what is there. An Earth satellite sends back measurements and sometimes photographs of the terrain below it, but a space probe must often carry out precise manoeuvres when at a vast distance from Earth. Probes have been sent to the Moon, Mars, Venus, Mercury, Jupiter and Saturn; they have orbited the Sun to investigate conditions in space on the far side of Earth's orbit, so keeping the whole Sun under surveillance; and they have given simultaneous measurements of space particles at several points widely distributed through the solar system.

Design The design of any space probe depends basically upon its destination and what it has to do. Almost all space probes and satellites have a fairly elementary framework around which can be added the various experiments and systems, and all probes of a particular class, such as the Mariners which were sent to Mercury, Venus and Mars, have basic similarities.

Near the Sun there are few problems of power supply; panels of solar cells will provide adequate supplies of electricity for several years. In the case of probes intended for Mars and beyond, however, the solar intensity is much lower and power from solar panels is limited.

In the case of the Pioneer craft destined for Jupiter and Saturn, no solar panels at all were used, since the solar intensity near Saturn is almost one hundredth of that near the Earth. Instead, four thermonuclear generators, producing heat from the radioactive decay of plutonium-238 were used to provide 130 watts of power. This output deteriorates with time, and it seems likely that only about 80 watts will be available when Pioneer 11 encounters Saturn in 1979.

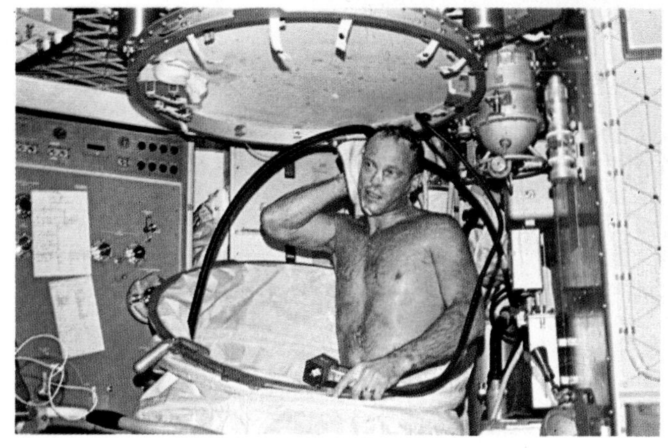

Above, top: an astronaut at the Skylab telescope controls. His shoes are locked in a grid to keep him in place; behind him is the airlock.
Above: drying up after a space shower. The astronaut uses a hand-held spray and a vacuum hose to suck up the droplets.

To protect the scientific instruments from radiation from these generators, the power units are located on booms pointing away from the instrument packages. Booms are often used on spacecraft where some particular instruments for collecting data, such as a magnetometer, may be affected by the others.

Another dominant feature of a space probe's design is the telemetry antennae. The transmitter power is usually very low, of the order of a few watts, so to make the best use of it the signals must be beamed back to Earth by a parabolic dish reflector, which gives a narrow beam. In the case of Pioneer 10 and 11, this dish dominates the whole craft, being nine feet (2.7 m) in diameter. A low-gain 'spike' antenna is also provided for transmission at a low rate if the main beam is not exactly aligned with Earth. Very large parabolic dishes on Earth, such as NASA's Deep-Space Network of dishes up to 210 feet (64 m) diameter, are needed to detect the very weak signals.

In order that the craft can point its antennae and experiments in the chosen direction, it must be stabilized in some way. The two methods available for any space vehicle are spin stabilization and three axis stabilization. In the former, the craft is set spinning by a platform on the launch vehicle just before it is sent on its way into space. Like a gyroscope, it will tend to stay spinning in the same direction in space. Where rapid pictures are to be taken and where a number of experiments have to be pointed in different directions, this is unsuitable. For most space probes, therefore, three-axis stabilization is used in which the craft is kept in a particular orientation by means of attitude correctors. This means that the orientation of the craft must be known, and for this Sun and star sensors are employed to detect the direction of their chosen object. The Sun is easy to find as the brightest object in the sky, and one star only is needed to fix the orientation of the craft. The star Canopus, the second brightest in the sky, is generally used because of its brightness and its large angle from the Sun.

Minor effects, such as the gravitational pulls of the planets, may disturb the attitude. In this case, the attitude is corrected by means of small gas (usually nitrogen) jets. Another method, with the advantage that it does not deplete gas supplies, is to arrange a set of reaction wheels inside the spacecraft. By spinning one of these, the probe can be made to turn in the opposite direction at a rather slow rate.

The Pioneer probes, unlike most others, are spin stabilized. These craft are designed for long lifetimes: Pioneer 6, which became operational in 1965, has given many years service, and Pioneer 11, launched in 1973, is intended to encounter Saturn in 1979. The gas used in three-axis stabilization could be used up on lengthy missions.

As well as correcting the attitude of craft, their orbits or paths through space have to be corrected from time to time. A probe in space will follow an ellipse round the Sun as predicted by Kepler's Laws, unless its rocket is fired. This changes the orbit, and as soon as the rocket stops, the probe will continue along a new orbit, slightly different. These course-correction manoeuvres are carefully calculated when the orbit of the probe is known, and if carried out accurately only one correction will be needed per mission. The craft is

Above: Pioneer 6, launched in 1965, is one of a series of probes designed to measure field strengths and particles in space, giving information about the Sun as well.
Right: a photograph of Venus, taken by Mariner 10. Venus is completely cloud-covered; this ultra-violet photo revealed atmospheric circulation zones.

commanded to fire its rocket for a precise length of time at a certain instant. In this way extreme precision of aiming can be carried out.

A number of space probes are designed not simply to travel past or to go into orbit around the planet, but have to land. Some of the earlier probes made little or no attempt to slow their velocity before striking the surface, and were known as 'hard landers'. Some, such as the Ranger series, made no attempt to brake their progress and hit the Moon, taking television pictures as they went. Others, such as Luna 9 and Luna 13, had more of a 'rough landing' since their retro-rockets fired until they were very close to the lunar surface, then cut off leaving the device to drop. The instruments were protected by being inside a ball-shaped capsule which was ejected from the main craft before impact to bounce and roll across the surface, coming to rest some distance away from the rocket. Four petals then opened out after an interval in such a way that the probe was forced upright and the instruments were revealed.

An improved technique was used by the Surveyor craft, five of which landed on the Moon in 1966 and 1967. These were equipped with controlled vernier rockets to keep the spacecraft attitude correct while the main retro-rocket slowed the craft down. This was then ejected and the vernier rockets took the spacecraft down to a soft landing. An essential feature of this technique is an automatic on-board controller linked to a radar altimeter and velocity sensor working on the Doppler principle, which enables the craft to know how high above the surface it is and how fast it is moving. (Guidance from Earth would be difficult because of the $2\frac{1}{2}$ second delay in the round trip of a radio signal between Moon, Earth and Moon). This system is now standard for soft-landing space probes. Subsequent probes, such as Luna 16, carried drills which took samples of the lunar surface. The upper part of the probe then used the lower part as a launching pad.

Another device is the unmanned lunar rover, the Lunokhod series. These are carried aboard soft landing craft, and move about the Moon's surface under the control of a driver on Earth. The vehicles are equipped with such instruments as X-ray spectrometers for soil analysis, X-ray telescopes and television cameras which can show the driver where the rover is going and also return higher resolution pictures.

The Moon has no appreciable atmosphere, but the planets Mars and Venus have thin and thick atmospheres respectively. By using the atmospheres to slow down the craft, less fuel is needed for the descent. The Viking craft due to land on Mars in 1976, for example, descends on a parachute after preliminary rocket braking, and then carries out a powered landing using rockets with a large number of nozzles to spread the exhaust in an attempt to reduce soil erosion—of importance since the probe is designed to examine the Martian soil for possible traces of life.

Above: Mariner 10 with its solar panels folded prior to launch. Its orbit passed both Venus and Mercury; it made its third encounter with Mercury in March, 1975.
Left: this photo of the moon's surface was taken from Lunar Orbiter 3 in 1967, which had an orbital altitude of less than 35 miles at its lowest point.

Above, top: a photograph of Jupiter showing convection zones on this gaseous planet. All of the planets out to Jupiter have been photographed close-up by space probes.
Directly above: a Soyuz command module, for a crew of three.

The very thick atmosphere of Venus presents particular problems and no completely successful landing has so far been carried out. The atmospheric pressure at the surface of Venus is about 90 times that on Earth, the temperature is about 475°C (887°F) and there appear to be very strong convection currents. Succeeding craft have been made to withstand greater and greater pressures as previous craft failed before reaching the surface; Venera 7 apparently reached the surface after a very rough parachute ride, and transmitted very faint signals for some 23 minutes. No rocket braking was used, and the craft seems to have struck the surface rather hard in the grip of a thermal current.

Instruments The point of sending space probes to the planets is to obtain details of conditions on the way and on arrival. The photographs sent back are the most spectacular results from probes, but scientific measurements of temperature, atmospheric pressure and composition, soil composition, magnetic field and particle densities are also made and sent back to Earth in the form of telemetry.

The cameras used for photography are not simple television cameras. The Moon probes Ranger and Surveyor had television cameras of special design, though neither returned the sort of continuous signals that normal broadcasting stations send out. Instead each picture took a short while to 'read' from the face of the camera tube. One reason for this is the handwidth of the telemetry channel used: not enough information could be sent in the time needed.

The problems of insufficient bandwidth, coupled with the distance of the probe from Earth and low power supplies, have led to a variety of methods being used. The highest quality pictures ever sent back from space were from the Lunar Orbiter craft which orbited the Moon prior to the manned Apollo landings. True photographs were taken by a pair of cameras, and the film was then processed on board the spacecraft by bringing it into contact with a chemical-impregnated film. The pictures were then read out by a 'flying spot' of light, produced on the face of a cathode ray tube which allowed the area of the photograph to be scanned as with a TV tube. The varying amount of light passing through the various parts of the film was picked up by a photomultiplier tube and turned into an electrical signal for return to Earth. The resulting photographs, when reconstituted, were sharp enough to reveal the grain structure of the original emulsion, and almost the whole lunar surface was mapped this way to a high resolution.

The Mariner craft sent to Mars, Venus and Mercury were equipped with tape recorders to record the camera output for later transmission at a slow rate. Mariner 9, for example, sent back a total of 7329 pictures of Mars over a period of a year from orbit. Two cameras, effectively small astronomical telescopes, were used with a variety of filters to measure colour and polarization of the surface. In addition, an infra-red radiometer or detector was aimed in the same direction as the cameras to measure the surface temperature.Similarly, a spectrometer observed the infra-red absorption resulting from carbon dioxide, the main component of the thin Martian atmosphere. Areas where greatest carbon dioxide was observed and hence where the atmosphere was deepest, could therefore be linked with the photographs taken to decide the height of the surface at each particular point.

In the case of the spin-stabilized Pioneers 10 and 11, sent to Jupiter, the spinning of the spacecraft would have made normal photography difficult. Instead, an imaging device was set to view a mirror which slowly turned, reflecting the scene around the craft into the detector. As the probe spun on its axis, so a complete picture was built up in lines, like a TV picture. No cathode ray tubes or other television equipment were used, however. Each picture took between 25 and 110 minutes to build up, but in view of the great distance over which signals had to be sent, it would not have been possible to transmit pictures more rapidly anyway.

The spacesuit

Our atmosphere on Earth gives us air, helps us to keep our body temperature constant and blocks out organically dangerous radiation, and a human being venturing into the vacuum of space must take such an environment with him. This is part of the function of any manned space vehicle; reduce the spaceship to a personal level and you have the spacesuit. Such has been the approach taken to the design of suits worn by both Russian and American astronauts for all activities outside their spacecraft and for back-up protection within it. Spacesuits are sometimes described as 'man-powered spacecraft'.

In spite of the absolute cold of space, almost −273°C (−459.4°F), solar rays are likely to overheat the spacesuit. Cooling is therefore highly important and the circulation of water through a network of tiny tubes has proved the most effective means of achieving this. The suit must also contain materials that make it completely airtight, flexible yet extremely tough, and resistant to solar radiation and micrometeoroids.

If, on an extravehicular excursion or 'spacewalk', the occupant is likely to remain close to the mother craft, then air and water can be circulated through an umbilical hose—like supplying a deep-sea diver through an air pipe. The umbilical also acts as a tether to stop the space-walker drifting away, and carries biomedical instrument links (to monitor such things as pulse rate) and communication links, plus electric power if the circulation pumps are attached to the suit rather than being in the mother craft.

For more remote activity such as lunar surface exploration, an independent supply plus associated machinery to operate it must all be carried by the wearer—paralleling the scuba diver—with the attendant limitation in endurance. Basic pressure garments, plugged into main systems, have also been worn for security during critical phases of flight such as lift-off and re-entry by both Russians and Americans, the suit being totally or only partially worn according to variations in contemporary views of the risk. On one Russian mission, Voskhod 1, the three crewmen took no suits at all with them. After the accidental depressurization just before re-entry of Soyuz 11, which killed the three crewmen, Russia reinstated fully suited re-entries for a cautionary period.

Early suits Spacesuit technology has naturally developed along with that of other space hardware. Spacesuits are all made to measure, as befitting such complex garments, often incorporating minor preferences of the wearer such as the positions of the pockets. Each man has, in fact, three suits, including one for reserve and one for ground training. The relatively simple pressure suit worn by Alan Shepard on the first US suborbital flight, however, bears little resemblance to the highly complex equipment worn by the lunar explorers.

Surveyor 3 landed on the moon in 1967 and returned more than 6000 television pictures of its surroundings. In 1969 the manned Apollo module landed near Surveyor so that the state of the craft after 2½ years in the lunar environment could be examined.

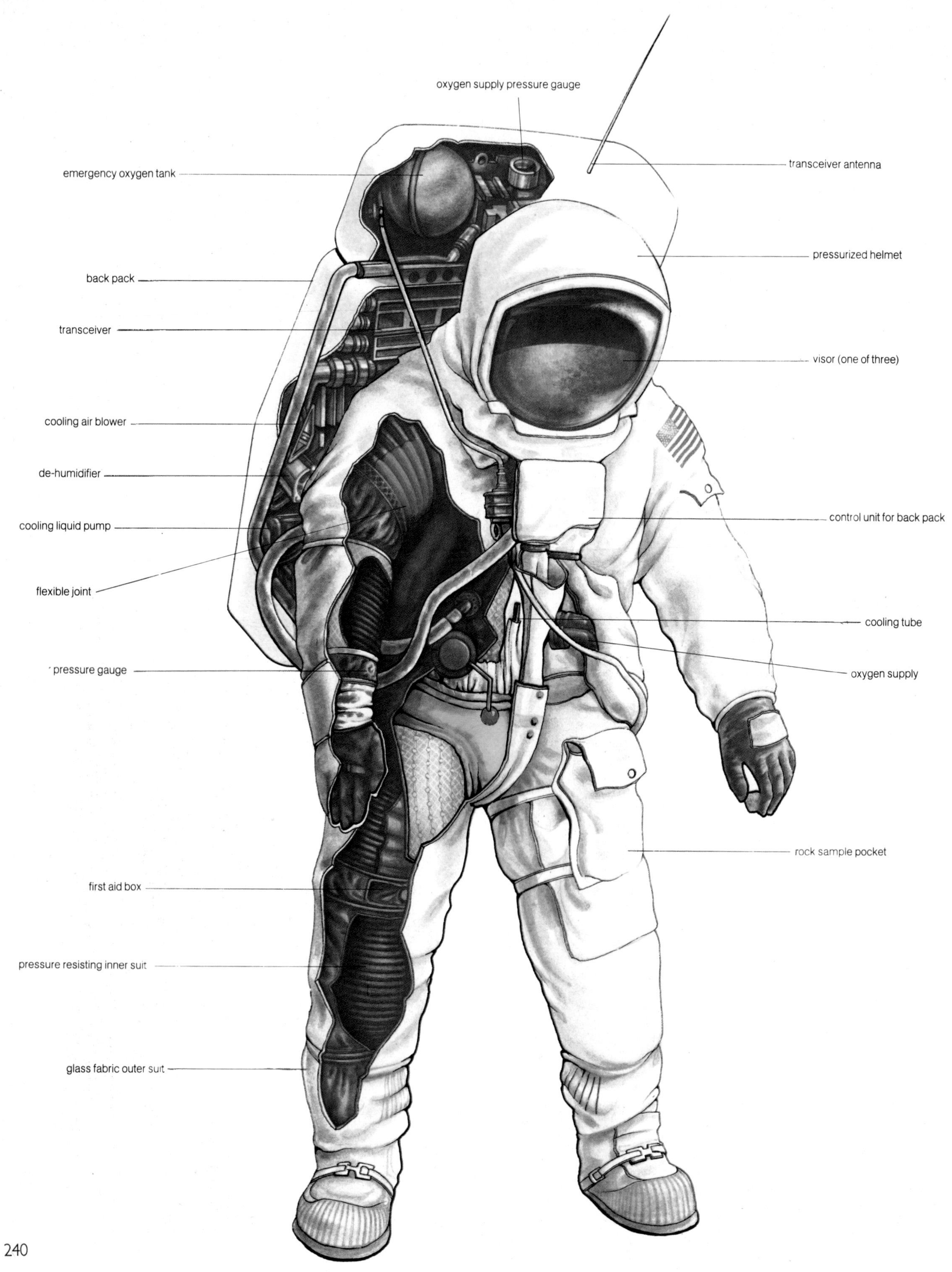
oxygen supply pressure gauge
emergency oxygen tank
transceiver antenna
pressurized helmet
back pack
transceiver
visor (one of three)
cooling air blower
de-humidifier
control unit for back pack
cooling liquid pump
flexible joint
cooling tube
pressure gauge
oxygen supply
rock sample pocket
first aid box
pressure resisting inner suit
glass fabric outer suit

For example, the crews of America's early two-man Gemini flights wore pressure suits against cabin depressurization but not designed for extravehicular activity. These suits had five layers, beginning with white cotton underwear with attachments for biomedical instruments. Next came a blue nylon layer, purely for comfort, and then a pressure garment of black neoprene-coated nylon, which pressurized at 3.5 psi (0.24 bar) of pure oxygen if cabin pressure failed. Over this was a link-net layer of Dacron-coated Teflon to hold the suit's shape when pressurized, and it was topped by a white HT-1 nylon layer to guard against accidental damage and reflected sunshine.

A new 16 lb (7.3 kg) lightweight suit was introduced just for Gemini 7, composed simply of a neoprene-nylon inner layer with HT-1 nylon on top, plus an ordinary pilot's helmet under a soft pressure hood.

Gemini space walks required extra protection which resulted in a seven-layer garment. Two extra layers, aluminized Mylar for thermal protection and a micrometeoroid protective layer, were inserted under the top layer of the basic five-piece suit. The total weight was 33 lb (15 kg)—of little significance in gravity-free orbit.

Apollo suits Apollo suits were even more complicated. When not in spacesuits, Apollo crew wore only light two-piece Teflon fabric overalls. The Command Module Pilot, who remained in the Apollo CM throughout the flight except for a brief spacewalk, required a simple five-layer pressure garment. But the lunar surface walkers wore two additional garments, plus a life-support backpack, all adding up to an Earth weight of 57 lb (26 kg). A modified version of this suit (without backpack) was also used by Skylab crews for outdoor activities.

The undergarment resembles long underwear and contains the cooling-water circulation tubing. Then comes the neoprene-coated pressure garment. On top, however, is an 18-layer 'integrated thermal meteoroid suit'. This begins with rubber-coated nylon and then non-woven Dacron, then aluminized Mylar film and Beta marquisette (a woven fabric) layers to inhibit heat radiation. On top is a layer of non-flammable Teflon-coated Beta cloth and finally an abrasion-resistant layer of Teflon fabric. The boots contain an even more diverse assemblage of materials. The helmet has two visors which together shield it against heat, ultra-violet and infra-red radiation and micrometeoroids. A tube leads into the helmet from a 2 pint (1.14 litre) water bag inside the suit neck ring. Finally, there is the self-descriptive Personal Life Support System, or backpack—a highly engineered unit which gives the wearer autonomy and security. In the terminology of NASA, the US space agency, it is then no longer a spacesuit, it is an Extravehicular Mobility Unit.

Opposite page: the type of suit worn on the Apollo missions has no provision for the disposal of solid body wastes, but urine can be disposed of by means of a tube leading to a 'fitted receptacle', a small tank in the right leg of the suit.
Below: on the left, the ease of motion in the improved Apollo suit is demonstrated. On the right is the Soviet suit used on the Soyuz flights.

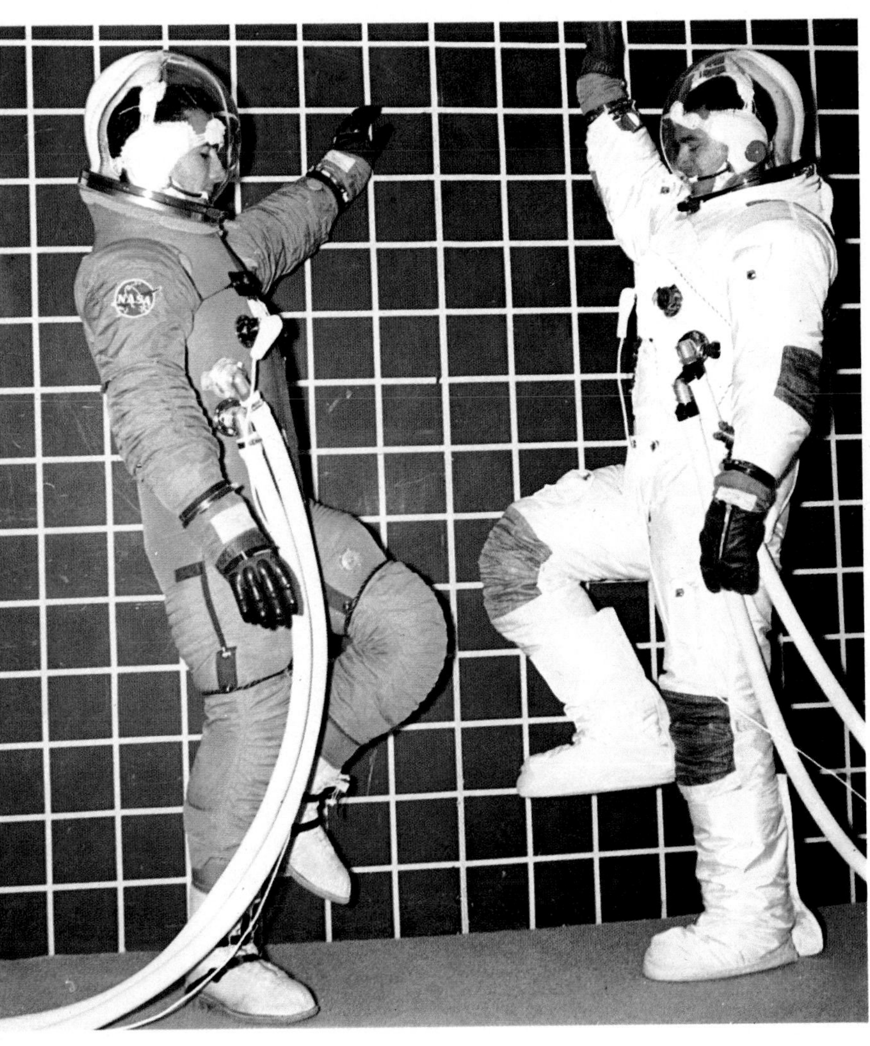

Space vehicles

The comparative novelty of space travel means that both American and Russian manned space activities have so far merely subsisted, relatively speaking, on makeshift equipment devised to master the basic knowledge and technology and to achieve certain limited objectives such as short one-stop visits to the Moon. In most respects a manned spacecraft compares in description with any unmanned satellite or space probe, but on a larger scale. Once basic technology is established, man's role becomes that of a far more efficient trouble-shooter and robot operator than we are yet capable of devising. His presence, however, demands additional expensive safety measures such as back-up systems and extra testing as well as large-scale hardware and complex life-support systems. The argument still rages as to whether the advantages merit the expense. The policy of NASA, the US space agency, is that manned and unmanned craft can be complementary and that both capabilities must be developed, but costed carefully.

Any manned space vehicle has certain fundamental requirements. It must have a powerful main rocket engine, preferably controllable, to perform course corrections, manoeuvres in Earth or lunar orbit and deorbiting retro-fire, plus sufficient reserve propellant. Today's fuels would be prohibitively bulky in the quantities required for long-duration missions and revolutionary rocket techniques using nuclear energy or ionized gases are under development for the spaceships of tomorrow.

It must also have an assembly of attitude-control thrusters (small rocket motors) to perform the same task as the ailerons, rudder and elevator of an aircraft. Usually duplicated for safety, thrusters are paired diametrically in opposition to apply torque (twisting force) round the appropiate axis. Each set of thrusters has its own fuel tanks and all are controlled by a joystick which electronically selects thrusters to control the craft in the same sense as an aircraft.

Systems Most systems depend on electric power, which was provided in early craft by short-life batteries. NASA introduced fuel cells in Gemini and Apollo, generating power by conversion of oxygen and hydrogen into drinkable water. Russia's Soyuz depends on solar generator panels of older technological vintage. Miniature nuclear generators, already used on some probes, are now becoming practicable for very lengthy manned missions.

Temperature control is important because, in space, the craft is cooked on one side by the Sun and chilled on the other by space. The loss of a solar panel on Skylab meant insufficient power to operate cooling systems properly, hence the need to erect a sunshield. During trans-lunar flight, the Apollo craft distributes heat by spinning slowly in what is termed the 'barbecue mode'.

Pressurization depends on sufficient air supplies to cope with normal consumption plus any cabin repressurizations required for space excursions. Air, water and food present a severe weight penalty and future flights of any length will probably have to recycle such consumables either mechanically or ecologically (using plants). The Russians have consistently used an Earth-type atmosphere of oxygen and nitrogen at near sea-level pressure. US craft have all relied on pure oxygen at only 5 psi (0.345 bar), except that Apollos have used an Earth atmosphere for launch and then converted to oxygen in orbit. This is a precaution taken after pure oxygen fuelled a fire that broke out in the spacecraft cabin during launchpad tests in preparation for the first manned Apollo mission, killing astronauts Gus Grissom, Ed White and Roger Chaffee.

Communications are the lifeline with Earth. Future spaceships will have considerable computer capacity to monitor their own systems, and future technology will have earned greater confidence. But today, a very large number of systems has to be monitored constantly from Earth, including crew health. In addition, voice and television channels must be maintained reliably and at adequate strength with limited on-board power, so there is a need for highly directional aerials operating on microwave frequencies.

Finally, the manned craft must have a means of returning its occupants to Earth. The severe conditions of atmospheric re-entry have dictated a modular approach to vehicles so far, in order to minimize the amount of vehicle requiring a special shape and covering to protect it from the 3000°F (1650°C) generated by the friction of the air at 17,000 mph (27,360 km/h). Capsules have been conical to let most of the surface hide behind the heat shield, or slightly curved as with Soyuz to improve aerodynamic lift and ease deceleration loads. The blunt heat shield is necessary to brake the craft to parachute-deployment speed. Heat shields have depended on ablation techniques whereby a thick plastic coating is allowed to burn off and dissipate the heat. A thinner coating is applied to all other surfaces.

Early vehicles In present day terms, 'spaceship' means Mercury, Gemini and Apollo on the US side and Vostok (East), Voskhod (Sunrise) and Soyuz (Union) on the Soviet side. As comparative evolutionary steps, the three can be matched up only in numbers of crewmen, having more or less leapfrogged each other in terms of technical advance.

Above: the Apollo 11 lunar module, which landed men on the moon in July 1969. Opposite page: an Apollo 17 astronaut taking samples of soil from the moon. The backpack is his Personal Life Support System, which enables him to work away from the lunar module.

The lift-off of the Apollo 11 moon mission in July 1969. The small rocket on top of the command module is to pull the module away from the Saturn 5 launch vehicle in an emergency.

As the space age dawned, the true potential of the unmanned satellite had yet to reveal itself, and as Man saw himself as the key to space exploration, research on manned spaceflight began early. Russia's second Sputnik carried a dog ('Laika') irretrievably into orbit. Five dog-carrying craft preceded the first ever manned space flight when, on April 12 1961, Yuri Gagarin made a single orbit of the Earth. Soviet experience was extended over the next two years by five more Vostok flights, the last carrying Valentina Tereshkova, the first and only woman in space by 1975.

Vostok comprised a spherical re-entry capsule and a cylindrical service module containing the retro-engine and more subsystems. Designed purely to achieve basic techniques of orbiting and recovery of manned craft, it carried ten days' battery power and was intended for full ground control, although equipped for basic pilot control.

America used chimpanzees for preparatory work, including a suborbital Mercury trial preceding each of two manned suborbital flights. A chimp was then orbited before John Glenn made America's first orbital flight on February 20 1962. Three more manned flights completed the programme.

Mercury was a simple bell-shaped craft, double-walled like all US manned vehicles, the outer wall being made of nickel alloy and the inner of titanium. Ceramic fibre insulated the walls while the heat shield was covered in a glass-fibre and resin mixture. All subsystems were located within the pressurized capsule itself, minimizing size and weight for the limited US launch capability of the time. Retro-thrust was provided by a package of three solid-fuel rockets strapped to the heat shield.

A major difference between the two craft, and one that set the pattern for all subsequent vehicles, was that the Russians opted for land recovery while the Americans preferred water, in spite of the massive recovery operations necessary (John Glenn received the services of 24 ships and 126 aircraft). Techniques had been perfected enough by the later Apollo flights to reduce this to a handful of ships and aircraft. Without the cushion of water, Vostok touchdowns were possibly quite heavy because, except for the first (Gagarin's) mission, the crewmen always used an ejection seat at 23,000 ft (7 km) and landed separately.

The next step involved two Russian flights called Voskhod, the first carrying three men, the second only two. General evidence plus a lack of pictures suggests that these were modified Vostoks designed (successfully) to upstage the Americans. In Voskhod 1, space for the two extra seats meant omitting ejection seats and spacesuits. With its spacewalk objective, Voskhod 2 required spacesuits and an airlock, hence the two-man crew.

Planning for the Apollo Moon landing programme had already begun even before the first Mercury flight, but the two-year Gemini programme of 10 two-man flights was introduced to develop essential rendezvous and docking, long duration and space-walking expertise. Basically an enlarged and improved Mercury, Gemini comprised a main re-entry module plus the launch vehicle adapter section used as an unpressurized bay for life-support and power-supply equipment, and was constructed largely of titanium and magnesium. It was the first craft to carry a computer, enabling variations in flight plan to be calculated. Ejection seats substituted for a launch escape tower (a small rocket fitted on top of the command module to pull it away from the launch vehicle in case of emergency), hence separate hatches were needed for each crewman.

Soyuz It is presumed that lunar applications were intended for the three man Soyuz which first flew in April 1967, but unfortunately crash-landed, killing the single crewman, Vladimir Komarov. Objectives seem to have veered towards space stations in 1969 when it became apparent that an American lunar landing was imminent. Several subsequent flights, expanding on Gemini-type activities, culminated in the launch and manning of the first Salyut orbital laboratory in April 1971. Soyuz 11, which docked with Salyut in June 1971, ended fatally for the crew, Georgi Dobrovolsky, Viktor Patsayev and Vladislav Volkov, when their ship accidentally depressurized during re-entry. Since then, Soyuz flights have continued in connection with follow-on Salyuts and towards the joint US-USSR rendezvous and docking mission.

Soyuz consists of three sections: a spherical orbital module in the nose, behind which is an igloo-shaped re-entry compartment, and a cylindrical service module in the rear to which a pair of 12 ft (3.7 m) long solar panels are attached. The orbital module is a general utility compartment containing working, eating and sleeping facilities. It has a main entry hatch, and a connecting hatch to the re-entry module so that it also serves as an airlock for space excursions.

The re-entry module has its three crew couches arranged fanwise to fit the confined space, and its heat shield contains a small solid-fuel rocket which fires at 3 ft (1 m) altitude to cushion touchdown. The service module contains most of the subsystems, including two liquid fuel engines (one spare) and fuel for manoeuvring up to 800 miles (1287 km) altitude.

Apollo Apollo was basically designed, in conjunction with the lunar module, to take three men to the Moon and back and to achieve little more. The conical command module, made of stainless steel sandwich containing stainless steel honeycomb, uses a phenolic epoxy resin on its ablative heat shield. A docking probe in the nose is removable (along with the lunar module's corresponding drogue) after connections are secure, to provide a clear tunnel between the two craft. The cylindrical service module contains the 20,500 lb (9300 kg) thrust main engine used for all manoeuvres. It also carries propellants for main engine and thrusters, plus oxygen and hydrogen for fuel cells and life support.

The Apollo lunar module (LM) was the first true spaceship, in being devoid of any aerodynamic characteristics. An octagonal descent stage carried the main engine and fuel, air and water supplies and general equipment for the surface stay. As the descent stage was not needed for the return flight, it was abandoned, serving as a launch pad for the multi-faceted ascent stage which contained the pressurized two man cabin and all systems of a spacecraft in its own right.

The last three Apollo missions used much-modified craft, enormously extending the mission capabilities. The mother ship carried a complex package of scientific instruments, exposed from one section of the service module for orbital study of the Moon, plus more supplies for extended flight time. The LM was similarly improved to allow not only for a three-day surface stay but also for three complete cabin repressurizations as well. In addition it carried a lunar roving vehicle (LRV).

The LRV, or 'rover', was a fold-up, two-seat car in which the astronauts were able to explore extensively the landing zone. As such, it was the forerunner of what will be an essential adjunct of future manned exploration of the planets. Its two 36 V batteries provided one hp from motors located on each of the four wheels for up to 57 miles (92 km) at up to 8.7 mph (14 km/h). It was equipped with inertial navigation and carried an Earth-controlled TV camera so that controllers could direct their own observations.

The future The next step will be NASA's space shuttle, a reusable manned freighter plying regularly between Earth and low orbit on a wide variety of missions. Evidence suggests that the Russians are also developing a similar type of vehicle. The shuttle is significant because it represents the beginnings of a practical maturity in manned space activities.

The shuttle will be similar to a conventional aircraft launched vertically and returning unpowered to a runway landing. Instead of ablative protection it will be covered with a new type of ceramic tile designed for repeated use. It will lift off on its three liquid-fuelled engines, fed from a huge

Lift-off is the climax of days of complex preparation. The first stage of Saturn 5 used liquid oxygen and kerosene as fuel; first-stage burn-out occurs 30 to 50 miles up.

external belly tank to which a pair of solid fuel boosters are attached, providing the bulk of lift-off thrust. The boosters will be jettisoned at altitude and recovered from the ocean for reuse. The fuel tank will be discarded from orbit for destruction in the atmosphere. Two smaller engines will provide orbital manoeuvring. The shuttle is mostly cargo bay —60 ft by 15 ft (18.3 m by 4.6 m) in size and capable of taking up to 65,000 lb (29,500 kg) into orbit.

The next step will be the permanent space station, assembled and serviced with the shuttle. A robot tug is being built to augment the shuttle by boosting payloads to orbits higher than the shuttle's limits. This could well evolve into a manned space tug which would be the first proper spaceship, operating permanently in space. When money permits further lunar exploration or building a lunar surface station, it will become a translunar ferry, connecting with the shuttle or space station in Earth orbit.

The interplanetary spaceship is already a theoretical reality. A manned Mars mission, now shelved, had been planned by NASA to begin in 1986. The two-year round trip was to have involved two identical six man vessels (for mutual support) assembled in orbit. As conceived, each ship was to consist of four huge cylinders, three in a row as propellant tanks for three nuclear engines, plus a crew section projecting end on from the central tank, its forward end slightly splayed to accommodate landing craft. Its length was 270 ft (82.3 m), its weight 1.6 million lb (0.73 million kg).

The two outer engines would have provided initial boost before being detached and returned to Earth orbit. The third was to power the remainder of the voyage. The ships were to remain in Mars orbit for 80 days, with half of each crew spending 30 days on the surface. Fuel limitations meant a 290-day return journey, crossing Earth's orbit in order to use the gravity of Venus to swing the craft back to Earth.

Navigation in space Journeys into space are based on very thorough 'flight planning' so that a Moon landing or a splashdown may be predicted with extraordinary precision many days or even months in advance. This is possible because there are no winds or tides, and the gravitational fields of Sun, planets and moons are known very exactly, so that the 'dead reckoning' can be remarkably precise.

For launch inertial navigation is used, but once in space, the craft and its crew are weightless and there is no unique vertical. Although position cannot be established, the platform is essential in order to keep the spacecraft on a steady alignment, which is set according to the brighter stars, while the motors are fired to change the velocity and amend the orbit. The accelerometers then measure the change of velocity and, when the correct value is reached, the thrust is cut off automatically.

Astro-navigation on Earth works only because the navigator needs just to fix his position relative to the Earth's surface. In space, there is no such reference and the stars can only give orientation in two dimensions. The third dimension of position in space is established by radar from the Earth supported by computing, though a modified form of astro-navigation can be used close to a planet or a moon where a reference surface exists. Doppler from the Earth measures velocity with extraordinary accuracy but the spacecraft is also fitted with Doppler navigation aids for landing purposes. Indeed, its on-board computing makes possible a safe return to Earth should the radio links break down.

Right: astronaut James Irwin with the Apollo 15 Lunar Rover Vehicle. On the moon, the vehicle was driven a total of 17.3 miles (28km), and about 168 pounds (76kg) of lunar rock and soil were collected. Opposite page: the Trifid Nebula, a glowing mass of gas, mostly hydrogen, 'excited' by the stars within it. The dark lanes are composed of gas which is not excited. The study of such nebulae leads to important discoveries in physics; stars such as our Sun are formed from such clouds of gas.

The Trifid Nebula is 3500 light-years from Earth, which means that at the speed of light it would take 3500 years to get there. Will mankind ever be able to travel that far? Perhaps the story of transport is only just beginning.

INDEX

Bold type indicates picture

005152

Allday Aluminium Co 49TR Alphabet & Image 15; 195CR; (NASA) 243 Aspect 241 Autair Ltd 207 Bell Aerospace Division/Textron Canada Ltd 78 Bettman Archive 27 Borg Warner Ltd 144/5 Paul Brierley 32 BOAC 221B British Aircraft Corporation 198; 205T; (Military Division) 202/3 British Airport Authority 221T British Railways Board/C F Clapper 116; 117; 118 British Transport Films 113; 119 British Waterways Board Museum 94/5 Tom Browne 106; 111 Brunel Society 105 Brunel University Library 59 Central Office of Information 60T; 73T Chubb Ltd 172 Mike Cornwell 165T Cranfield Institute of Technology/Brierley 39;/Goldblatt 214 Colorific (John Moss) 212; 228; 244; (NASA) 245 Daily Telegraph Colour Library 96T S. Davies & Sons Ltd 218T John Day 127 Daymark 166 Decca Radar Ltd 222/3 Douglas Dickens 50; 51 Mary Evans Picture Library 10CB; 120; 128 Fairey Marine Ltd 48 Ferranti 215; 217 Fiat 136B Fodens 168 (2, 3, 4); 169 Ford Motor Co 142; 152B; 179 GES Germany 179CR General Motors 28; 38; 136T; 140/1 John Goldblatt 199 Sue Gooders/Guildhall 10CT Hale Observatories 247 Robert Harding Associates 93TL Michael Holford (British Museum) 8; (Science Museum) 157R D. V. Hoskins 205B Illustrated London News 104 Imperial War Museum 40/1; 188/9; 189B Information Canada Photothèque 57TR International Harvester 177CR International Labour Office 46/7 Keystone 138/9 Courtesy London Transport Executive 1 Lucas/St. Maur Sheil 23B Archibald Maclean 223 William McQuitty 49BL Mansell Collection 123 Marconi Radar 219 Massey Ferguson 175 Ministry of Defence 74 Model & Allied Publications 19B NASA 3R; 235; 236R; 237L; 238T; 239; 246 National Motor Museum 25; 134T; 134/5 Novosti 128R Photo Library of Australia 167B Photri 43; 60B; 68; 135R; 148B; 194B; 195BL; 203; 211; 229; 232; 236L; 237R; 242 Picturepoint 29; 93BR; 110/1B; 112; 122; 124; 168T Ponts et Chaussees France 97 Porsche 24B Popperfoto 2/3; 4/5; 52L; 53; 56TL; 58; 69; 102/3; 130/1; 141; 190B Port Authority of New York & New Jersey 220/1 Radio Times Hulton Library 152T; 178 Raleigh 11TR Rickett Encyclopedia of Slides 9 Ian Ridpath 238B; 241R courtesy Rolls Royce 138/9 Ronan Picture Library 8B; 10T Royal Aeronautical Society 186/7 (Baden Powell Collection); 206 Royal Navy 90 Royal Scottish Museum/Syntax Films Ltd 13 Science Museum 16; (Hoffman) 36; 132; (Barker) 145; 174; 184/5; 194T Short's 210 Smith's Industries 159; 161B; 216 Harry Sowden 44/5 Spectrum 61; 65; 98; 115; 156; 177BL Sperry Vickers 88 Staatsbibliothek Berlin 133; 164; 165B; 180 Stone Manganese Marine 84; 85 Swiss National Tourist Office 121B Technisches Museum Germany 135TL Union Pacific Archives 107B UPI 190T; 226/7 US Navy 3L; 70/1; 76L; 79 (inset) Roger Viollet 193 Vickers Ltd 195TR John Watney 56BR; 57BL; 79; 91T; 100/1; 161T Western Americana 107T Anastasios White 167T Zefa 92; 99; (Pictor) 162; 181

Illustrations

Allard Graphic Arts 91 (Decca Navigator Co.); 96; 114/5; 129; 218 Arka Graphics 137; 191; 192B Jim Bamber 197B John Batchelor 6/7; 34/5 John Bishop 14; 26; 37; 42; 82/3; 143; 170/1; 172/3 (Lansing Bagnell); 182/3; 196; 197; 200/1; 233; 234 Roy Castle 224/5 Terry Hand 189 Eric Jewell Associates 192T Frank Kennard 22; 30; 31; 33; 154 (Girling); 155; 156; 157 Ken Lewis 80/1 Tom McArthur 10; 11; 12; 17; 18; 19; 20; 21; 24; 144; 146; 147; 148; 149; 153; 158; 160 Osborne/Marks 23; 52; 64; 72/3; 86/7; 89; 101; 108/9; 121; 125; 126; 150; 151; 163; 176; 213; 230/1; 240 Peter Sarson/Tony Bryan 54/5; 62/3; 66/7; 74/5; 76/7; 110/1; 208; 209 Saxon Artists 47 Typographics 134L

G R E Y H O U N D L I N E S